Self-Organization in Complex Ecosystems

MONOGRAPHS IN POPULATION BIOLOGY

EDITED BY SIMON A. LEVIN AND HENRY S. HORN

A complete series list follows the index.

Self-Organization in Complex Ecosystems

RICARD V. SOLÉ

AND

JORDI BASCOMPTE

PRINCETON UNIVERSITY PRESS
PRINCETON AND OXFORD

Published by Princeton University Press,
41 William Street, Princeton, New Jersey 08540

In the United Kingdom: Princeton University Press,
3 Market Place, Woodstock, Oxfordshire OX20 1SY

Library of Congress Cataloging-in-Publication Data

Solé, Ricard V.
Self-organization in complex ecosystems / Ricard V. Solé and Jordi Bascompte.
p. cm.—(Monographs in population biology ; 42)
Includes bibliographical references and index.
ISBN-13: 978-0-691-07039-1 (cl : alk. paper)
ISBN-10: 0-691-07039-3 (cl : alk. paper)
ISBN-13: 978-0-691-07040-7 (pbk. : alk. paper)
ISBN-10: 0-691-07040-7 (pbk. : alk. paper)
1. Ecology—Mathematical models. 2. Self-organizing systems. I. Bascompte, Jordi, 1967- II. Title. III. Series.

QH541.15.M3. S65 2006
577/.015118–dc22 2005047632

British Library Cataloging-in-Publication Data is available

This book has been composed in Baskerville BT

Printed on acid-free paper.∞

pup.princeton.edu

Printed in the United States of America

10 9 8 7 6 5 4 3 2 1

Dedicated to
Elisabetta (RS) and Evita (JB)

Contents

List of Figures and Tables

Figures

Tables

Acknowledgments

The following people supplied figures from their own work: P. Turchin, W. Schaffer, R. Desharnais, K. Higgins, A. Liebhold, E. Ranta, M.A. Rodríguez, J. Allen, M.E.J. Newman, T. Keitt, L.A.N. Amaral, E. Meron, and O. Ovaskainen.

We thank all our coauthors. Robert M. May and an anonymous referee made extremely valuable suggestions to a previous draft.

RS wishes to thank Stuart Kauffman, Patrick Marcos Nikolaus, Brian Goodwin, José Montoya and the members of the Complex Systems Lab at UPF. Special thanks to the Santa Fe Institute, where many ideas presented in this book were born and many parts written.

JB wishes to thank P. Jordano, C. J. Melián, M. A. Fortuna and the rest of the Integrative Ecology Group. The staff of the Estación Biológica de Doñana, especially Fernando Hiraldo, Carlos Soler, and Juan Calderón, smoothed things during his transition to the Spanish scientific system. The staff of the National Center for Ecological Analysis and Synthesis, especially Jim Reichman and Sandy Andelman, provided stimulating opportunities for scientific visits during which many parts of this book were written.

We are very thankful to Pere Alberch, Per Bak, and Ramon Margalef, three outstanding scientists whose contributions had a tremendous influence on our work. Pere Alberch's thinking shaped many of our evolutionary views, largely changing our perspective on the role of selection and self-organization in nature. Per Bak was an example of a scientist delighted in prodding others to look for simple explanations of how nature might work. He was looking for a general view of complexity that was personal but far reaching and the influence of his ideas keeps growing. Ramon Margalef was a great thinker who pioneered the search for unifying principles and the use of simple, conceptual models. We had many unforgettable discussions with him while starting our long journey into ecological complexity. He was always critic but careful, rigorous but funny, teacher and friend. All three passed away and left a powerful legacy behind. It was a privilege crossing paths with such extraordinary people. This book is also a modest tribute to them.

Self-Organization in Complex Ecosystems

CHAPTER ONE

Complexity in Ecological Systems

Ecology has been eminently a descriptive science despite some pioneering work by theoreticians such as Lotka, Volterra, Nicholson, and others. Description is a first step toward understanding a system. However, such a first step needs to be accompanied by the development of a theoretical framework in order to achieve real insight and, whenever possible, predictive power. Ecologists are increasingly facing the challenge of predicting the consequences of human-induced changes in the biosphere. For example, we need to better understand how biodiversity declines as more habitat is destroyed, or how harvested populations are driven to extinction as harvesting rates are increased. Toward that end, it is necessary to integrate field and experimental ecology with theory (Odum, 1977). This integration is much more common in the physical sciences, where theory and experimental data have always marched together (May, 2004).

This book describes a theoretical view of ecosystems based on how they self-organize to produce complex patterns. It focuses on very simple models that, despite their simplicity, encapsulate fundamental properties of how ecosystems work. They are based on the nonlinear interactions observed in nature and predict the existence of thresholds and discontinuities that can challenge the usual linear way of thinking. These models have shown the possibility of ecosystem self-organization at several scales. Simple nonlinear rules are able to generate complex patterns. This view contrasts with traditional approaches to ecological complexity based on extrinsic explanations. Thus, population cycles were said to be complex because multiple variables and external influences such as the climate are involved; spatial patterns in the distribution of a species were said to be nonhomogeneous because the spatial distribution of nutrients or perturbations is also heterogeneous; communities were said to be complex because there are hundreds of species interacting in a somewhat random way. This book presents theoretical evidence of the potential of nonlinear ecological interactions to generate nonrandom, self-organized patterns at all levels. In time, nonlinear density-dependence can generate complex dynamics such as deterministic chaos (chapter 2); in space, the combination of local nonlinear growth and short-range dispersal can generate a myriad of spatiotemporal patterns (chapter 3); at the community level, simple buildup mechanisms such as the preferential attachment of species to generalists can generate complex, invariant ecological networks (chapter 6).

By reviewing general properties of ecological systems, we emphasize processes that are equivalent to physical systems. General predictions can be made about ecosystem dynamics and their response to human-induced perturbations. Although there are several excellent books on parts of this plethora of processes (population ecology, spatial ecology, community ecology), none integrates them into a single framework. What is to be gained from this exercise is a comprehensive view of how simple processes act to build the patterns we observe in ecology at different levels and how predictions based on linear thinking may be misleading. Another characteristic of this book is to reveal such common mechanisms by using a suite of tools from physics and mathematics. These tools are used to emphasize their connection to ecological systems. They help in relating ecological systems to other physical systems.

THE NEWTONIAN PARADIGM IN PHYSICS

The Hardy-Weinberg law in population genetics can be thought of as an equivalent evolutionary version of Newton's first law stating that gene frequencies in a population remain fixed in absence of any force applied to them (e.g., migration and selection) (May, 2004). Despite this analogy, however, most ecologists would probably agree that there is little relation between the complexity of natural ecosystems and the simplicity displayed by any example derived from Newtonian physics. The standard physical systems we are familiar with involve either a single or a small number of particles. Of course understanding such "simple" systems might not be so simple, especially when dealing with the sophisticated mathematical framework required. But beyond the difficulties of the microscopic dynamics of atoms and elementary objects, macroscopic systems composed by a few or even many units (such as the solar system) are essentially well understood in terms of fundamental laws.

Several basic examples provide a perfect illustration for these laws. A perfect pendulum, for example, is a deterministic system described by only two variables: its position and its velocity (fig. 1.1). It can be shown that two deterministic equations describe its motion, defining how the position (angle ϕ) and the velocity v change through time (Arnold, 1978; Strogatz, 1994):

$$\frac{d\phi}{dt} = v \tag{1.1}$$

$$\frac{dv}{dt} = -\mu v - \frac{g}{L} \sin \phi \tag{1.2}$$

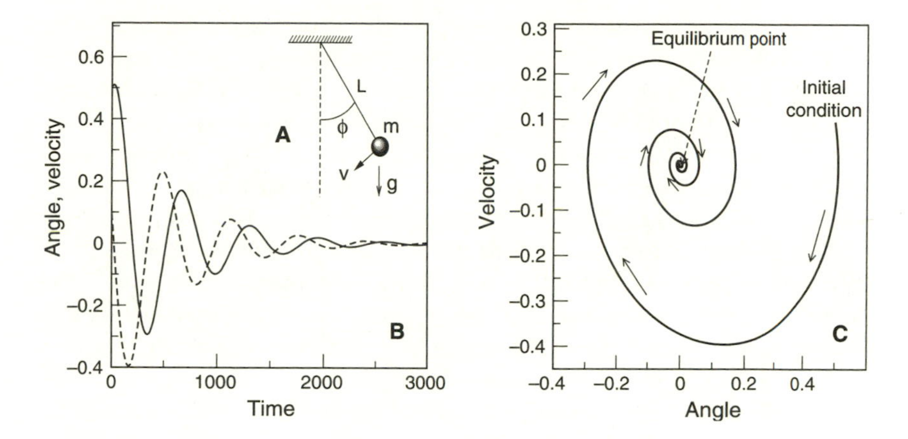

FIGURE 1.1. (A) One of the simplest examples of predictable dynamics: the pendulum. Here a ball of a given mass moves in a deterministic fashion and its state is fully characterized by only two variables: its position ϕ and its velocity v. Their evolution (for a given set of parameters) is shown in (B): Damped oscillations drive the system toward its equilibrium state. In (C) the same trajectories are represented by using the phase space (ϕ, v).

The motion takes place under the action of the gravitatory field g acting upon a mass m and a friction force proportional to the pendulum's speed. Consistent with our experience, the system will approach a state of repose in which $\phi^* = 0$ and $v^* = 0$. This dynamical evolution is displayed in figure 1.1b. Both the position (continuous line) and the velocity (dashed line) present dampened oscillations. Such an equilibrium state will be obtained starting from any possible initial condition, and we call it the global "attractor." It can be visualized by using an appropriate *phase space* Γ, here defined by the two variables required, that is, $\Gamma = (\phi, v)$. As shown in figure 1.1c, the orbits displayed by the pendulum converge to a single point attractor $(0, 0) \in \Gamma$.

Not surprisingly, the mathematical framework required in order to describe the motion of such a regular system is simple. And since simplicity in the underlying model and simple motion seem naturally related, one might conclude that: (a) simple dynamic patterns will be describable by means of simple mathematical models, and conversely (b) simple mathematical models will display simple, predictable dynamics. Although the first is typically true, the second turned out to be false (chapter 2).

For a long time, predictability has been the hallmark of the Newtonian success: the new visit of a given comet can be predicted with the highest accuracy and no one is surprised by such achievement. The success of classical mechanics has been a reference point for many scientists. Is there any hope of formulating such types of general laws for ecosystems? Probably not. One obstacle for a physics-oriented theoretical ecology is the fact that physical and biological systems strongly differ in some essential ways. Biological structures involve features such as functionality that have no equivalent within physics (Hopfield, 1994). Biological entities (at least at some scales) involve reproduction and evolution, and although some physical systems can "replicate," the implications of such event in terms of information are completely different. And of course there is a strong historical component to be added to the whole picture: species evolve, they change their interactions, and eventually the whole ecosystem changes. Ecosystems as such cannot be considered units of selection, but one may well ask if universal laws are shaping their structure at least on some scales.

The challenge posed by the previous issue is enormous, and one might ask first if physics can be an appropriate area to compare with ecological science. One possible answer is that most methods used in statistical physics are actually generic for any system composed by many parts. At its most extreme, econophysics offers a clever example (Mantegna and Stanley, 2000). Here the elementary units are so-called agents. These agents are humans, computer programs, or a mixture of both. Humans introduce an enormous level of potential complexity, and it makes no sense to write down the "microscopic" equations for them. In other words, the accurate description that would please a reductionist approach is simply forbidden. However, by assuming that these entities are not so complex *in making their decisions*, it is possible to understand, on a quantitative basis, many fundamental processes that define a market. Perhaps not surprisingly, several components of the key types of interactions that emerge in the economy are common to ecological dynamics. Actually, these similarities have been explicitly introduced within the framework of so-called computational ecologies, where networks of computers and agents display dynamic patterns not different from those recognizable as competition, cooperation, or symbiosis (Kephart et al., 1989).

Hierarchies and Levels of Description

Ecosystems can be understood and analyzed at very different, nested scales (fig. 1.2). At the most fundamental level, single species (or descriptors of some of their relevant features) define the smaller scale. Beyond this basic units, we have a level of description in which interactions with other species must be considered. At this level, new

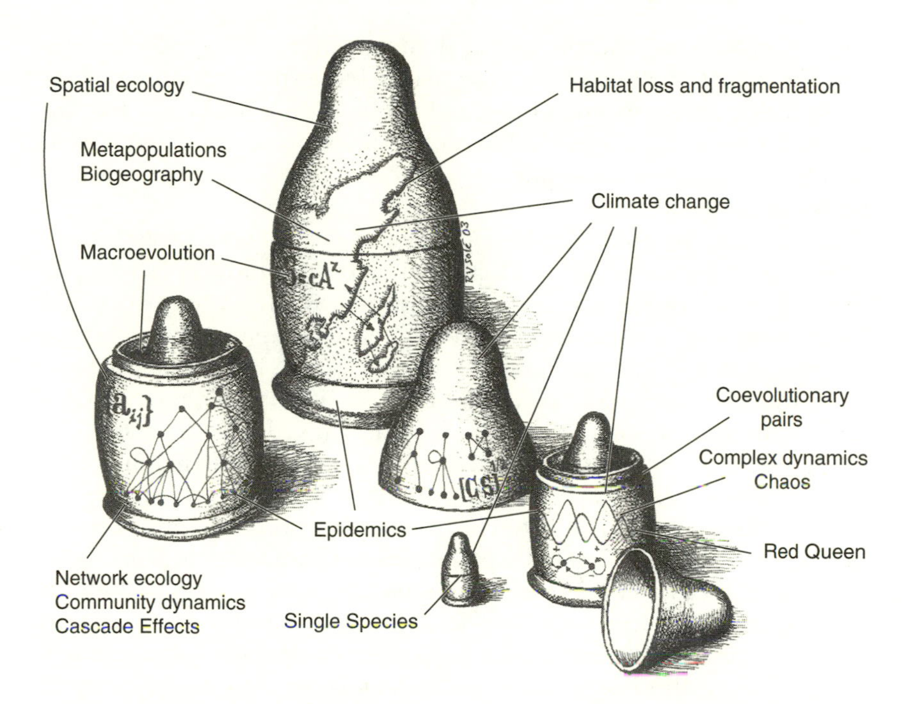

FIGURE 1.2. Levels of organization in complex ecosystems. A nested hierarchy of complexity levels can be defined, from single species to biogeographic patterns. Different properties and different problems can be identified at several scales, and some large-scale patterns cannot be reduced to the inferior levels of the hierarchy.

phenomena such as population cycles (chapter 2) can only be understood after coupling among species has been introduced. Evolutionary paths also need to consider the biotic environment, and some theories of evolutionary ecology, such as the Red Queen hypothesis (chapter 7), emerge as the natural explanatory framework. Under this scenario, new phenomena, such as the appearance of parasites and epidemics have to be taken into account. Beyond this point, the next step in the hierarchy involves the community level, in which the networks of interactions between species provide a unifying picture. At the community level, different regularities can be observed suggesting the presence of universal principles of community organization (chapter 6), but historical events implicit in the assembly process have to be taken into account. How external and internal changes propagate through the web of interactions among species and affect its stability is one key example of how ecological complexity is influenced by structural properties. At the higher level, the spatial context, the variability associated to local climate influences, and the global patterns of biotic organization define the top of our hierarchy. Our planet has experienced numerous large-scale changes through time, and the biosphere has been a leading actor. Through the history of life on Earth, extinction and diversification have been interacting and the first cannot be understood without the second (chapter 7).

But it would be misleading to think about the previous nested structure as a fully hierarchical one. Besides, the understanding of the upper parts of the hierarchy are somewhat, but not totally, decoupled of the lower members. The influences between each component are likely to be bidirectional. Climatic changes influence biotas, but the response of them to such change can further increase the trends (chapter 7). Habitat fragmentation (chapter 5) is a leading cause of biodiversity loss, although its effects must be understood in terms of their impact on the web of interacting species. Evolutionary responses are shaped by climate, but some overall patterns of ecosystem organization seem to largely stem from fundamental principles of community organization.

Although ecosystems are biological entities and as such include in their description concepts such as selection, adaptation, or information on a large scale, they are under the same physical laws as any other system. They also involve memory, information, and nonlinearity as essential features, and their interactions often behave in all-or-nothing ways, thus leaving little chance for the underlying details to effectively matter (chapter 4).

DYNAMICS AND THERMODYNAMICS

In a search for general, physics-inspired laws of ecological organization, the first stop is thermodynamics and statistical mechanics (Ulanowicz, 1997). Since ecosystems must fulfill the three laws of thermodynamics, perhaps some of the patterns observed in complex biotas are a consequence of such laws. Energy is a quantity common to all processes. It flows, is stored, and is transformed. When dealing with energy flows, the components and the boundaries of the system must be properly specified (Odum, 1983). One standard, classical approach is to make use of energy-flow diagrams (Odum, 1983; Ulanowicz, 1997). Based on early developments in general systems theory (von Bertalanffy, 1951, 1968), energy diagrams provide holistic descriptions of the ecosystem where all parts, including the environment, are considered (fig. 1.3).

Many physical systems are made up of multiple components (in interaction or in isolation). The simplest example is provided by an ideal gas, where a huge number of atoms (or molecules) occupy a given, fixed

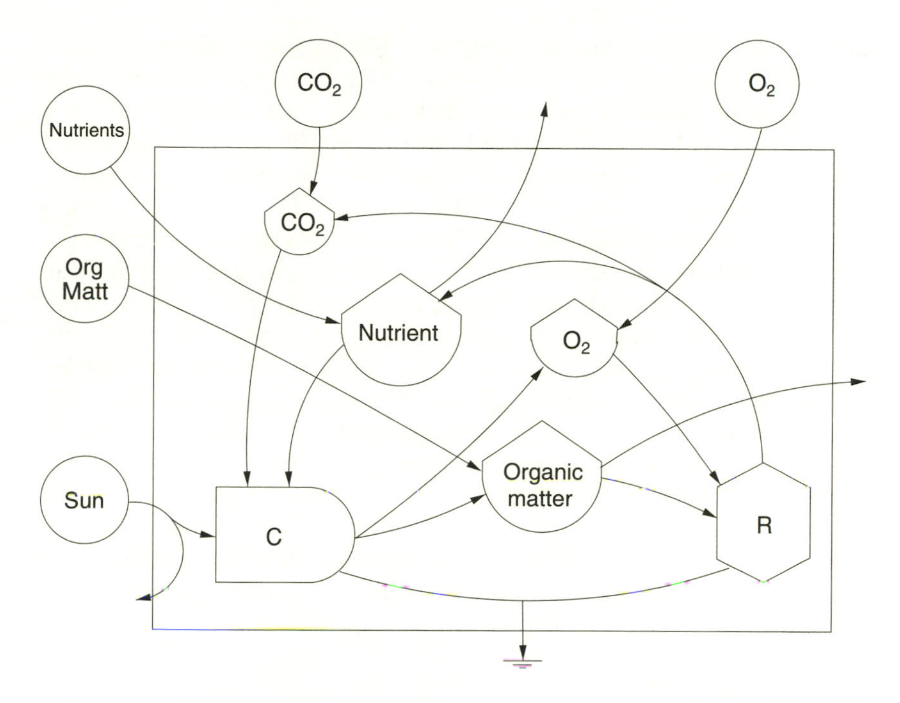

FIGURE 1.3. Energy diagram displaying the main processes involved in production, consumption, gas exchanges, and recycling in a typical aquatic ecosystem (redrawn from Odum, 1983). Here a very simple producer-consumer (C-R) model is shown. The arrows indicate energy flows. C refers to photosynthetic production and R to respiration of the consumer parts of the system.

volume of space. These molecules do not interact (or they interact very weakly) and thus can be considered independent of each other. This assumption, together with conservation in the number of particles and their energy, allows the construction of a well-defined body of theory in which we can extrapolate from single units to the whole. In other words: the average behavior of single particles provides full insight into the global behavior. For a standard physical system composed by N particles of mass m_i, an energy function can be defined as follows:

$$E(r_i, \ldots, r_N; v_1, \ldots, v_N) = \frac{1}{2}\left\{\sum_{i=1}^{N} m_i v_i^2 + U(r_{ij})\right\} \tag{1.3}$$

where the first part in the sum correspond to kinetic energies $T_i = m_i v_i^2/2$ and the second to potential energies, which in general will depend on the relative distance among particles, $r_{ij} = |r_i - r_j|$, through some functional dependence. When this term can be ignored or is too small compared with kinetic energy, we have a system that is basically described in terms of separated, representative items. However, once the interactions become relevant (and we talk about strongly interacting systems), collective phenomena start to modify expectations based on understanding single units. In particular, and this is an especially relevant point that will be discussed through the book, sudden changes will occur under small changes of the system's parameters. A given parameter such as temperature, precipitation rate, or infection rate can increase or decrease without having a detectable effect on a system's properties and suddenly a totally new qualitative property might appear. Within the ecological context, this corresponds to the so-called catastrophic shifts (chapter 2). Typically, these shifts end up in a new state with little resemblance to the original one, and very often there is no way back unless significant parameter changes occur. Irreversibility thus appears to be a rather common feature of ecological complexity. Such a property does not match comfortably within standard thermodynamics, where reversibility is a natural constraint. This problem will be discussed in some detail in chapter 4 in relation to the emergence of scaling laws and phase transitions in ecological systems.

Order and Disorder

Since purely thermodynamical arguments might fail to help understand the origins of ecological patterns, we should turn our attention toward a more dynamical view. Several questions emerge when dealing with ecological dynamics. What is the nature of the processes underlying the observed patterns? Is it possible to capture their main ingredients within simple mathematical equations? Are they predictable? Answering these questions forces us to look into the spectrum of expected patterns generated from deterministic and stochastic processes.

At one extreme in the spectrum of possible dynamical patterns we have randomness. Here events take place in a stochastic manner, and the appropriate description of the system must be cast in terms of probabilistic measures. Chance is here the main actor, and the coin toss the appropriate icon. Tossing a coin involves a simple mechanism but somehow it deals with a very large number of degrees of freedom. The same situation applies for a pinball machine: although it should be possible *in principle* to know how the dynamics will develop in time, it is practically impossible to do so. A coin toss, in spite of its simple definition, pervades the unpredictable character of random systems involving an infinite number of degrees of freedom.

At the other extreme in the spectrum, deterministic, simple dynamical systems offer (in principle) the other side of the coin. A deterministic system would provide a perfect predictability: once the initial condition is known, the future is available. This is the case of the example discussed at the beginning of this chapter.

But of course, things aren't necessarily so easy. A perfect illustration is offered by nonlinear discrete maps (to be explored in chapter 2). Consider a population of size x_n at a given year n (here $n = 1, 2, \ldots$). As will be shown in chapter 2, this time-discrete approach to population dynamics is appropriate for a number of ecological systems, such as insect populations with non-overlapping generations. The population at year $n+1$ will be obtained from the population at the previous year through some mathematical function. One of the simplest models is given by the following map:

$$x_{n+1} = \mu x_n (1 - x_n) \tag{1.4}$$

Equation (4) is called the *logistic map* and exhibits very complex dynamics in spite of its obvious simplicity. Since this complexity and its consequences will be of relevance in the next chapter, here we only highlight the random character of some specific properties of a large class of simple, deterministic dynamic systems. In particular, one class of dynamics that can be displayed by the previous model involves the generation of a sequence of values $\{x_i\}$ that never repeats itself. An example is shown in figure 1.4 for $\mu = 4$. The black dots represent the actual states, and the dotted line helps illustrate how the consecutive values are connected. By looking at the black dots, it is difficult to appreciate any regular pattern (although some correlations seem to be present).

Let us consider a given situation where we only have access to a coarse-grained description of the system's state. To be more precise, let us consider a binary space state obtained from a so-called Markov partition, where each value x_i has an associated bit S_i:

$$S_i = \begin{cases} 0 & : \quad 0 \le x_i < 1/2 \\ 1 & : \quad 1/2 \le x_i \le 1 \end{cases} \tag{1.5}$$

This mapping creates a string of ones and zeros that only retain a very rough part of the underlying dynamics. The surprise comes when the structure of this string of bits is analyzed: although the original dynamical system is completely deterministic, the binary sequence generated from the Markov partition is not distinguishable from a coin-toss process. Randomness is obtained from determinism.

The previous result illustrates the fact that the boundaries between determinism and randomness are not so well defined as one would first suspect. This emergence of complex dynamics will interact, within

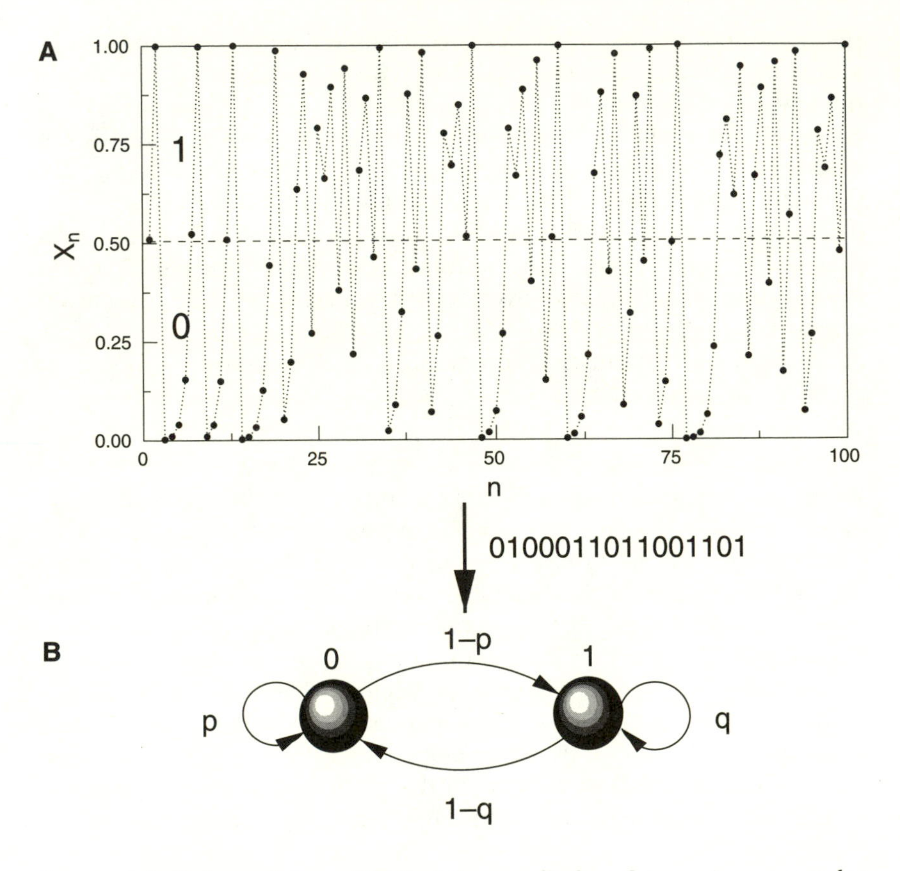

FIGURE 1.4. Order and chaos. The logistic map displays, for $\mu = 4$, very complex dynamics, here indicated as a sequence of black circles (A). If a binary partition is used in order to generate a sequence of ones and zeros, a stochastic sequence is obtained, which can be summarized in a transition diagram (B). In this picture p and q are probabilities, which for the logistic map give $p = q = 1/2$, as for a fair coin toss.

the ecological context, with other fundamental factors such as spatial degrees of freedom and evolutionary dynamics. As we will see, the results are often unexpected.

EMERGENT PROPERTIES

One key ingredient to be introduced in this book is the presence of emergent phenomena. Often, when looking at the macroscopic features of a given system, scientists have tried to find the origin of these properties by looking at the structure of their component parts. Such a view roughly defined the reductionist approach (Wilson, 1998). But the fact that the properties of the individual units cannot always explain the whole has been known from the earliest times of science. In this context, it is often said that the whole is more than the sum of the parts, meaning that the global behavior exhibited by a given system will display different features from those associated to its individual components. A more appropriate statement would be that "the whole is *something else* than the sum of its parts" since in most cases completely different properties arise from the interactions among components. As an example, the properties of water that make this molecule so unique for life cannot be explained in terms of the separate properties of hydrogen and oxygen, even though we can understand them in detail from quantum mechanical principles. The same limitations apply to biology: Some properties such as memory in the brain cannot be reduced to the understanding of single neurons (Solé and Goodwin, 2001). Life itself is a good example: nucleic acids, proteins, or lipids are not "alive" by themselves. It is the cooperation among different sets that actually creates a self-sustained, evolvable pattern called life. Over the last decades of the twentieth century the shortcomings of the reductionist approach had become increasingly apparent, and at some point a new type of view, known as integrative biology began to emerge.

Is reductionism the appropriate way of exploring complex systems? This is a debated issue. Within physics, extreme reductionists are not difficult to find. Advocates of reductionism, such as E. O. Wilson, have pointed out that modern science has been successful largely because of the analytic approach to reality (Wilson, 1998). But this view is being abandoned and replaced by a more global view of reality that takes into account the emergence of new properties. Let us consider one example.

Imagine a system composed by a set of n interacting species $S_1, \dots, S_n$ whose populations are indicated as $x_1, \dots, x_n$. The interactions are schematically described in figure 1.5a for $n = 6$. The diagram indicates that each species requires the help of another one for its growth to occur: species S_2 needs S_1, species S_3 needs S_2, et cetera. This model was introduced within the context of prebiotic evolution: molecular species would cooperate to enhance their success by means of this so-called hypercycle (Eigen and Schuster, 1979). Species abundances can reach steady states, oscillate, or behave in very complex, unpredictable ways. The cooperation requires the presence of all components: the lack of a single species destroys the hypercycle.

In spite of its simplicity, the hypercycle displays an enormous set of dynamical patterns, and its relevance goes far beyond molecular repli-

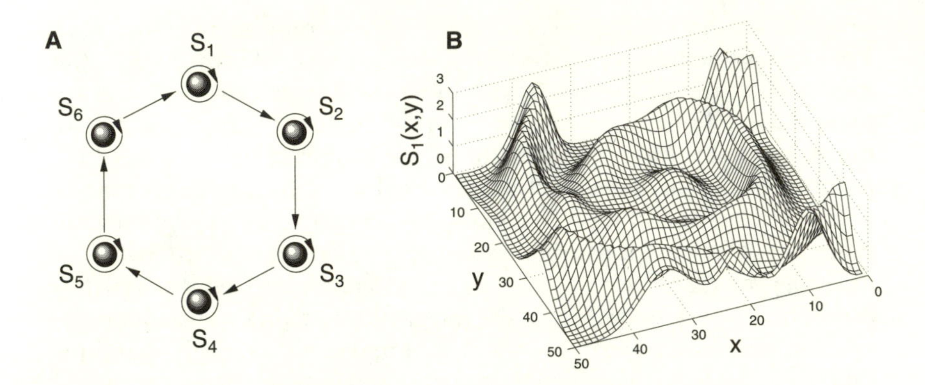

FIGURE 1.5. Emergence of long-range order in a set of cooperating species. In (A) the basic scheme of species interactions is shown. Here the reproduction of each species requires the help of another one. Such a dependence is closed, forming a catalytic cycle (the so-called hypercycle). When these populations interact within a spatially explicit, local context, complex structures, such as spiral waves (B) emerge. Here the population size of the first species $S_1(x, y)$ is shown.

cators (Maynard Smith and Szathmáry, 1995; Cronhjort and Blomberg, 1997). Consider now a simple extension of the previous model. The units (molecules or organisms) move now in a two-dimensional spatial domain where initially we place random amounts of individuals of each species at each site. The domain can be a discrete square lattice of patches where each patch is linked to its four nearest ones. Movement takes place locally, and interactions (as they occur in reality) are also localized in space. Cooperation thus takes place in a well-defined spatial context where units diffuse toward the nearest locations, and interactions are limited to available molecules in a neighborhood. The surprise comes from simulating the previous model (Boerlijst and Hogeweg, 1991; Chacon and Nuño, 1995; Cronhjort, 1995) on a lattice. An example of this is shown in figure 1.5b, where the local concentration of the first species is shown for different points on a 50×50 lattice. In spite of the fact that interactions only occur inside a given patch and that molecules or individuals can only move step by step between nearest lattice sites, a large-scale structure has been formed. The size of this structure has a length that is similar to the whole spatial domain and thus its origins cannot be explained from the local nature of the interactions. The spiral wave is a common motif of patterns emerging in models of interacting populations (chapter 3) as a result of a collective synchronization phenomenon.

Spiral waves emerge from a large number of nonlinear models with local interactions. The origin of the spatial structure can be properly explained once the transfer of information introduced by diffusion is taken into account. In other words, once local *interactions* among nearest sites are considered. These spiral waves cannot be reduced to the knowledge of the interactions among components and represent an emergent phenomenon with little relation to the species-specific features of individuals. Only interaction and diffusion rates are required in order to obtain a complete picture of the emerging structure. The impact of these structures on evolutionary responses has been explored by a number of authors (Boerlijst and Hogeweg, 1991; Boerlijst et al., 1993). Spiral waves in the invasion of parasites, and in a more general context spatially extended patterns, pervade ecological complexity and affect how ecosystems react against external stresses (chapters 3 and 5). Biodiversity patterns are also mediated by the fact that interactions are spatially localized. To a large extent, the limitations imposed by spatial interactions have a creative impact on ecological structures, allowing many different strategies to emerge (chapter 6).

ECOSYSTEMS AS COMPLEX ADAPTIVE SYSTEMS

Many macroscopic features of complex ecosystems emerge from interactions among their components. These patterns may, in turn, influence the further development of the interactions. As indicated by Simon Levin, ecosystems actually belong to a class of far-from-equilibrium systems known as complex adaptive systems (CAS). This type of system is characterized by several properties (Arthur et al., 1997; Levin, 1998), none totally independent among them. These include, together with far from equilibrium dynamics:

1. Localized interactions: Units in CAS typically interact with a limited set of neighbors. The constraints imposed by local interactions are responsible for the emergence of a large number of characteristic features, such as spatial patterns and coexistence among competing species (chapters 2 and 3).
2. Absence of well-defined top-down control: Complex patterns are obtained from local interactions and they largely influence the subsequent evolution of the system dynamics. Although top-down control appears to be present in ecosystem organization (chapter 6), it is seldom a strict one, often counterbalanced by bottom-up forces.

3. Heterogeneity in network organization: Complex ecosystems have a network structure that pervades their behavior and response against perturbations (chapter 6). Such networks have a complex structure with a nested hierarchy of interactions. Evolution and adaptation are strongly influenced by these networks.
4. Adaptation: Ecological systems display adaptation at different levels. Species adapt to given external conditions (such as temperature fluctuations) and properly respond to them. At a larger scale, food webs may adapt to a particular regime of perturbations.
5. Evolvability, that is, the presence of mechanisms allowing new features to emerge: This is an intrinsic feature of many complex systems, from cellular to economic webs. The capacity to respond to changing conditions on an evolutionary time scale is obvious from microevolution to macroevolution (chapter 7).

Ultimately, the structure and universal properties of ecosystems are linked to evolutionary forces operating at different scales. But these forces operate within an ecological context and thus both terms cannot be separated. What are the components of evolution that shape ecological patterns?

Evolution and the Ecological Theater

Ecosystems are the result of historical processes. The building of an ecosystem involves, on short time scales, path-dependent processes defining ecological succession. On larger time scales, species themselves change and coevolutionary dynamics arise. Succession is illustrated by the progressive colonization of an abandoned field, eventually ending up in the building of a mature forest. Since succession is historical, different communities are obtained starting from a given regional species pool. Contingency thus plays a role. But what is the impact of path dependence on the overall ecosystem's structure? The analysis of mature communities reveals that, beyond the specific composition of the final community, the same macroscopic traits of community organization are universal. Examples are provided by the network structure of the food and mutualistic webs or the patterns of species abundance (chapter 6).

Such universality reminds us of a different perspective of evolutionary change emphasizing the role of fundamental constraints (Kauffman, 1993; Goodwin, 1994; Solé and Goodwin, 2001). These theories suggest that basic, universal laws of organization shape the large-scale architecture of biological systems. Some of these basic laws and principles are presented in chapters 2 and 4. They result from a limited spectrum of dynamical patterns, the presence of inherent multiplicative processes (a species with a higher population is more likely to leave more offspring, see chapter 7), and to conflicts arising from competitive forces. The latter is the case in the levels of diversity allowed in a given habitat and how species turnover proceeds: There is a conflict between forces increasing diversity (immigration and speciation) and those reducing it (due to interactions). Such a conflict (chapter 4) can explain the emergence of universal patterns of community organization (chapters 6 and 7). As it occurs in some physical systems, conflicting constraints often end up in a very small repertoire of patterns of organization, which we recognize as universals.

The previous observations actually connect with the suggestion that natural structures result from a process of tinkering (Jacob, 1977; Solé et al., 2002b). The possible and the actual would differ strongly due to constraints intrinsic to the universe of potential structures. In this context, the dynamics and evolution of complex ecosystems would be shaped not only by selection and history but also by fundamental laws and intrinsic constraints.

CHAPTER TWO

Nonlinear Dynamics

THE BALANCE OF NATURE?

A popular view assumes a balance of nature. According to this view, species fluctuate until reaching a stationary abundance, a specific number given by the energetic constraints of their habitat. Of course, climatic changes and other perturbations may move populations away from their steady state, but after enough time, they will reach the same equilibrium values. As it happened with early Newtonian physics, early ecological theories were strongly tied to the concept (and goal) of stability. As discussed in the previous chapter, no other example of regularity, predictability, and equilibrium is better than the simple pendulum. By means of the power of mathematical analysis, its behavior can be understood and predicted with the greatest accuracy (Arnold, 1978).

The ecological "pendulum" is well represented by the logistic equation (Begon and Mortimer, 1986). If x indicates the population size of a given isolated, basal species, its time evolution can be described by means of the following deterministic equation:

$$\frac{dx}{dt} = \phi_{\mu}(x) = \mu x \left(1 - \frac{x}{K}\right) \tag{2.1}$$

where μ and K are the growth rate and carrying capacity of the population, respectively. This model has been used in a variety of contexts (DeAngelis and Waterhouse, 1987) and has been validated by controlled experiments (Begon and Mortimer, 1986). It can be used as an illustration of the concepts of stability, attractor, and predictability. The previous equation can be solved analytically. If $x = x_0$ at $t = 0$, it can be shown that

$$x(t) = \frac{K}{1 + \left[\frac{K - x_0}{x_0}\right] e^{-\mu t}} \tag{2.2}$$

and thus $x(t) \to x^* = K$ as $t \to \infty$. In other words, any initial condition $x_0 > 0$ eventually converges to the carrying capacity $x^* = K$, consistently with our intuition. Since all trajectories starting from $x_0 > 0$ become "attracted" by x^*, we call it an attractor of the dynamics. This is also the case for other models, in which stable points describe the long-term behavior of the solutions. But since most dynamical systems cannot be

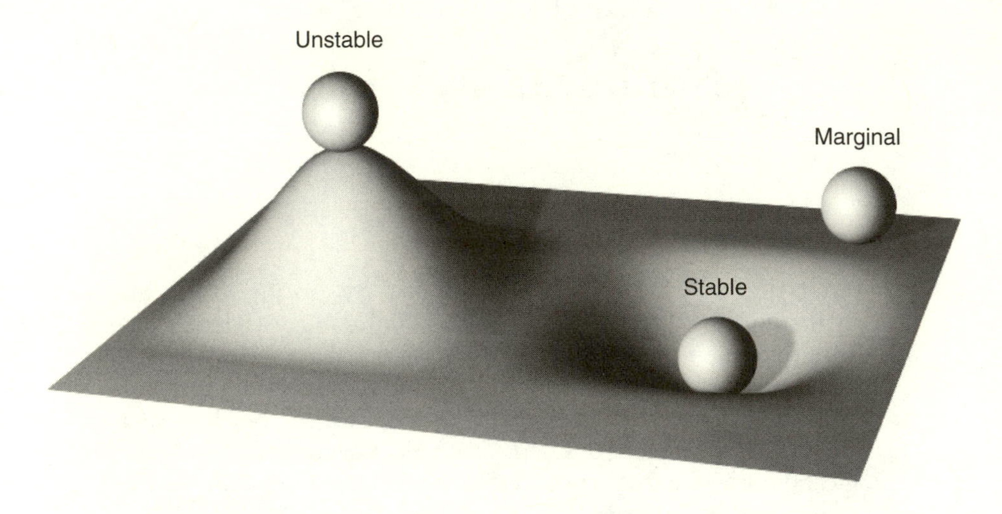

FIGURE 2.1. The three types of (linear) stability, through a simple mechanical analogy. Here three marbles are at equilibrium at three different points in the landscape.

easily solved analytically (or at all), a qualitative approach can be taken: the so-called linear stability analysis (LSA).

The first step in LSA is to identify the set π_μ of equilibrium (fixed) points. It is defined as:

$$\pi_\mu = \{x^* \mid \phi_\mu(x^*) = 0\} \tag{2.3}$$

The members of this set are those such that the derivative of the dynamic system is zero and so they are equilibrium points. If $x(0) = x^*$, it is true that $x(\infty) = x^*$. But obvious differences appear when comparing the behavior of different fixed points. The basic idea is illustrated in figure 2.1 by means of a mechanical analog. Three marbles are shown at three different locations on a landscape. The three of them are at equilibrium. However, a small displacement from their equilibrium points will modify their states in completely different ways. The *stable* equilibrium point will be recovered after some transient time (as it occurs with a pendulum). The *unstable* one will never recover its original position: the initial perturbation will be amplified and the marble will roll down and away. A third, apparently less interesting possibility is the so-called marginal state: on a totally flat surface, a marble displaced to another close position will just stay there.

The nature of the different points $x^* \in \pi_\mu$ is easily established by considering how a small perturbation will evolve in time (Strogatz, 1994). Consider $x = x^* + y$, where y is a small change (such as adding a small amount of individuals to the equilibrium population). Assuming that the one-dimensional dynamics can be described by means of a system $dx/dt = \phi_\mu(x)$, and since x^* is a constant, the dynamic of the perturbation is obtained from:

$$\frac{dy}{dt} = \phi_\mu(x^* + y) \tag{2.4}$$

Assuming that y is small, we can perform a Taylor expansion of the previous function:

$$\frac{dy}{dt} = \phi_\mu(x^*) + \left[\frac{d\phi_\mu}{dx}\right]_{x^*} y + \left[\frac{d^2\phi_\mu}{dx^2}\right]_{x^*} y^2 + \dots \tag{2.5}$$

Since $\phi_\mu(x^*) = 0$ (by definition) and if the first derivative is non-zero, we can use the linear approximation:

$$\frac{dy}{dt} = \lambda_\mu y \tag{2.6}$$

where we indicate:

$$\lambda_\mu = \left[\frac{d\phi_\mu}{dx}\right]_{x^*} \tag{2.7}$$

The linear system is exactly solvable, with an exponential solution:

$$y(t) = y(0)e^{\lambda_\mu t} \tag{2.8}$$

This solution immediately provides a criterion for stability: if $\lambda_\mu < 0$, then $y \to 0$ and thus the system returns to its *stable* equilibrium. If $\lambda_\mu > 0$, the perturbations grow and thus the point is *unstable*. Both possibilities are separated by a *marginal* state defined from $\lambda_\mu^c = 0$.

For the logistic model, we have $\pi_\mu = \{0, K\}$. It is easy to prove that $x^* = 0$ is unstable and $x^* = K$ is stable, consistently with our intuition. Since limited resources always constrain real ecologies from unbounded fluctuations, and assuming that nature works under simple laws, one may conclude that ecosystems should be predictable, perhaps even in some stable equilibrium state. However, real populations tend to display complex fluctuations, and the understanding of their nontrivial nature changes considerably our view of ecosystems.

POPULATION CYCLES

From the early years of ecology as a science, some authors have pointed out the tendency of some populations to show well-defined cycles. The Canadian lynx (*Lynx canadensis*) and its main prey, the snowshoe hare (*Lepus americanus*), are a classical example. Fur records from the Hudson Bay Company, a fur-trading company with hunters trapping in the forests of Canada, provide one of the largest temporal series in ecology (fig. 2.2a). As observed, lynx cycle with a frequency of about 10 years (Wolff, 1980). The complete web of interactions for this system is shown in figure 2.2b. Similarly, the four-year cycle of microtine rodents in boreal and arctic regions was first described by Elton (1942) and has been widely studied since then (Hanski, 1987; Hanski et al., 1991; Hanski et al., 1993; Erlinge, 1987; Turchin and Hanski, 1997; Turchin et al., 2000). These examples have provided empirical support for the predictions made by the theoretical work of Lotka and Volterra (see also next chapter), who showed the propensity of simple prey-predator models to generate cycles.

A first theoretical explanation of the origins of population cycles was provided by the periodic orbits displayed by some nonlinear population models. The consumer (C) and the resource (R) or, in other words, predators and prey populations, follow a two-dimensional dynamical system

$$\frac{dR}{dt} = \phi_1(R, C), \tag{2.9}$$

$$\frac{dC}{dt} = \phi_2(R, C). \tag{2.10}$$

We can ask under what conditions intrinsic oscillations can develop. A first step in such analysis requires, as with one-dimensional systems, the identification of the fixed points $P^* = (R^*, C^*)$. Now $P^* \in \pi_\mu$ if $\phi_1(R^*, C^*) = \phi_2(R^*, C^*) = 0$. The linear stability analysis proceeds in a qualitatively similar fashion, but now the so-called Jacobi matrix $\mathbf{L}_\mu$ needs to be calculated:

$$\mathbf{L}_\mu(P^*) = \begin{pmatrix} \dfrac{\partial \phi_1}{\partial R} & \dfrac{\partial \phi_1}{\partial C} \\ \dfrac{\partial \phi_2}{\partial R} & \dfrac{\partial \phi_2}{\partial C} \end{pmatrix}_{P^*} \tag{2.11}$$

This matrix actually contains the first derivatives of the Taylor expansion near P^* and the stability of each fixed point is evaluated from $\mathbf{L}_\mu(P^*)$ by analyzing the eigenvalues $\{\lambda_1, \lambda_2\}$ (see appendix 1). If they are both real and negative, then P^* is stable. If they are real and at least one is positive, the point will be unstable. But an additional possibility is provided by complex eigenvalues, that is, $\lambda_{1,2} = \alpha \pm \beta i$. Now the fixed point will be stable if the real part is $\alpha < 0$ and unstable for $\alpha > 0$. In this case, the trajectories inward or outward the fixed point will be spirals.

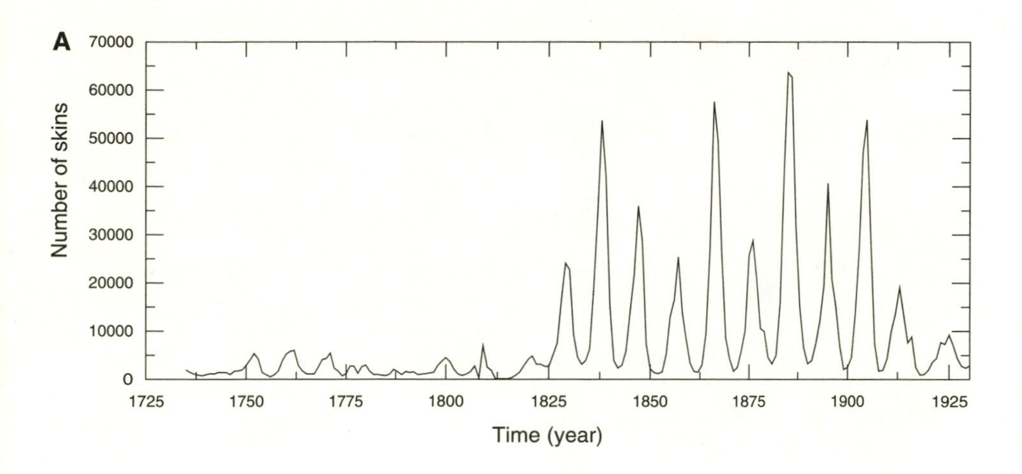

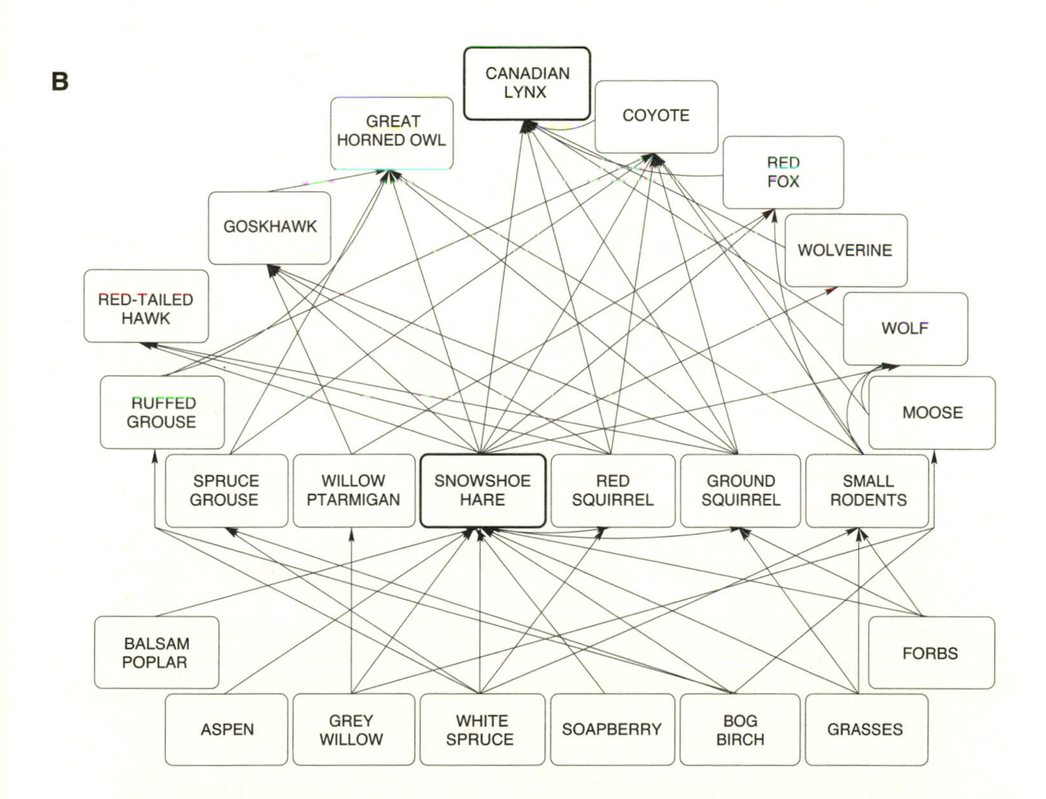

FIGURE 2.2. (A) Fluctuations in the population of Canadian *Lynx canadensis*, measured in terms of annual number of furs trapped by the Hudson Bay Company. (B) The underling trophic structure of the food web in which the lynx is embedded.

An example is given by the following predator-prey model (Roughgarden, 1979):

$$\frac{dR}{dt} = \mu R \left(1 - \frac{R}{K}\right) - \Phi(R, C)C \tag{2.12}$$

$$\frac{dC}{dt} = \beta \Phi(R, C)C - \delta C \tag{2.13}$$

where $\Phi(R, C)$ describes the explicit form of the functional response involving predator satiation when prey are very common. Here

$$\Phi(R, C) = c\left[1 - e^{-\alpha R/c}\right], \tag{2.14}$$

where c and d are two parameters. The qualitative behavior of this and related systems can be systematically analyzed through the linear analysis (May, 1973; Murray, 1989; Case, 2000). In figure 2.3 we show three different examples of the dynamics displayed by the model. In this case, the carrying capacity is used as a bifurcation parameter. Three characteristic situations are shown, involving both stable points (a, b) and periodic orbits (c, d). As noted, there are two types of representations in figure 2.3. On one hand, a temporal series where prey and predator densities are plotted through time. This is the standard representation of a dynamical system. However, sometimes it is more useful to plot the so-called phase space. A phase space is a plot in which each axis represents one variable of our system. In this case, prey and predator abundance. At each time step, the system can be defined as a point in this phase space, a particular density of prey and predator. The sequence of points defines the trajectory, and after a large enough number of iterations, the system enters its attractor, that is, the time-invariant solution reached independently of the initial condition. A system evolving to an equilibrium would have a point attractor in phase space as shown in figure 2.3a and b. A periodic system would be represented by a limit cycle, a loop that is repeated again and again (fig. 2.3d).

As discussed in Roughgarden (1979), the oscillations occur because predator satiation is a destabilizing factor. At low K, the stabilizing effect of density dependence by prey prevents them from the rapid increase necessary for escaping predator control. But once K is large enough, predation is unable to check the increase, and prey approach their carrying capacity, triggering a rapid increase in predator numbers after some characteristic delay.

The qualitative change that takes place from a stable point attractor into a limit cycle is called a Hopf bifurcation (Guckenheimer and Holmes, 1983; Strogatz, 1994; Kuztnesov, 1995). In general, qualitative changes in the dynamics are called bifurcations. The parameter

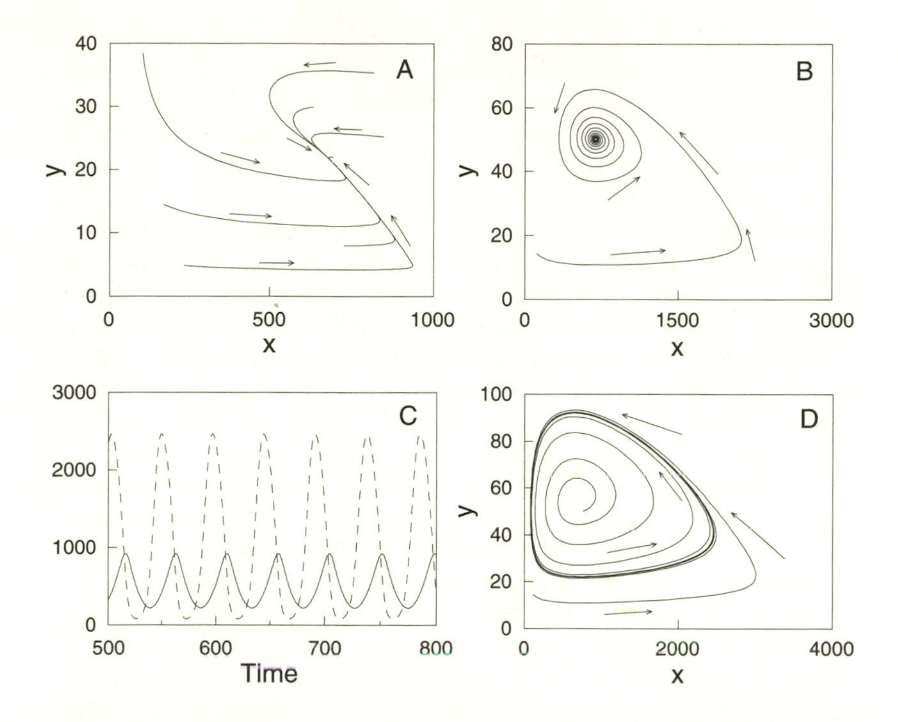

FIGURE 2.3. Nonlinear dynamics of a simple Lotka-Volterra model with predator satiation (based on Roughgarden 1979). X and Y represent resource and consumer density, respectively. Here $\mu = 0.5, \alpha = 0.01, \beta = 0.02, \delta = 0.1$, and $c = 10$. In (A) and (B) we have stable points, obtained for $K = 1000$ and $K = 2500$, respectively. In (C–D) an example of the limit cycles displayed by this model for $K = 3500$ is shown. Trajectories in the phase space are shown in (a), (b), and (d), where prey and predator density at each time step are represented in the x and y axis, respectively. In (A) and (B) the attractor is fix point, while in (D) it is a limit cycle. In (C) the temporal series of both species is shown here indicated with continuous (predator) and dashed (prey) lines. Predator populations have been rescaled ($10y$). In (D) the corresponding limit cycle is shown. The arrows indicate the movement of trajectories in the phase space. These simple models illustrate the origin of cycles in prey-predator systems.

values at which bifurcations occur are called bifurcation points. The Hopf bifurcation is only one example and there is a well-defined set of conditions under which it takes place (see, for example, Strogatz, 1994). Several types of bifurcations will be discussed in this book, and they play an important role in understanding biocomplexity. The presence of bifurcation points provides one of the clearest illustrations of nonlinearity: Small changes in key (bifurcation) parameters can lead to totally different types of systems behaviors.

There has been a long tradition of studying the population cycles of small mammals, since they may provide evidence for density-dependence and population regulation. Despite this tradition, the question about the relative importance of density-dependent regulation *versus* density-independent (e.g., climatic fluctuations) factors in shaping population variation is still open (Hassell, 1985; Turchin, 1990). Several factors have been suggested as responsible for the cycles, as food supply, dispersal, genetic mechanisms, predators, parasites, and diseases (see Krebs et al., 1973 for a review). From all these suggested causes, the most often quoted are interaction with food or interaction with predators (bottom-up or top-down mechanism). A very different kind of hypothesis is that of the maternal effect, which has been adduced to explain population cycles of forest lepidoptera (Ginzburg and Taneyhill, 1994). The key assumption is that delayed density dependence is caused by transmission of quality through different generations via the maternal effect. Low-dimensional, difference models relating averaging quality of individuals to patterns of abundance produce cycles that in some cases can give better fit to real data than other types of models. Furthermore, these models predict the period of the oscillations of the species studied (Ginzburg and Taneyhill, 1994).

For the case of the snowshoe hare, Sinclair et al. (1988) gives support to the hypothesis of a time lag in the interaction with its predator. Keith (1983), on the other hand, suggests that the origin of the cycle is a delayed response of the hares to available food. Often however, different mechanisms are not exclusive, and the reason for the cycles could be the synergetic effect of various mechanisms. This is the hypothesis defined by Wolff (1980) and Krebs et al. (1995). The last result is interesting in relation to the present book, for it suggests a criticism to the widely used practice in ecology of disentangling the different processes behind an observed pattern by breaking them down into the components, and studying these elements in isolation. If there is a nonlinear, nonadditive effect between different factors, a pattern can only be understood under an integrated perspective.

In this context, Peter Turchin and colleagues devised a way to test for the contributions of the above two causes for cycles. Turchin et al. (2000) ask themselves whether it is possible to tell from the lemming time series if they act as the resource or as the consumer. To do this, they use theoretical predictions according to which the peaks in the resource cycles will be rounded, while they will be sharp in the consumer cycles (fig. 2.4). By applying statistical tests to check this idea on time series of both voles and lemmings, they conclude that different causal mechanisms explain the cycles of voles and lemmings. Lemmings show very sharp peaks, which suggests that they act as functional predators,

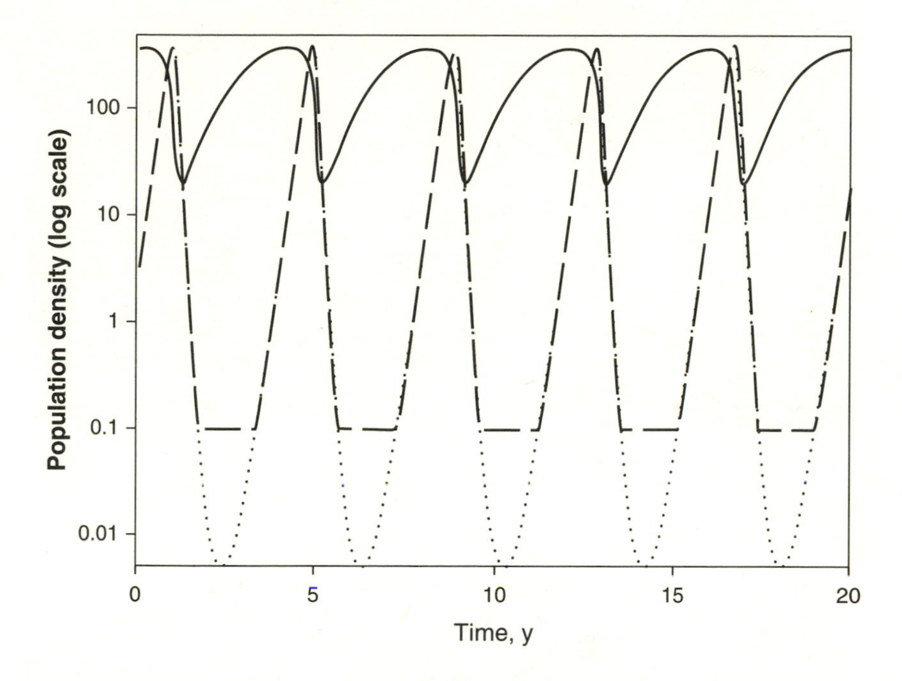

FIGURE 2.4. Population cycles of a resource and a consumer species, according to a general prey-predator model (from Turchin et al., 2000). Solid and broken curves show prey and predator density, respectively. As noted, peaks are rounded for the resource species, while they are sharp for the consumer species. Reproduced with permission of *Nature*. Figure courtesy P. Turchin.

that is, their cycles are driven by the interaction with the plants they eat. On the other hand, voles show rounded peaks. This suggests that they act as resources. Their cycles are driven by interactions with their predators (Turchin et al., 2000; see fig. 2.5).

In summary, the issue of population cycles is both a key and an open question. The reason is that as with many other examples, it is almost impossible to perform clear experiments in the field due to the large spatial and temporal scales involved. In this case, simulations can provide an opportunity to "test" the plausibility of different mechanisms. Since a lot of biological information exists for the classical examples, this information can be introduced into an individual-based simulation (IBS). There has been an increase in the use of IBS in ecology during the last years, due to both the increase in computational power and to the recognition that local interactions and individuality are important

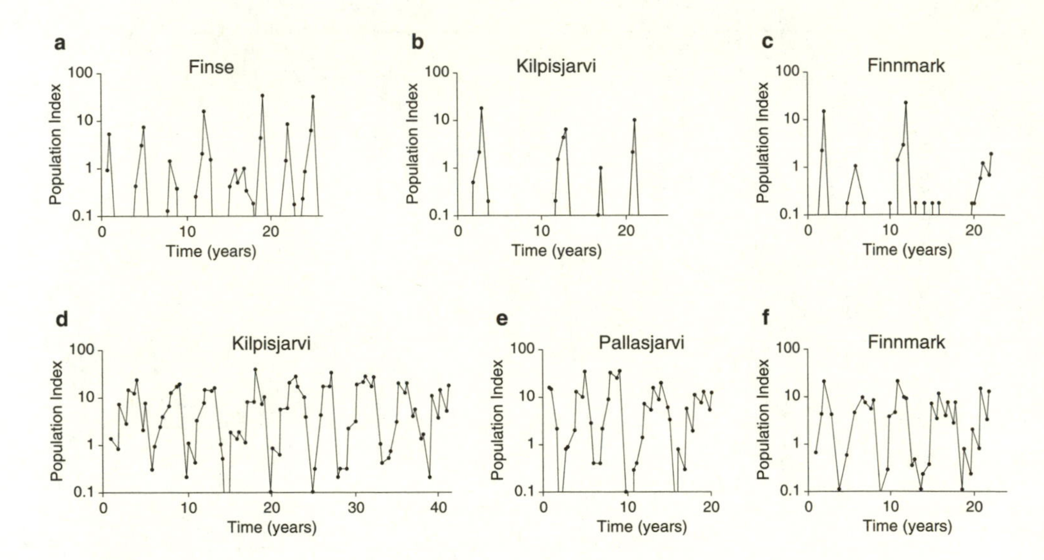

FIGURE 2.5. Population cycles for lemmings (top) and voles (bottom) at different localities. Voles show rounded peaks, which (in comparison to prey-predator models such as the one plotted in the previous figure) suggests that their cycles are driven by interactions with their predators. Lemmings show sharp peaks, suggesting that their cycles are driven by vegetation-consumer interactions. (a) Lemmings at Finse; (b) lemmings at Kilpisjarvi; (c) lemmings at Finmark; (d) voles at Kilpisjarvi; (e) voles at Pallasjarvi; (f) voles at Finnmark. Based on Turchin et al. (2000). Reproduced with permission of *Nature*. Figure courtesy P. Turchin.

(Caswell and Jonhs, 1992; De Angelis and Gross, 1992; Metz and de Roos, 1992; Solé and Valls, 1992; Wilson et al., 1993; Judson, 1994).

IBS treats each individual as unique, and its behavior is followed through a set of probabilistic rules. That is, the individual has a probability of moving, of finding food, of encountering a predator, of mating, et cetera. Population-level properties, such as population cycles, are then an emergent property of the sum of behaviors of lower-level entities (individuals). This contrasts with analytical models in which the level of description is the population, that is, an equation describing population-density changes through time. For cases where biological information is available, and the goal is to provide answers related, for example, to the conservation of an endangered species, IBS is a useful approach. For the case of the snowshoe hare, Bascompte et al. (1997) tested the plausibility of several hypothesis to explain the ten-year cycle.

Since the beginning of population studies, it was clear that some populations fluctuate in an erratic way. This was assumed to reflect stochastic variations in environmental variables. Density-dependent population regulation could show either steady states or cycles and should change smoothly under smooth external changes. And that was all, or not?

CATASTROPHES AND BREAKPOINTS

An especially important feature of most complex systems is the presence of multiple attractors. Depending on the initial condition, the system will flow toward different final stationary states. Such multiplicity of states arises in simple, low-dimensional systems and has great consequences in ecology (May, 1977; Scheffer et al., 2001). A remarkable feature of this multiplicity is that, under some conditions, as a given parameter smoothly changes, an abrupt transition in a system's state can occur. The best-known example of this behavior is offered by the outbreaks of spruce budworm (*Choristoneura fumiferana*) populations. This is one of the most destructive native insects in the northern spruce and fir forests of the eastern United States and Canada. The spruce budworm consumes the leaves of coniferous trees and is eaten primarily by birds. The outbreaks occur approximately every forty years and lead to devastation of the pineries. The loss of timber is an economical disaster for the industry.

Other ecological systems seem to exhibit a similar two-phase behavior on different time scales, together with sudden shifts between the two alternative states. This situation has been particularly well documented from the analysis of lake ecosystems (Carpenter and Kitchell, 1996). The external event that triggers the shift can be climate, nutrient input, changes in water availability, habitat fragmentation (see chapter 5), or harvesting. Beyond the diverse nature of the potential events that can trigger the shifts, all these systems share some common features related to the presence of several types of nonlinearities.

The nonlinear dynamics exhibited by a prey-predator system such as the spruce budworm can be approximately described by:

$$\frac{dN}{dt} = \mu N \left(1 - \frac{N}{K}\right) - \Phi(N), \tag{2.15}$$

where $\Phi(N)$ introduces the effect of predation. Here a so-called type-III functional response describing how predators consume changes with prey availability was used:

$$\Phi(N) = \frac{\beta P N^2}{\alpha^2 + N^2}, \tag{2.16}$$

where β and α are two parameters and P represents a constant pool of predator birds. The functional form of $\Phi(N)$ introduces satiation in predator response. It can be shown that $\Phi(N)$ is very small at small N but rapidly increases as $N > \alpha$, with a saturation $\Phi(N) \to \beta P$ at large N. A similar equation results when vegetation and a constant population of grazers are considered.

This model exhibits sudden population changes as the parameters are tuned. In order to see how the multiple attractors appear, and their transitions, let us further simplify the model to:

$$\frac{dx}{dt} = x(1-x) - \frac{\gamma x^2}{\alpha^2 + x^2} \tag{2.17}$$

by using the parameter transformations $x = N/K, \gamma = \beta P/rK$ and a time rescaling $t \to \mu t$. The set of nontrivial fixed points is:

$$\pi_\mu = \left\{ x \mid x(1-x) = \frac{\gamma x}{\alpha^2 + x^2} \right\}. \tag{2.18}$$

We can graphically illustrate the presence of this multiplicity of fixed points by plotting together the two contributions to dN/dt (logistic growth and predation), as shown in figure 2.6a. Here, three different levels of predation are considered. They are represented by the dashed lines, corresponding to high ($\gamma = 0.35$), intermediate ($\gamma = 0.22$), and low ($\gamma = 0.10$), following May's (1977) analysis. High and low predation rates provide two very different steady values of N. But at intermediate levels three possible states are present (here indicated by arrows). It can be shown that two of them are stable and the middle one unstable.

We can see that, by increasing γ (predation rate) and assuming that the system starts at a high density (large N), the steady state will initially smoothly decrease with P. But eventually this state disappears and is replaced by the low-density state through a sudden jump. This is known as a *catastrophe* (Zeeman, 1976; Poston and Stewart, 1977).

The underlying topology responsible for the shifts is schematically shown in figure 2.6b. This surface corresponds to the set of fixed points $x^* \in \pi_\mu$ close to the three-states domain. Here two arbitrary parameters, indicated as (P_1, P_2) define the parameter space. As one of them (P_1) is smoothly increased, the system's state (here indicated as a marble) moves from left to right. The presence of a fold in the surface (a) is responsible for the jump to a new state (b).

An important aspect of the catastrophic shifts is a key feature of complex systems: the presence of *hysteresis*. This property is illustrated in figure 2.6b. When P_1 returns (again smoothly) to its previous value, the system's state now moves on the lower branch of the surface ($b \to c$) until a new switch point is reached (c) and the system moves to the upper branch. In other words, the system now returns to its original

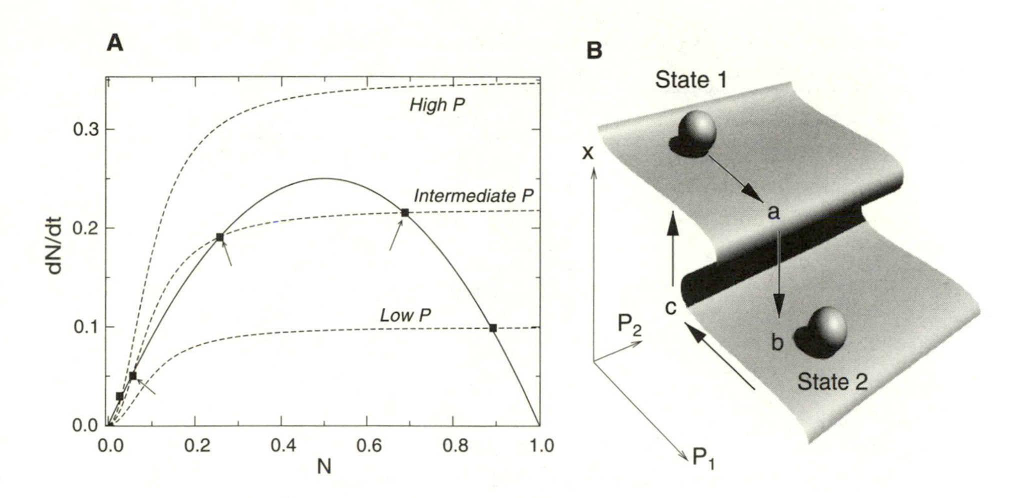

FIGURE 2.6. Breakpoints and catastrophes. (A) Multiple attractors in the spruce (normalized) budworm model for $\alpha = 0.1$. Here the rate of population change is shown as a function of N in terms of the logistic curve (continuous line) and the predation term (dashed lines). Three levels of predation (high, intermediate, and low, corresponding to $\gamma = 0.35, 0.22, 0.10$) are indicated. In (B) the basic idea of the catastrophe point is illustrated by means of a mechanical analog. The surface indicates the set of stationary states (fixed points) available for each parameter combination (P_1, P_2). As one of the parameters is smoothly changed (here P_1), the state changes smoothly until the folded region is reached (a): Now a sudden transition (a catastrophe) takes place and a totally different state (b) is attained. If the parameter moves back to its original value, a new shift will occur at a *different* point (c). In other words, the same initial conditions are recovered after a different path has been followed.

state, but through a different pathway: the previous history of the system is thus a relevant aspect in its dynamics. The degree of hysteresis is variable from one ecosystem to another (Scheffer et al., 2001). As an example, shallow lakes can have a pronounced hysteresis in response to nutrient load.

Systems exhibiting two alternative states have been identified in a wide range of situations. Shallow lakes are particularly well studied (Carpenter and Kitchell, 1996), and rapid shifts are actually well known to occur under human-induced eutrophication. The two states here are: (a) clear water with rich, submerged vegetation and (b) a turbid state with largely absent vegetation and high phytoplankton density. Although we could conjecture that the original, clear-water state can be recovered by decreasing nutrient concentrations to the previous values

TABLE 2.1. Examples of Systems Exhibiting Alternative States, Together with Examples of Events That Can Trigger Shifts Among Them

SYSTEM	STATE I	STATE II	I → II	II → I
Lakes	Clear water	Turbid	Plant decline	Fish decline
Coral reefs	Corals	Macroalgae	Sea urchin decline	Unknown
Woodlands	Herbaceous cover	Woodlands	Fires/Tree cutting	Grazer decline
Deserts	Perennials	Bare soil	Climate change	Climate change

Based on Scheffer et al., 2001. Here "decline" labels different sources of population decline, from natural to human based.

before the shift, it actually occurs that substantially lower nutrient levels are required in order to return to the original situation. Dedicated whole-lake experiments have shown that temporary reductions in fish populations can bring the lake to the clear state. The reason is that fishes control *Daphnia* populations, which are key phytoplankton grazers (Carpenter and Kitchell, 1996). An appropriate reduction of fish stocks triggers reduction of phytoplankton biomass and turbidity, which enhances vegetation recovery. Vegetation increases water clarity, thereby enhancing its own growing conditions.

Other examples of multiple states are indicated in table 2.1, together with the nature of the underlying shifts. In spite of their widely different features, they share obvious similarities. First, the contrast among the alternative states seems to be due to the dominance, at each phase, of species involving different life forms. Second, shifts are usually triggered by stochastic events (such as pathogens, fires, or climate extremes). Finally, feedbacks that stabilize different states involve both biotic and nonbiotic mechanisms.

It is worth mentioning that these shifts seem to be operating in large-scale climate changes. It has been shown that the (nonlinear) coupling of atmosphere and ocean can lead to sudden shifts between alternative states, with a rapid drop of temperatures over a short time scale (of about one decade). This observation is based on models of ocean circulation (Rahmstorf, 1996) but is also supported by extensive analysis of paleoclimatic data (Taylor, 1999). The current, steady global warming could increase the inflow of freshwater into the North Atlantic, which is required to power the oceanic current that transports warm water to western Europe and eastern North America. Data from ice cores and ocean sediments indicate that 11,650 years ago the climate in Greenland switched from ice-age conditions to the current relatively warm conditions (a warming of 5 to 10 degrees Celsius on average) in only forty years.

DETERMINISTIC CHAOS

In the early 1970s, the theoretician Robert M. May made an outstanding contribution with implications far beyond ecology (May, 1974, 1976). He studied simple, discrete-time, single-species models such as the logistic equation, a discrete-time version of model (2.1):

$$x_{t+1} = F_\mu(x_t) = \mu x_t(1 - x_t), \tag{2.19}$$

where x_t represents population size at generation t, and μ is the growth rate. The variable x is normalized, that is, we have scaled the variable by dividing population size by the maximum value. As opposed to continuous-time models such as (2.1), discrete-time models are used to describe populations with discrete, non-overlapping generations such as arthropods. The previous equation is probably the simplest model we can write to describe the dynamics of an ideal population with density-dependence. Note that it is a deterministic model. There are no random terms. If we know the population value at one generation, we know the whole story of the system, we can iterate and estimate population values at successive generations. So we should not expect too much complexity from such a simple model. If $\mu \leq 1$, the population does not grow fast enough, and it goes extinct after a few generations. If $\mu > 1$ there is a positive solution given by $x^* = 1 - 1/\mu$. Is this solution stable? To study the stability of a solution, in analogy what we did for the continuous models, we have to see how small deviations (ξ) from the solution (x^*) evolve through time as we have already illustrated for continuous-time models. The solution will be stable if small deviations decrease, and unstable otherwise. To check this we expand model (2.19) around the equilibrium through Taylor expansions. Thus, if ξ_t is a small perturbation near x^* (i.e., $\xi_t = x_t - x^*$), by using the definition of equilibrium, we have:

$$\xi_{t+1} = x_{t+1} - x^* = F_\mu(x_t) - x^*. \tag{2.20}$$

We can Taylor-expand F_μ around the equilibrium value (x^*) as follows:

$$\xi_{t+1} = F_\mu(x^*) + \frac{\partial F_\mu(x^*)}{\partial x^*}(x_t - x^*) - x^* = \frac{\partial F_\mu(x^*)}{\partial x^*}(x_t - x^*). \tag{2.21}$$

Now we have a linear equation describing the dynamics of the perturbation, that is, how the distance to the equilibrium $\xi_t = x_t - x^*$ changes through time:

$$\xi_{t+1} = \frac{\partial F_\mu(x^*)}{\partial x^*}\xi_t \equiv \lambda \xi_t. \tag{2.22}$$

As easily seen from the previous equation, the perturbation will grow ($\xi_{t+1} > \xi_t$) if $|\lambda| > 1$. Note that for continuous-time models, the perturbation will grow if $\lambda > 0$ as illustrated at the beginning of this chapter. Thus, we have a criterion for studying the stability of solutions. Following the example of the logistic equation (2.19), we can say that the equilibrium point ($x^* = 1 - 1/\mu$) will be stable if:

$$\left| \frac{\partial F_\mu(x^*)}{\partial x^*} \right| = |2 - \mu| < 1, \tag{2.23}$$

by means of which, the solution will be stable in the domain $\mu \in (1, 3)$.

Let us consider again the dynamics of the population described by model (2.19). The key point here is that such dynamics are very dependent on μ. We have already said that if the growth rate is lower than one, the population will go extinct. Let us consider a larger μ-value (for example $\mu = 2.95$), and imagine we start from an initial condition $x_0 = 0.25$. That is, the initial population size is a fourth of its maximum value. We can then iterate model (2.19) and plot the temporal series (fig. 2.7a). As can be seen, population density grows through time until it reaches an equilibrium value of $x^* = 0.661$. The iterations before arriving at the time-invariant value are called transients (this concept, so innocuous here, will become of tremendous importance in the next chapter when we consider spatially extended systems). If we had started from different initial conditions, transients would be different but would ultimately drive the population to the same steady state. So far, no surprises; a simple model generates simple dynamics: The population evolves toward a constant value, a steady state. From then on, nothing new happens. But what happens if we try a largest value for μ? Imagine that our ideal population grows faster; let us say that μ is now 3.2. Now we can see that the population does not follow into a time-invariant density, but oscillates in time between two values. One year the population has a low value, the next year a high value, and so on (fig. 2.7b). We have encountered cycles, in this case of period two. What happens is that as we increase continuously μ, we encounter a critical value μ_{c_1} at which a bifurcation takes place. At such a critical point, the previously stable solution x^* loses its stability. We have already shown that the equilibrium solution is stable in the domain (1,3). For $\mu > 3$, the solution becomes unstable, and a small perturbation will drive the orbit far from x^*. The new stable solution is in fact a period-two cycle. In the manner we have performed for the equilibrium solution, it can be shown that there is a domain of stability for this period-two orbit.

Suppose we increase again μ, for example we set it to a value of 3.45. Now the solution is a period-four cycle, the population repeats its value every four years. By further increasing μ we found additional

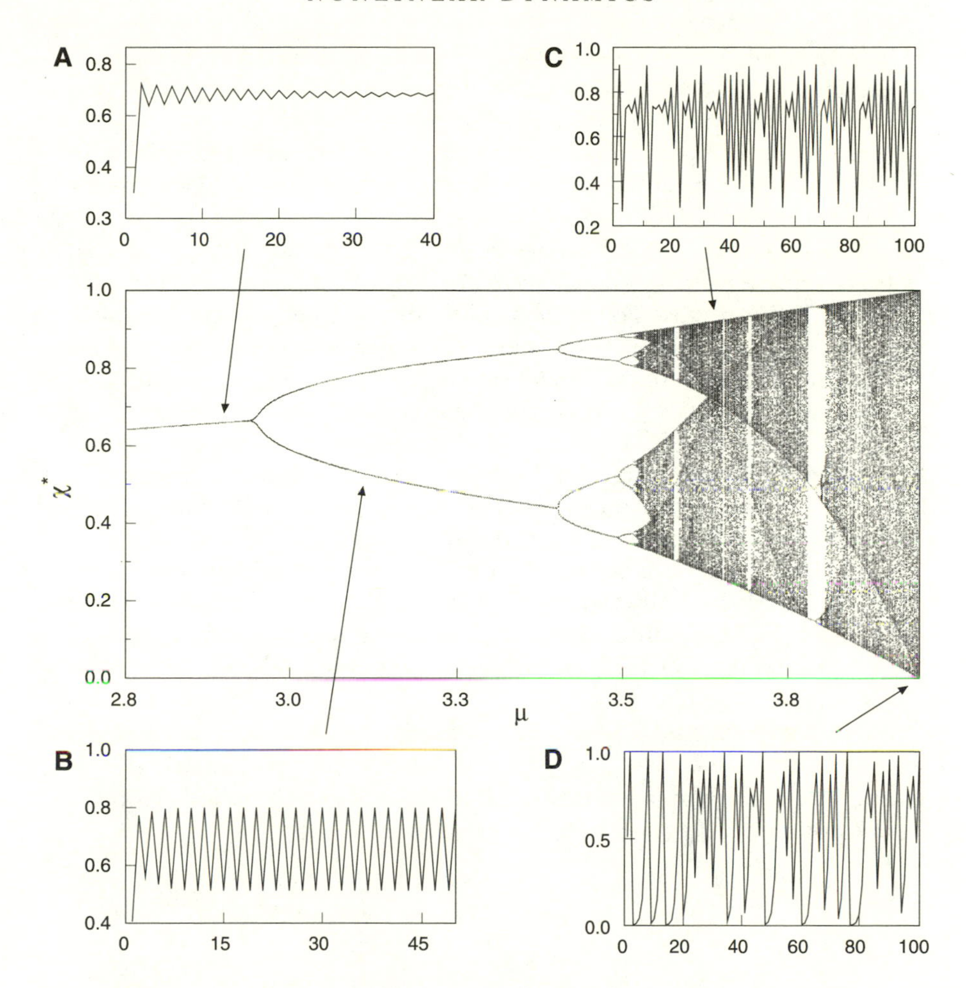

FIGURE 2.7. Chaos in the logistic map. Here four examples of the time series are shown, for (A) $\mu = 2.95$ (point attractor); (B) $\mu = 3.2$ (periodic orbit) and two chaotic orbits, for (C) $\mu = 3.7$, and (D) $\mu = 4.0$. The central picture shows the bifurcation diagram.

duplications of period. The dynamic becomes a bit more complicated but remains in a constant cycle after all. Until here nothing is new. Discrete-time models can generate steady states and cycles as illustrated in previous sections for continuous-time models. But beyond a critical μ, May (1974, 1976) found a completely different and unexpected kind of dynamics. Here figures 2.7c and 2.7d correspond to $\mu = 3.7$ and $\mu = 4.0$, respectively. Now no order seems to be present. The population oscillates in a seemingly random way, without periodicity.

The dynamic never repeats itself no matter for how long we iterate the model. This kind of dynamic was called deterministic chaos (the word "chaos" was coined as a mathematical term by Tien Yien Li and James A. Yorke [1975] in their classic paper "Period Three Implies Chaos"). The concept of deterministic chaos has deep implications for our understanding of complex time series.

The suite of dynamical behaviors shown by model (2.19) can be seen from a single picture called a bifurcation diagram (fig. 2.7); μ is represented in the x-axis. For each μ-value, the long-term dynamics (after transients are discarded) is plotted. Thus, if $\mu = 2.95$ as in figure 2.7a, after iterating the model for, let us say 200 iterations, we plot the following 100 iterations. As this μ-value corresponds to a steady state ($x^* = 0.661$), we will be plotting this single point 100 times. For the μ-values corresponding to a 2-period cycle we will plot a high and a low value, and so on. For the chaotic region, since the dynamic is aperiodic, we will be plotting a set of different points. This plot is called a bifurcation diagram because by tuning a relevant parameter (the so-called bifurcation parameter, μ) we generate a period doubling route to chaos. A very similar diagram is obtained using different types of one-hump maps, such as the so-called Ricker map, which is defined as follows:

$$x_{t+1} = x_t e^{\mu(1-x_t)}, \tag{2.24}$$

and which also displays a transition to chaos through period-doubling bifurcations.

EVIDENCE OF BIFURCATIONS IN NATURE

Bifurcations in the Lab: The Case of Tribolium

One key concept from the last section is the existence of bifurcation scenarios leading to chaos. Progressive increases in a given parameter promote successive bifurcations eventually leading to irregular, but deterministic, behavior. If there is some relationship between the bifurcation parameter and any environmental variable, then the environment can force populations to behave in different ways. For example, external forcing (input of energy) in the system may act as a bifurcation parameter. When its value is small, the population will stay on a steady state, but when the external forcing is larger, the population will show cycles and chaos (Kuramoto, 1984; Kot et al., 1992; Solé and Valls, 1992).

Similarly, the individual-based simulation described above for the snowshoe hares identified key variables that acted as bifurcation pa-

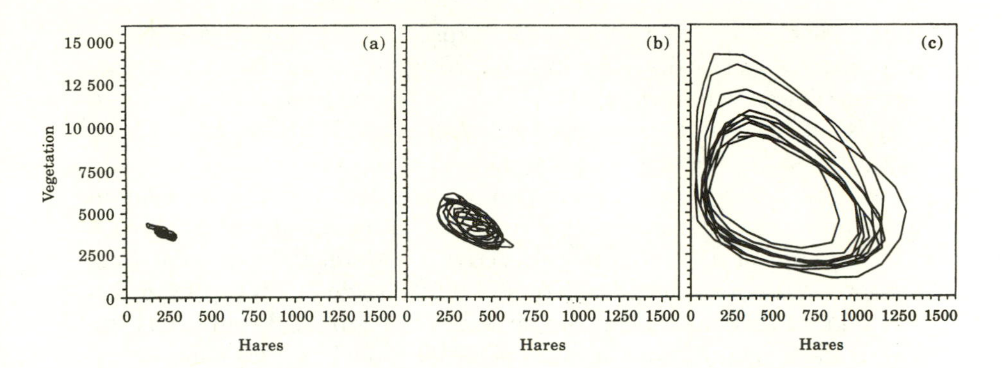

FIGURE 2.8. Bifurcations in a hare-vegetation Individual Based Simulation. Each plot represents a phase space, with hare density on the x-axis and vegetation density on the y-axis. As the maximum amount of vegetation is increased, dynamics shifts from a noisy steady state to a noisy limit cycle. Modified from Bascompte, et al., 1997.

rameters. One example is the maximum vegetation that can be accumulated at a spatial point. This variable has deep implications for the dynamics of hares. As the maximum vegetation is increased, the dynamic shifts from a steady state to limit cycles (fig. 2.8). Since maximum vegetation can simulate the addition of extra food, different hypothesis concerning the origin and maintenance of cycles can be tested by means of this approximation. One can carry out "experiments" that would be much more complicated to perform in nature.

However, the best example of the period-doubling route to chaos in population dynamics is provided by the flour beetle *Tribolium* growing in the lab. Costantino et al. (1995, 1997) used a smart design to experimentally induce transitions in the population dynamics of *Tribolium* growing in milk bottles. The basis of their contribution is the interaction between a modeling and an experimental approach. This interaction is particularly successful in the present case due to the knowledge of the biology of the flour beetle. Cannibalism is frequent in this species, which introduces a strong nonlinear density-dependent mechanism.

If L_t, P_t, and A_t represent the number of feeding larvae, nonfeeding larvae, and callow adults, respectively, at time t, the following system of difference equations can be written (Costantino et al., 1995):

$$L_{t+1} = bA_t \, exp \, (-c_{ea}A_t - c_{el}L_t), \tag{2.25}$$

$$P_{t+1} = L_t(1 - \mu_1), \tag{2.26}$$

$$A_{t+1} = P_t \, exp \, (-c_{pa}A_t) + A_t(1 - \mu_a), \tag{2.27}$$

where $b > 0$ is the number of larval recruits per adult per unit of time in the absence of cannibalism; μ_1 and μ_a are the larval and adult probabilities, respectively, of dying from other causes besides cannibalism. Here, cannibalistic interactions are introduced by means of the exponential terms. Thus, $exp\ (-c_{ea}A_t)$ and $exp\ (-c_{el}L_t)$ are the probabilities than an egg is not eaten in the presence of A_t adults and L_t larvae. In a similar way, $exp(-c_{pa}A_t)$ is the probability than a pupa survives in the presence of A_t adults. The unit of time is two weeks, which is the feeding larval maturation interval; after a time unit a larvae either dies or pupates.

Costantino et al. (1995) first proceeded by finding the maximum likelihood estimates of the parameters in model (2.25–2.27) by fitting this model into their data from the laboratory. Figure 2.9 represents the bifurcation diagram for the parametrized model as a function of c_{pa}, the attack rate of pupae by adults. Next, Costantino and colleagues repeated the previous procedure with a stochastic version of model (2.25–2.27), where noise was added on a logarithmic scale. Once parameters were estimated by least squares techniques, they fixed adult mortality and manipulated recruitment rate into the adult stage ($exp[-c_{pa}A_t]$) by adding or removing young adults at the time of a census at the end of each time step (two weeks). Their results are conclusive. The observed dynamic for each c_{pa}-value is consistent with the dynamic predicted by the model with the corresponding c_{pa}-value. The dynamic shifts from a steady state, to a limit cycle, and to chaos as c_{pa} is increased. The nonlinear deterministic behavior was clearly observed despite the unavoidable superimposed experimental noise. One cannot "prove" the presence of chaos from such short time series, but one can determine chaos in the model for the parameters estimated from the lab populations. What it is more important, the observed time series is consistent with the model output for each value of adult mortality. A similar conclusion is obtained by looking at the reconstructed attractors in the adult-larvae phase. As Costantino et al. (1995) concluded, "this rigorous verification of the predicted shifts in dynamical behavior provides convincing evidence for the relevance of nonlinear mathematics in population biology."

One important message for biological control and management of natural populations arises. Human control (e.g., changing a recruitment rate or a death rate) can have unexpected results by suddenly changing the dynamics of the population under consideration. This may be the case for the Canadian lynx once hunting pressure crossed a given threshold (Gamarra and Solé, 2001), as illustrated in the next section.

Bifurcations in the Field: Lynx Returns Revisited

One might argue that the previous results are due to the special features displayed by systems with intrinsic discrete dynamics. Most of our

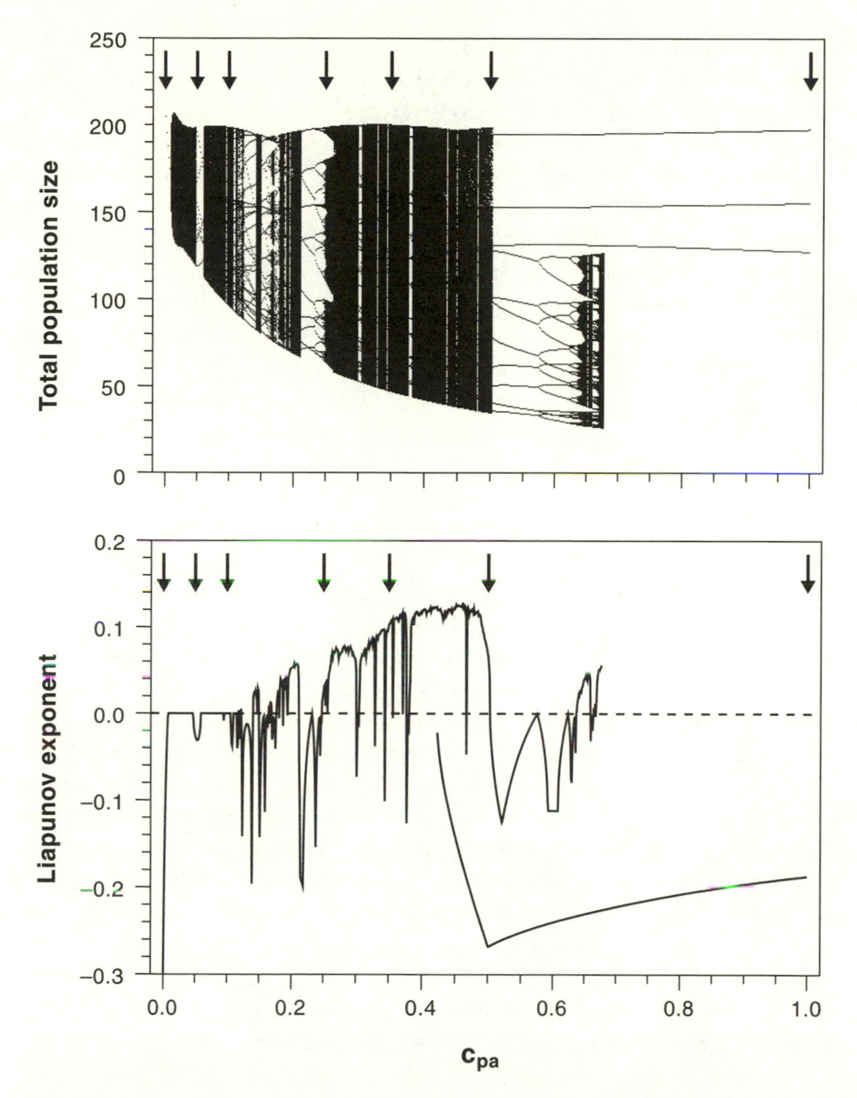

FIGURE 2.9. Top: Bifurcation diagram for Tribolium dynamics model, using parameter values based on experimental data of the flour beetle *Tribolium*. The bifurcation parameter is adult mortality, which is experimentally manipulated. Bottom: Lyapunov exponent of the same system. Arrows indicate the experimental treatments. Based on Costantino et al. (1997). Reprinted with permission from *Science*. Copyright (1997) AAAS. Figure courtesy R. Desharnais.

daily experience deals with continuous time, and many populations have overlapped generations. Is chaos also likely to occur in continuous population models? Examples rapidly accumulated for systems involving three interacting species. For continuous, autonomous dynamical systems, no chaos can occur below dimension three. But once three variables are at work, the topological constraints imposed by lower dimensions vanish (Guckenheimer and Holmes, 1983; Nicolis, 1995).

Consider the following three-trophic level model, which was proposed as a model of plankton dynamics (Scheffer, 1991). The three species correspond to planktonic algae (A), herbivorous zooplankton (Z), and carnivorous zooplankton (C). The model is given by the set:

$$\frac{dA}{dt} = \mu A - cA^2 - \gamma \frac{A}{A + h_a} Z, \tag{2.28}$$

$$\frac{dZ}{dt} = \epsilon\gamma \frac{A}{A + h_a} Z - \delta_z Z - \rho \frac{Z}{Z + h_z} C, \tag{2.29}$$

$$\frac{dC}{dt} = \epsilon\rho \frac{Z}{Z + h_z} C - \delta_c C, \tag{2.30}$$

where a type-II functional response has been used. All the parameters can be roughly estimated (Scheffer, 1991). By tuning the algal growth rate μ (while maintaining all other parameters fixed), a sequence of bifurcations is observable, as shown in figure 2.10a–c. As different critical μ-values are crossed, new qualitative behaviors appear, defining a period-doubling bifurcation scenario, eventually leading to chaos. Although the sequence of bifurcations is somewhat analogous to the one displayed by the logistic map, the similarity becomes obvious once the appropriate bifurcation scenario is plotted. One way of doing this is to take one of the variables (say C) and just keep the maxima through time. This procedure gives a discrete sequence $M_\mu = \{C_1, C_2, \ldots, C_n, C_{n+1}, \ldots\}$ that is then used exactly as with the discrete map (fig 2.10g). A period-doubling bifurcation scenario is obtained, with all the features displayed by one-dimensional maps with a single maximum (Feigenbaum, 1978). Of course another question immediately emerges: how can we determine if the observed fluctuations are chaotic? Can chaos be distinguished from noise? Are bifurcations observable in the field? The first two questions will be answered in later sections. The last might have been shown to have an affirmative answer from the study of the lynx time series (Gamarra and Solé, 2001).

Detecting a bifurcation in a nonexperimental data set of this type might be a difficult task. First, a long time series is needed. Second, some known change in a key parameter must take place in order for the bifurcation to occur. Although most researchers have studied only the high-amplitude phase of the time series (from 1821 to 1909) it has actually three well-defined phases, as we can see in figure 2.2a. The first is a

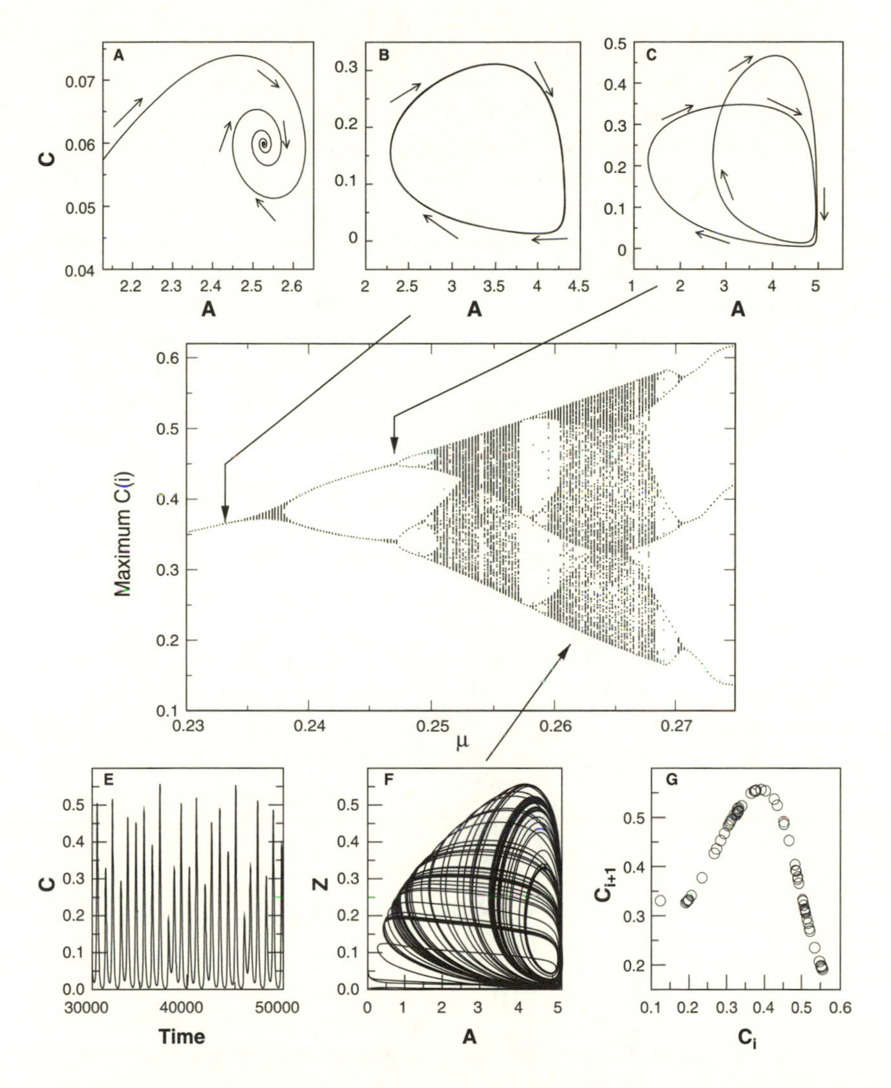

FIGURE 2.10. Bifurcations and chaos in a model of plankton dynamics (Scheffer, 1991). The central picture shows the bifurcation scenario obtained for increasing levels of the algal growth rate μ. For each μ, the maxima attained by the population of carnivorous zooplankton (C) are shown. Three examples of the different attractors are indicated in (A–C) involving fixed points, limit cycles, and a period-two orbit. In (E–F) two views of a chaotic dynamical pattern are shown. When the corresponding set of maxima $\{C_i\}$ for this pattern is plotted in a return diagram (C_i, C_{i+1}), a nearly one-dimensional shape is recovered (G), consistent with the Feigenbaum scenario displayed by simple discrete maps. Variables A, Z, and C represent abundances of algae, herbivorous zooplankton, and carnivorous zooplankton, respectively.

low-amplitude phase, starting in 1735. In this phase, the same periodic component is observable. The third phase, starting in 1910 reveals a decline in the amplitude of the fluctuations. One possible explanation, suggested by Schaffer, is the presence of a bifurcation phenomenon. Since fur trade increased through time, it might have happened that some critical threshold in hunting (and thus predator death rate) was reached in the lynx-hare-vegetation system. Such a threshold might have generated a bifurcation from a periodic, low-amplitude dynamic to a chaotic, strange attractor (Schaffer, 1984). However, this explanation goes against our intuition. Since increased hunting means increased predator rates, we should expect a *decrease* in the amplitude of the fluctuations. This is actually what occurs, for example, in the previous model of plankton dynamics. But again, our intuition was wrong. Gamarra and Solé (2001) analyzed the available evidence for an external change in hunting pressure, looking for a rapid shift near the end of the first phase. The evidence was found: The number of trading posts increased exponentially at the beginning of the nineteenth century. Actually, the number of employees from the Hudson Bay company peaked near 1820 and remained high afterward (fig. 2.11b). This is an obvious source of parameter change, but needs to be validated by an appropriate model. In order to test the conjecture that increased hunting generated a bifurcation in the system, a simple model (Blasius et al., 1999) was used. The following set of equations describes the dynamic:

$$\frac{dv}{dt} = av - \alpha_1 f_1(v,h), \tag{2.31}$$

$$\frac{dh}{dt} = -bh + \alpha_1 f_1(v,h) - \alpha_2 f_2(h,l), \tag{2.32}$$

$$\frac{dl}{dt} = -c(l - l^*) + \alpha_2 f_2(h,l), \tag{2.33}$$

where v, h, and l indicate populations of vegetation, hares, and lynx, respectively. Here $f_1(v,h) = vh/(1 + k_1 v)$ (Holling type II) and $f_2(h,l) = hl$ (Lotka-Volterra linear response). One ingredient in this model is new: the presence of the parameter l^* describing the minimum lynx population that triggers a switch toward alternative preys in the lynx diet (fig. 2.2b). This is an elegant way of introducing a realistic mechanism to avoid extinction under a highly fluctuating dynamic. The model displays the same period-doubling bifurcations shown in Scheffer's model. Actually, as death rate is increased, the system shows successive bifurcations and increasing amplitudes in population fluctuations. An example of the strange attractors obtained in the chaotic domain is shown in figure 2.11a.

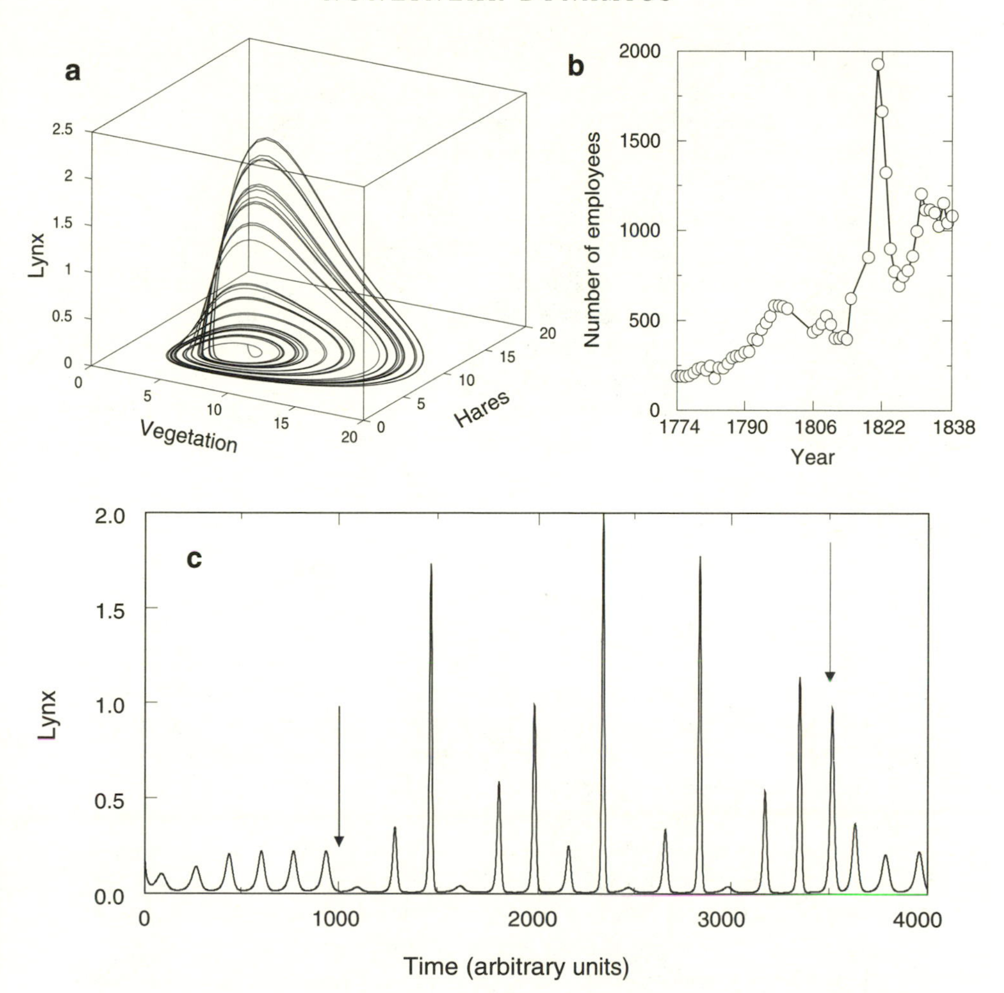

FIGURE 2.11. (a) Strange attractor in the lynx-hare-vegetation model. Here $a = b = 1, c = 10, \alpha_1 = 0.2, \alpha_2 = 1, k_1 = 0.05, l^* = 0.006$ and $v_{max} = 150$. (b) Number of employees from the Hudson Bay Company. (c) Temporal series generated by the model. Arrows indicate the shifts in phase as a consequence of changing hunting pressure (see Gamarra and Solé, 2001).

By starting from a low-amplitude, periodic attractor (first phase, fig. 2.11c), the effect of a rapid increase in hunting pressure can be studied by using a time-dependent mortality rate $c = c(t)$. The simplest choice is to consider a stepwise dependence, that is, $c(t) = c_0 + \delta\theta(t - T)$, where $\theta(z) = z$ for $z > 0$ and zero otherwise. In this way, starting at some value c_0 in the domain of limit cycles, a shift to $c = c_0 + \delta$ occurs at $t = T$. We can see in figure 2.11c that the introduction of this change (here indicated with an arrow at $T = 1000$) displaces the system toward a high-amplitude phase, as observed from the real time series (fig. 2.2a). A second shift is also indicated (at $T = 3500$) in which the system returns to its previous hunting rate, and the dynamic shifts again to low-amplitude fluctuations. This is likely to be the origin of the pattern observed in the third part of the fur-return time series (Gamarra and Solé, 2001). The reason for this behavior can be understood by taking into account the nonlinear responses implicit in the predator-prey interactions. Although it is true that increased hunting promotes the decrease of lynx populations, it is also true that it triggers rapid bursts of hare populations in response to decreased predation. This in turn makes possible a very rapid, almost explosive regrowth of lynx numbers, which can actually reach much larger sizes. Thus the nonlinear response of the interacting populations (provided that $l^* > 0$ and no lynx crash is allowed) naturally explains the pattern observed in field data.

UNPREDICTABILITY AND FORECASTING

The concept of deterministic chaos was not only discovered in models of population dynamics. Rather, the first conceptual ideas arose at the beginning of the twentieth century in relation to cosmology. The great mathematician Henri Poincaré discovered that the attraction among three celestial bodies generates complex and aperiodic orbits. That is, the Newtonian paradigm already breaks down when moving from two celestial bodies (e.g., the earth and the moon) to three celestial bodies. Since the complexities of the calculations were impossible to find (no computers at that time), the subject had to wait several decades.

In 1963 the meteorologist Edward Lorenz published an influential paper (Lorenz, 1963). Lorenz was interested in the issue of weather forecast. He built a simple model of the weather with three coupled, nonlinear differential equations:

$$\frac{dx}{dt} = a(y - x), \tag{2.34}$$

$$\frac{dy}{dt} = x(b - z) - y, \tag{2.35}$$

$$\frac{dz}{dt} = xy - cz. \tag{2.36}$$

Lorenz's system is derived from the laws of hydrodynamics. The three variables in his model can be interpreted as the vertical component of the speed, and the vertical and horizontal variations in temperature; a, b, and c are three parameters related to viscosity and thermic diffusion (a is the so-called Prandtl number and b is related to the Rayleigh number). For our purposes, we have only to remember that model (2.34–2.36) describes the dynamics of a fluid layer with a difference of temperature between the superior and the inferior surface. As is well known, if the temperature difference is less than a critical threshold, heat is transported by conduction. But after a critical difference, heat is transported by convection, and turbulence arises for an even higher difference. For a particular parameter combination corresponding to the turbulent regime, Lorenz started with an initial condition and he iterated the model to the equivalent in real time of several weeks. What he found was a surprise. Despite the system's being completely deterministic, it shows aperiodic dynamics, the type of dynamics that one would expect for a noisy system (fig. 2.12b).

We can plot the system's phase space, which in this case will contain three variables. We had already seen that trajectories in phase space are captured by the attractor, and we had seen two types of attractors, that is, fixed points corresponding to stationary states, and limit cycles corresponding to cycles. As in the two previous examples of lynx and plankton dynamics, now we have encountered deterministic chaos. Figure 2.12a represents the projection on the $z - x$ axis of the 3-D attractor of Lorenz's system. Far from the random cloud of points we may have expected by looking at the temporal series, we find a beautiful object. Order is hidden within chaos. This attractor is something much more complex than a limit cycle. The term *strange attractor* has been coined to describe the attractors corresponding to chaotic systems. Since the system is deterministic, the attractor has a well-defined shape, and all trajectories will ultimately become trapped there. But two different trajectories will never cross (repeat) each other, which means that the dimension of this object is fractal, a concept fully explained in chapter 4.

The previous result was of extraordinary importance. Until then one assumed a relationship between knowledge and prediction, and between determinism and periodicity. Thus, if one knows the laws that drive a deterministic system, and can measure an initial condition, future values can be predicted. Besides, a deterministic system should display a simple behavior such as steady states and cycles. However, we are now facing a deterministic system that shows randomlike dynamics, that is, the kind of dynamics we would expect for a random variable. Lorenz noticed that the system was very dependent on initial conditions. There is always a mistake when estimating a given variable such as temperature, and the difference between the real value and

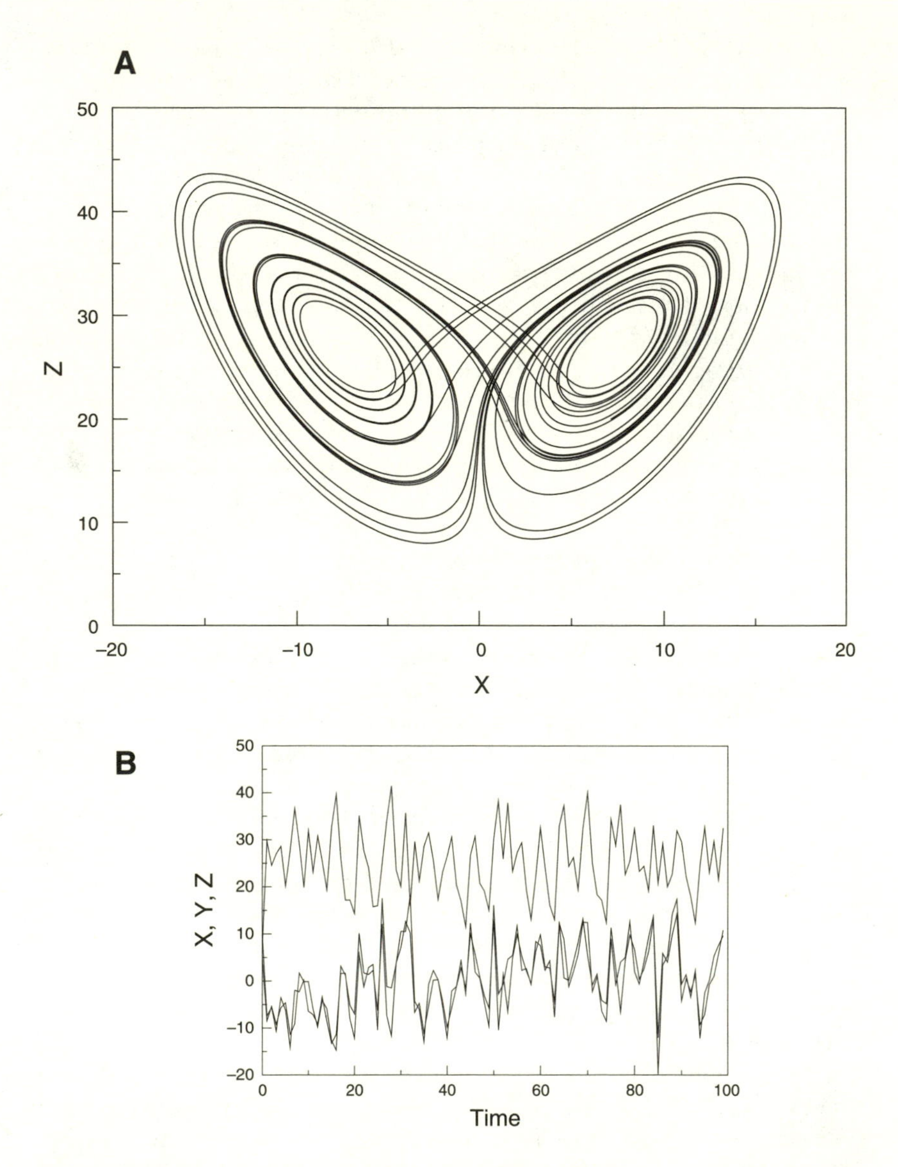

FIGURE 2.12. (A) Lorenz's strange attractor. This object lies in a phase space determined by three climatic variables, x, y, z. The figure plots a two-dimensional projection on the variables $x - y$. The system is described by a deterministic system of three nonlinear differential equations. Once a specific initial condition is given, the system can be iterated. After some iterations, the trajectory will be captured in this object, although it will never repeat again. (B) represents the temporal series for the system, where the randomlike, aperiodic dynamics of deterministic chaos can be observed. Parameters are $a = 10, b = 28$, and $c = 8/3$.

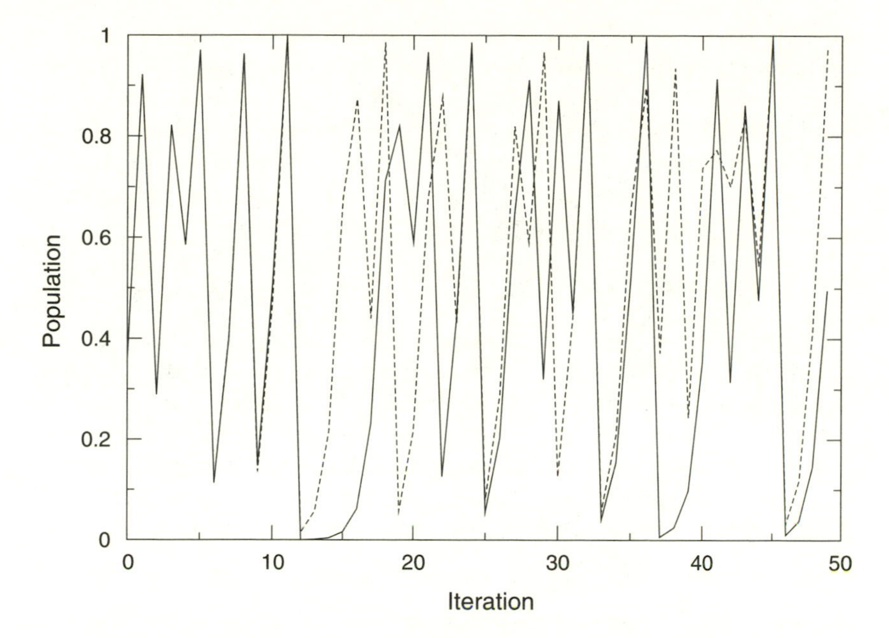

FIGURE 2.13. Chaos implies a strong dependence on initial conditions. The figure represents two iterations of the logistic map for $\mu = 4$ corresponding to the domain of deterministic chaos. The initial conditions are: $x_0 = 0.10$ (continuous line) and $x_0 = 0.10001$ (broken line). That is, there is a difference (error) of only 0.001%. At the beginning, both trajectories coincide, but after a certain time horizon, they start to diverge and end up completely uncorrelated. Long-term forecast is thus forbidden for nonlinear systems.

the estimate will not remain small as it would for linear systems. Due to the nonlinearities in the model, the error grows exponentially with time. There is a time horizon beyond which prediction has nothing to do with reality. To increase the time horizon, let us say by two, one should have to increase the precision several (more than two) times. In figure 2.13 we illustrate the phenomenon of dependence to initial conditions, which is the hallmark of deterministic chaos. Here two time series generated for the logistic with $\mu = 4$ and slightly different initial conditions are shown. We can imagine that the continuous line corresponds to an estimate of initial condition by a scientist. This initial condition is $x_0 = 0.1$. The broken line corresponds to a time series starting from a value $x_0 = 0.10001$, which we can assume corresponds to the "real" value. The difference or error in estimating the initial condition is as small as 0.001%. We could assume that this error is negligible as would be the case for a linear system in which small errors remain small through time. But what we observe is different. Even when at the beginning both trajectories coincide, after a time horizon they start to diverge, and after a while both are completely uncorrelated. Systems exhibiting deterministic chaos can be predicted at short time windows (they are deterministic), but long-term prediction is impossible despite knowing the dynamics underlying our system and having a good (as good as one wants) estimation of the initial condition.

Dependence to Initial Conditions: The Lyapunov Exponent

Dependence to initial conditions can be estimated by using the so-called Lyapunov exponent. The Lyapunov exponent is a quantification of the degree of instability of the system, or, in other words, of how fast nearby trajectories diverge in the phase space. To understand this concept, let us start by considering two similar initial conditions x_0 and $x_0 + \epsilon$ separated a distance ϵ, where ϵ is very small ($\epsilon = 0.00001$ in fig. 2.13). After n iterations, the distance between these two points will be

$$\delta(\epsilon) = \epsilon e^{n\lambda(x_0)}, \tag{2.37}$$

where λ is the Lyapunov exponent. As one can see from the previous equation, the exponential term gives us the rate of stretching between the two close trajectories. We can find three different situations: The distance between the two trajectories decreases, remains constant, or increases through time. For the distance between the two initial conditions to decrease, ϵ has to be multiplied by a number smaller than one (note that these considerations are the same as we made for the stability analysis above; as noted above, the Lyapunov exponent characterizes the degree of instability of a dynamic system): $exp(n\lambda(x_0)) < 1$ if $\lambda(x_0) < 0$. That is to say, two close trajectories will converge into the same fixed point attractor if the Lyapunov exponent is negative. If $\lambda(x_0) = 0$, the initial difference will remain the same: The system is trapped in a limit cycle. Finally, chaotic systems have a positive Lyapunov exponent. In general, there is a Lyapunov exponent for each dimension of phase space. In this case, the characteristic Lyapunov exponent will be the largest one. A system will be chaotic if such an exponent is positive.

To find an expression for the Lyapunov exponent, we can write:

$$\epsilon e^{n\lambda(x_0)} = |F_\mu^n(x_0 + \epsilon) - F_\mu^n(x_0)|. \tag{2.38}$$

From the previous equation, the Lyapunov exponent can be defined as follows (see appendix 1):

$$\lambda(x_0) \approx \frac{1}{n} \sum_{i=0}^{n-1} log \left| \frac{\partial F_\mu(x_i)}{\partial x_i} \right|. \tag{2.39}$$

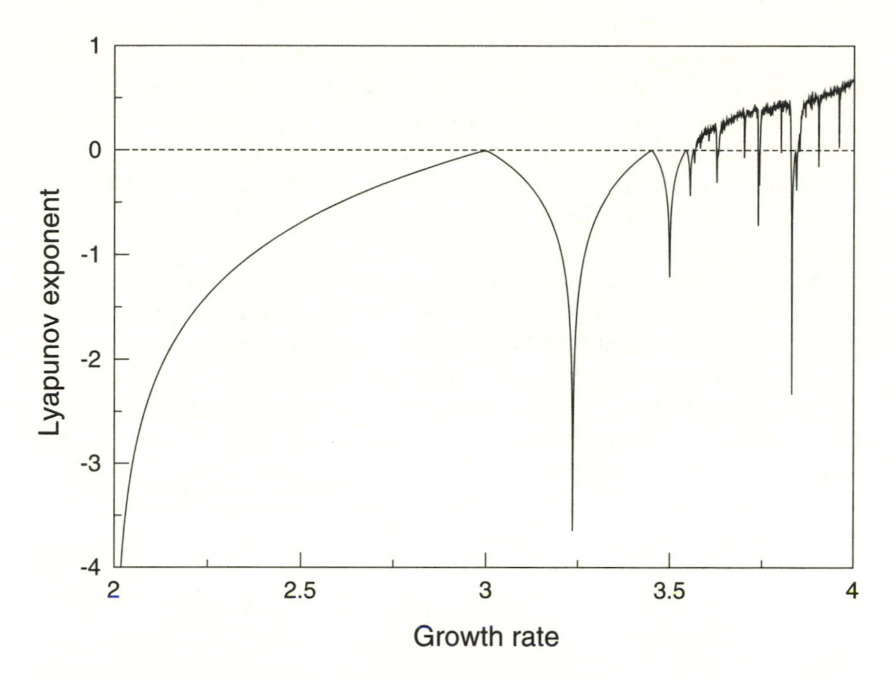

FIGURE 2.14. The Lyapunov exponent is plotted versus the growth rate (μ) for the logistic map. A positive Lyapunov exponent is typical for chaotic systems and characterizes the rate at which nearby trajectories diverge exponentially (a concept shown in previous figure). The drop in the Lyapunov exponent toward negative values within the chaotic domain correspond to periodic windows. Compare this figure with the bifurcation diagram in figure 2.7.

The previous equation determines the Lyapunov exponent for a dynamic system, and so the rate at which initial conditions diverge. The Lyapunov exponent is an unequivocal characterization of the type of dynamic exhibited by a system. Figure 2.14 plots the Lyapunov exponent according to (2.39) for the logistic map (equation 2.19). Compare this figure with figure 2.7, where the bifurcation diagram for the logistic map was represented. By comparing both figures, one can see how the Lyapunov exponent becomes positive in the chaotic domain, indicating that initial trajectories will diverge exponentially. Also, one can observe that there are regions within the chaotic domain in which the Lyapunov exponent becomes negative. This corresponds to the existence of periodic windows as already noted in the bifurcation diagram.

In summary, in the previous sections we have seen that simple, nonlinear deterministic systems can generate complicated dynamics. This has deep implications for our understanding of complex time series.

Before the discovery of deterministic chaos, one would have assumed that noise was the main force behind some of the series generated by chaotic systems. Now we see that the same complex series may be generated by a simple deterministic, low-dimensional system. If this is so, we can understand the mechanism behind such dynamics. We can even predict at shorter time scales, although prediction is impossible beyond a given time horizon.

The definition of the Lyapunov exponent seen in this section is based on a mathematical function. Of course we do not know precise equations behind ecological time series. All we have is a time series. Later, we will emphasize that there are different algorithms to approximate the largest Lyapunov exponent from a temporal series obtained without any information about the underlying function. Though other measures have been used to characterize a chaotic time series, we will not review these in detail, but will briefly mention them when considering the search for chaos in real ecosystems.

THE ECOLOGY OF UNIVERSALITY

The period-doubling route to chaos shown by the logistic map and similar maps has general properties. This was first noted by Mitchel Feigenbaum (1978), a physicist working at the time at Los Alamos National Laboratory. The discovery of universality in the path to chaos for a wide class of nonlinear systems was an advance comparable to the development of the physics of critical phenomena (see chapter 4). Let us review here his contribution.

Given a diagram displaying period-doubling bifurcations, one can mark the critical points of the bifurcation parameter μ_k at which a new bifurcation k takes place. For example at $\mu_1 = 3$, there is the first bifurcation, shifting from a steady state to a 2-period cycle. At μ_2 the dynamics shift to a 4-period cycle, and so on. Feigenbaum (1978) showed that for a large value of k, the following relationship holds:

$$\delta = \lim_{k \to \infty} \frac{\mu_k - \mu_{k-1}}{\mu_{k+1} - \mu_k} = 4.6692\ldots \tag{2.40}$$

The demonstration of the constant was done by Feigenbaum using the renormalization group technique, a celebrated technique in physics that is powerful in relation to several problems related to complexity. May and Oster (1976) used a different approach to prove Feigenbaum's constant based on stability of orbits of period k and Taylor expansions.

The existence of Feigenbaum's constant is important in that it points toward a clear structural pattern in a bifurcation diagram. It gives us an idea of the "distance" between two successive bifurcations in relation to the distance between the next two. What is most interesting from Feigenbaum's work is that this constant is universal. That is, it is independent of which particular map we use (we could have used a wide range of maps such as that for the logistic equation). Any nonlinear, one-dimensional map that has a single hump or maximum, despite differences in details, has the same constant. This is the reason δ is called a constant. This is an amazing result that we could not have anticipated. It means that there is some common structure behind the period-doubling route to chaos in different systems.

One could think that this is a curiosity, but it is far from that. If we deal with a continuous time system, we can plot the successive peaks. A model such as (2.19) may then be a characterization of the discrete version of the system. Discrete time models can be interesting in themselves as analogies of discrete time populations, but they can also characterize continuous time systems.

If one has data from a real system such as a chemical reaction, and a temporal series is recorded, one can plot the successive peaks as illustrated in fig. 2.10G. From the series of peaks, we plot a bifurcation diagram, provided a good bifurcation parameter is found. Interestingly, the resulting bifurcation diagram has the same constant, that is, it is not an artifact from simple models, but a property of real chaotic systems (May, 1976; Cvitanovic, 1984).

The implications of universality are that some basic mechanisms are general and so independent of details of specific systems. Some general properties are exhibited by a broad class of chaotic systems, those showing period-doubling bifurcations. This is an important concept and will be emphasized throughout this book: the search for dynamic properties that are independent of specific systems. Thus we can get basic and general information to characterize a wide spectrum of complex, nonlinear systems. This also has implications for the strategy of model building. As noted, despite their simplicity, models such as (2.19) have the key properties (nonlinearity, density dependence) that characterize real populations. Thus, despite the omission of many details from these simplistic models, we may expect that the relevant features are maintained. Simple models may be, after all, relevant when dealing with complex ecological systems.

Universality in its different aspects will be encountered throughout the book. One of them is percolation, a characteristic bifurcation point where a phase transition takes place. If the Feigenbaum constants are a temporal property of nonlinear systems, percolation is a spatial property. Imagine a spatial lattice such as a chessboard. If we randomly start introducing trees at each lattice site, clusters will come to appear as the density of trees is increased. The probability of moving from tree to tree, from one edge of the lattice to the other, however, does not increase linearly with tree density as one could imagine. Instead, the probability is almost zero below a critical density and almost one beyond it. This critical density is called the percolation point and again does not depend on further details. It is only determined by the dimensionality of the system (a two-dimensional lattice in our case and clusters defined as sites connected by one of their four sides). We will be considering the concept of percolation in chapter 4, and one application to the problem of habitat fragmentation in chapter 5.

EVIDENCE OF CHAOS IN NATURE

As already noted, the discovery of chaotic dynamics revealed that simple, nonlinear models can show complex dynamics. This contribution was relevant not only in ecology, but in science in general, since it was one of the fronts in which deterministic chaos was introduced. Thus, until now we have seen that chaos can be generated by ecological models, and some evidence indicates that it is one possible pattern found in natural systems. But is deterministic chaos a widespread type of behavior? We have already discussed the results by Costantino and colleagues who show the presence of chaos in the experimental design with *Tribolium*. Mixed data and modeling strongly supported the presence of a bifurcation toward chaos in the Canadian lynx dynamic. In this section we further explore the evidence found for chaos in real ecosystems.

The first work trying to address this question is the paper by Hassell et al. (1976). These authors collected temporal series on twenty-eight arthropod species, both from the field and the laboratory. They fitted the data to a model developed previously by Hassell (1975). This model was a discrete time model similar to (2.19), but more realistic, and it exhibited all the range of dynamic possibilities depending on the values of the parameters. The result was conclusive. From twenty-eight populations, twenty-six were fitted within the steady state domain, one within the domain of cycles and the remainder within the domain of chaos. The obvious conclusion was that chaos does not seem to be common in nature. This was a relevant result for various reasons, mainly because it was the first attempt to find chaos in the real world. As a consequence it stimulated a large number of subsequent papers that contributed to this area. However, as the same authors mentioned, one has to be cautious when interpreting this data. For example, the population that corresponded to the most stable set of parameters, the butterfly *Zeiraphera diniana*, is well known for exhibiting regular cycles. Morris (1990) has pointed out that it is difficult to derive the kind of dynamic exhibited by a population by fitting data to a model. Results may depend on the particular model used or on the procedure to fit the data. Similarly, besides biological variables considered in models of a single population, spatial effects may contribute to the dynamics (Bascompte and Solé, 1994), at least for continuous time systems (Murray, 1989; Pascual, 1993) or discrete time systems with age structure or specific interspecific interactions (Hastings, 1992).

Hassell et al.'s (1976) paper had such an influence in the ecological community, that the idea arose that deterministic chaos was not an important component in ecosystems. In the next section we will see in detail the reasons adduced. The same approach was used by other authors, arriving at similar conclusions (Thomas et al., 1980; Mueller and Ayala, 1981). These studies have proposed both group selection (the former) and selection at the level of the individual (the latter) to explain the dominance of stable dynamics.

Since fitting models to data can introduce some biases, the next step was to look at the possible existence of chaos in nature by using more appropriate tools. A first stage of papers, led by William Schaffer and Mark Kot, used techniques derived from the physical sciences such as attractor reconstruction, attractor's dimension, and the Lyapunov exponent. Other techniques were derived by ecologists working in specific ecological problems, such as nonlinear forecasting (Sugihara and May, 1990). We will not review in detail these procedures, but instead briefly mention them in the context of specific examples.

Schaffer and Kot (1986a) reviewed examples suggesting the presence of chaos and concluded that chaos is an important force in ecological systems. One of their examples was the lynx data mentioned in previous sections. Although eminently periodic, the lynx record has the hallmark of chaos (Schaffer, 1984): The frequency of the cycle is quite constant; the amplitude, however, is quite variable. To reach this conclusion, Schaffer (1984) looked at the lynx attractor. Since he knew neither the variables, nor how many of such variables were involved, he used a well-known technique, the so-called embedding method (Takens, 1981). This consists of defining new variables that are the same time series with some time lag. For example, one axis would be the density of lynx at year t ($x(t)$), another axis will be density at year $t + \tau$ ($x(t + \tau)$), and so on. It might seem an ad hoc way of using temporal data, but it actually makes sense once we realize that populations fluctuate as a consequence of their interactions with *all* species to whom they are linked. In some way, the degrees of freedom must be implicit in the time series of each population.

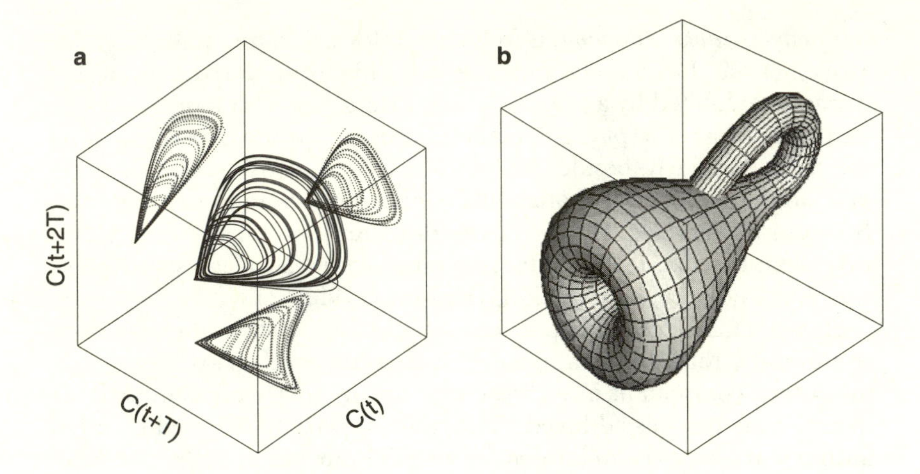

FIGURE 2.15. Attractor reconstruction. In (a) we show an example of a strange attractor (obtained from the time series of carnivorous zooplankton in Scheffer's model) reconstructed in a three-dimensional embedding space. In (b) a Klein's bottle is shown as a three-dimensional object. Because of its particular structure, it crosses itself in three dimensions and only becomes noncrossing in four dimensions.

An example is shown in figure 2.15a. Here the strange attractor of Scheffer's model of plankton dynamics is reconstructed from the time series of carnivorous zooplankton (compare with fig. 2.10F). The picture actually shows two different embeddings: The black plot is the reconstructed attractor in three dimensions, whereas the three projections (shown in gray) would be two-dimensional reconstructions. If the embedding dimension is smaller than the system's dimension, the resulting reconstructed attractor will be folded and trajectories will intersect (as happens with the projections in the previous example). Once we reach the appropriate embedding dimension, we will see that no intersections arise (well-defined quantitative methods are applied in order to determine this). In this way, we would say that the number of variables associated to Scheffer's model is three, since using a three-dimensional reconstruction provides a faithful representation of the dynamics. It is important to mention that sometimes additional dimensions must be added. Actually, Takens theorem shows that in order to guarantee that the attractor completely unfolds, the embedding dimension must be (at least) twice the system's dimension. This is illustrated in figure 2.15b for the so called Klein's bottle. This is a two-dimensional object (a surface), but it can only be embedded (with no self-crossings) in four dimensions.

Schaffer reconstructed the lynx attractor with this technique, finding evidence for the stretching and folding of trajectories, which is the hallmark of deterministic chaos. We have seen in a previous section that stretching is a component of chaos, the dependence on initial conditions. However, at the same time, chaotic systems, since deterministic, are constrained within an attractor, a strange attractor. This means that nearby trajectories cannot diverge forever. After diverging for a while they are "folded" again, so the dynamic within a strange attractor is a combination of successive stretchings and foldings. Another conclusion of Schaffer (1984) is that only three variables are necessary to describe the system, that is, the system is low dimensional. One could imagine that these variables are lynx, hares, and vegetation, which is suggested by another type of study related to the origin of the cycles (Krebs et al., 1995).

Other examples studied by the attractor reconstruction were the outbreaks of the insect *Thrips imaginis*, the cycles of the rodent *Clethrionomys glareolus*, and the record of measles in Baltimore (Schaffer and Kot, 1986a). Epidemiology has in fact provided some of the best time series in ecology, and some of the strongest evidences for chaos. Epidemiological systems are no more than simplified ecological systems, and some properties are therefore similar to other ecological systems. It is worth mentioning that the time series coming from epidemics are by far the best data sets in which to look for nonlinear dynamics and chaos in population dynamics, both in space and time (Anderson and May, 1991; Grenfell et al., 2001). Here we will just emphasize that while mumps and chicken pox seem to reflect a periodic time series (with a period of one year forced by the school calendar), data on measles before introduction of mass vaccination seems to reflect the existence of deterministic chaos.

Schaffer and Kot's (1986a) paper introduced the view that deterministic chaos may be relevant for real ecosystems. They concluded their review by noting that "the concept of chaos is both exhilarating and a bit threatening. On the one hand, it offers a deterministic alternative to the idea that population fluctuations are solely the consequence of external perturbations. At the same time, chaos could undermine much of the conceptual framework of contemporary ecology. In so doing, it forces us to grapple with realities that a generation of ecologists has ignored."

Other techniques to detect chaos are based on the estimation of the dimension of the reconstructed attractor. A strange attractor is a fractal object (trajectories are constrained on it but never cross). This means that a strange attractor has a non-integer dimension (a fractal dimension, a concept introduced in chapter 4). The *correlation dimension* of reconstructed attractors (Grassberger and Procacia, 1983) is a technique to estimate this dimension.

The Lyapunov exponent is, as we have noted in an earlier section, one of the most characteristic determinants for the presence of deterministic chaos. There is an algorithm that estimates the largest Lyapunov exponent from a temporal series, that is, without any explicit reference to a mathematical function (Wolf et al., 1985). This algorithm estimates the average rate of divergence between close trajectories in the attractor. If

$$\Theta(t,\tau) = \frac{\|x(t+\tau) - x'(t+\tau)\|}{\|x(t) - x'(t)\|} \tag{2.41}$$

denotes the ratio between the Euclidean distance of two initial close trajectories after a time lag τ, with regard to the initial distance, then the largest Lyapunov exponent λ_L can be estimated as:

$$\lambda_L = \frac{1}{r\tau} \sum_{t=1}^{r} ln(\Theta(t,\tau)). \tag{2.42}$$

Wolf et al.'s (1985) algorithm is useful in the sense that allows one to estimate the largest Lyapunov exponent from a temporal series without any information of the equation(s) governing the dynamics. However, problems still remain. Both Wolf et al.'s algorithm and the correlation dimension of a reconstructed attractor are techniques derived by physicists thinking about physical problems. The problem with this is that time series in ecology are far shorter and with higher levels of noise (both intrinsic and sampling error). The promise of the mid-1980s to use all these tools to characterize chaotic dynamics was soon weakened by noting that the tools were probably not very useful in ecology, although some interesting applications were done in other areas of biology such as physiology (Glass and Mackey, 1988). The problem was set for ecologists to derive their own measures for their own data.

Ellner and Turchin (1995) have applied nonlinear techniques to estimate the Lyapunov exponent. Their strategy is as follows. Since data contains large amounts of noise, they do not calculate the Lyapunov exponent directly from the data, but first estimate the map that best fits the data. This map determines population size at a year as a function of population sizes at previous years, that is $N(t) = f(N(t-1), N(t-2), \ldots, N(t-d), e_t)$, where d is the embedding dimension and e_t represents exogenous influences over some finite interval in the past. By nonlinear techniques they find the specific map that best describes the data, and then with the computer they can start a second trajectory as close as possible to the one observed in their data. In this way they estimate the Lyapunov exponent, with the advantage that it can be done with series that present high levels of noise.

A different approach is taken by nonlinear forecasting (Sugihara and May, 1990a; see also Kaplan and Glass, 1995). Roughly speaking, this approach consists of dividing the temporal series into two halves. The first half will be used as a source of known data, and the second half considered an unknown future. The method can determine the optimal embedding dimension, which will be the one that provides the highest correlation between the prediction and the observation. Let us say this embedding dimension is 3. One proceeds by considering one point of the second half of the data. For this point one finds the points of the first half, which are closer, defining the smallest possible vertex of a triangle (or n-dimensional polyhedron of triangular sides) that can be constructed with those points. For a given number of time steps, one looks at the position of these points (this is known as it belongs to the first part of the temporal series). They define a new triangle, and one can thus calculate its baricenter, which will be the prediction. One can then compare this prediction with the "real" value from the second half. One then repeats this procedure for all the points, calculating the correlation between the predicted and the observed pairs of points. One then increases the time lag (prediction time) and repeats the procedure for each new time lag. One can then plot correlation versus prediction time (Sugihara and May, 1990a) as done in figure 2.16.

Interestingly, nonlinear forecasting clearly distinguishes white noise from deterministic chaos. Since there is no correlation in white noise series, one finds that the correlation coefficient is as good (or as bad) from one time step to the next as after several time steps. In other words, we would find a flat line when plotting correlation coefficient versus prediction time (fig. 2.16). Nonetheless, if a time series is chaotic, correlation for small prediction times will be high (the system being deterministic), but such correlation will drop exponentially as the prediction time is increased.

Sugihara and May applied their nonlinear forecasting technique to several ecological time series, namely, the measles record from New York City already studied by Schaffer and Kot, the record of chicken pox in the same city, and a temporal series corresponding to the monthly record of diatoms present in samples of marine plankton collected at the Scripps Pier (San Diego) between 1920 and 1939. The conclusions are as follows: For the measles time series, the correlation decays as prediction time increases, indicating evidence for a chaotic system. This is in agreement with previous results by Schaffer and Kot (1986a) discussed above. However, the chicken pox record shows that correlation is low and does not change with prediction time; this is strong evidence for white noise. Finally, for the case of diatom abundance, correlation decreases again with prediction time, but in this case

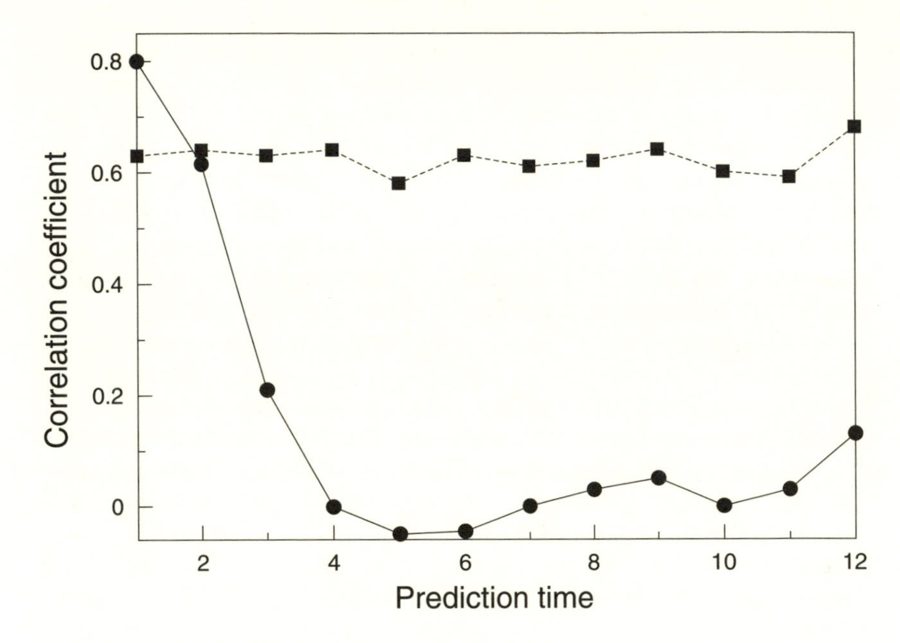

FIGURE 2.16. Nonlinear forecasting. The correlation coefficient between predicted and observed values is plotted versus prediction time. Continuous line corresponds to deterministic chaos; broken line corresponds to white noise. As noted, correlation is constant for random systems exhibiting white noise, while it decays exponentially for chaotic systems due to their dependence on initial conditions. From Sugihara and May (1990a).

the correlation is much lower than that for the measles. This means that there is evidence for a chaotic system with strong levels of noise superimposed.

Nonlinear forecasting can only deal with white noise. If noise is correlated, this can confound the results. However, the important point is that Sugihara and May developed an interesting approach to the problem of distinguishing chaos in real data, and that some of their results confirm those obtained by other approaches. At the same time, their pioneering approach stimulated other ecologists to look at additional data sets.

Additional evidence for chaos in nature was obtained from the dynamics of annual plants (Tilman and Wedin, 1991). Their approach was as follows. They plotted biomass density at one year t versus density at the next year $(t + 1)$. This is a *Poincaré map*. If the dynamic is stationary, one would find a single point (or a cloud of points clustered around this if there is external noise). For a two-periodic cycle we would find two points, and so on. Finally, for a chaotic system, we would find a quadratic curve. They plotted twenty-four data points coming from four annual intervals repeated in six different parcels. They fitted the observed points to a quadratic curve and calculated its maxima and its slope. This information was then compared to a model previously developed to describe the dynamics of annual plants. The model was a nonlinear, deterministic model that could exhibit the whole spectrum of dynamic behaviors. The observed maxima and slope corresponded to a set of parameter values in the model corresponding to chaos.

Note that this approach mixes temporal and spatial data. This is an interesting point, since temporal series are of a limited length, but often there are spatial replicates. Frank (1991) had already shown that data obtained in a single point through time was equivalent to data obtained at a single time at different spatial locations using a host-parasitoid model. Expanding on this idea, Solé and Bascompte (1995) developed a new class of Lyapunov exponent, which can be used mixing spatial and temporal data. For simple coupled map-lattice models (a concept that will be fully explained in the next chapter), they show that the estimate using as few as ten iterations from enough lattice sites was the same as using the standard Lyapunov exponent (using the Wolf et al.'s algorithm) with data from a single point through a large enough number of iterations (Solé and Bascompte, 1995; see also Nikolaus et al., 2002).

Another set of evidence for chaos comes from studies of the dynamics of voles and small rodents in northern Europe. Hanski et al. (1993) find that despite the strong periodic component of the cycles of rodent populations in Fennoscandia, these are chaotic. Furthermore, they suggested that the delayed density dependence to specialist predators (mustelids) is the mechanism generating chaos. These authors used a general prey-predator model, estimated the parameters using real time series, and found that the majority of model-predicted dynamics are chaotic. A complementary line of evidence comes from nonlinear time series analysis of the longest time series. The estimated Lyapunov exponent is positive in all the northern populations, which confirms the suggestion for chaos in these populations. The southern population has a negative Lyapunov exponent, which agrees with the latitudinal gradient in the stability of these populations, according to which southern populations are stable, and they become more unstable (cycles and chaos) as we move toward northern latitudes. Further studies (Turchin and Ellner, 2000) have shown that vole dynamics in northern Fennoscandia are characterized by alternating periods of order and irregularity. That is, the global Lyapunov exponent is close to zero, but in short periods, the dynamic is very unstable.

The above studies provide evidence for one or a small number of examples, but the next question is how common chaos is in nature. At this point, several researchers have made an intensive effort to analyze large data sets and determine the importance of chaotic dynamics. An example is the study of Turchin and Ellner of a wide spectrum of both lab and field time series, where nonlinear techniques were used to estimate the Lyapunov exponent mentioned above. There are two main results in this work that are worth mentioning here. First, they found a wide spectrum of populations, ranging from stability to chaos. Second, for both laboratory and field data sets, the frequency distribution of estimated Lyapunov exponents has a mode near or at zero. Thus, Turchin and Ellner's position is intermediate between the two extremes: Chaos is neither totally absent, nor the main mechanism to explain population fluctuations. Interestingly, the authors note that "If any, our results provide some support for the hypothesis, recently put forth by some theoreticians that interacting populations should coevolve to dynamics at the edge of chaos." However, further reanalysis allows one to interpret these results on the dynamics of Fennoscandian voles in the opposite direction since all possibilities (stability, chaos, quasi-periodicity) are observed, not only one. This point will be discussed later in this book in relation to a view of ecosystems as dynamic systems evolving toward a critical state (stable, but near instability).

As Hastings et al. (1993) have pointed out, there is evidence for chaos in nature, and probably one of the few things we can conclude is that not all populations are chaotic and at least one is. Reality lies somewhere in the middle. But even in the case in which chaos is not found in nature, studies on deterministic chaos are extremely important because they show us that density dependence can lead to types of dynamics other than cycles and steady states. Note that if a population is chaotic, erratic oscillations are not opposed to the existence of density-dependent regulation.

CRITICISMS OF CHAOS

We have already mentioned that the first search for chaos in natural populations suggested that chaos may not be common in nature (Hassell et al., 1976). Assuming that it was a possibility not realized in real ecosystems, some ecologists thought about mechanisms to avoid chaos. The absence of chaos from real populations would be a consequence of the structure of the bifurcation diagram (Berryman and Millstein, 1989a). Note that for low values of the growth rate (bifurcation parameter), the population value is high. As the successive bifurcations appear, the lower branch becomes closer to zero, that is, at some point in the cycle, population numbers are small. Finally, for the chaotic region, the population will be close to extinction in certain years. Berryman and Millstein (1989a) pointed out that in such circumstances, small environmental fluctuations could easily lead to population extinction. Thus, there was a strong selective force to avoid chaotic dynamics. Chaos would be maladaptive.

Selection at different levels has been considered the mechanism for avoiding chaos in nature. Thomas et al. (1980) proposed group selection, while Mueller and Ayala (1981) proposed selection at the level of the individual. Some discussion arose around this concept and the particular mechanism involved in avoiding chaos (Nisbet et al., 1984; Lomnicki, 1989; Mani, 1989; Berryman and Millstein, 1989b). We will stop at this point, since a key factor will be introduced in the next chapter. There, we will see that space introduces a different perspective. Even questions related to temporal properties cannot be fully answered without considering space.

The bifurcation diagram in figure 2.7 will be useful to illustrate another criticism to the presence of chaos in nature. As already noted, there is a nested structure of periodic windows inside the chaotic domain. The chaotic domain has so-called periodic windows, that is, regions in which the dynamic becomes periodic again. By slightly increasing μ, we suddenly move from a chaotic dynamic to a periodic one. The nesting of periodic windows within the chaotic domain is fractal, that is, it has a self-similar structure (fractals are definitively introduced in chapter 4). If we enlarge a periodic window, we will see again the period doubling route to chaos and new periodic windows within the chaotic domain. From a technical point of view, this means that deterministic chaos is *structurally unstable*. That is, even if we have a parameter combination corresponding to deterministic chaos, a small perturbation of the bifurcation parameter will move dynamics away from chaos. This is another criticism to chaos and once more we have to wait for the introduction of space.

Finally, it has been noted that the logistic equation (2.19) and similar one-dimensional, nonlinear maps generate chaos only for high values of the growth rate. One can think that these growth rates are unrealistically high. If one estimates the growth rate in a real population and this is lower than the minimum value generating cycles, this seems evidence enough to disregard the possibility of chaos underlying population fluctuations. Once more this consideration has to be put in a spatiotemporal context. Dispersal among patches may induce an otherwise stable population to become cyclic or periodic as we will see in the next chapter.

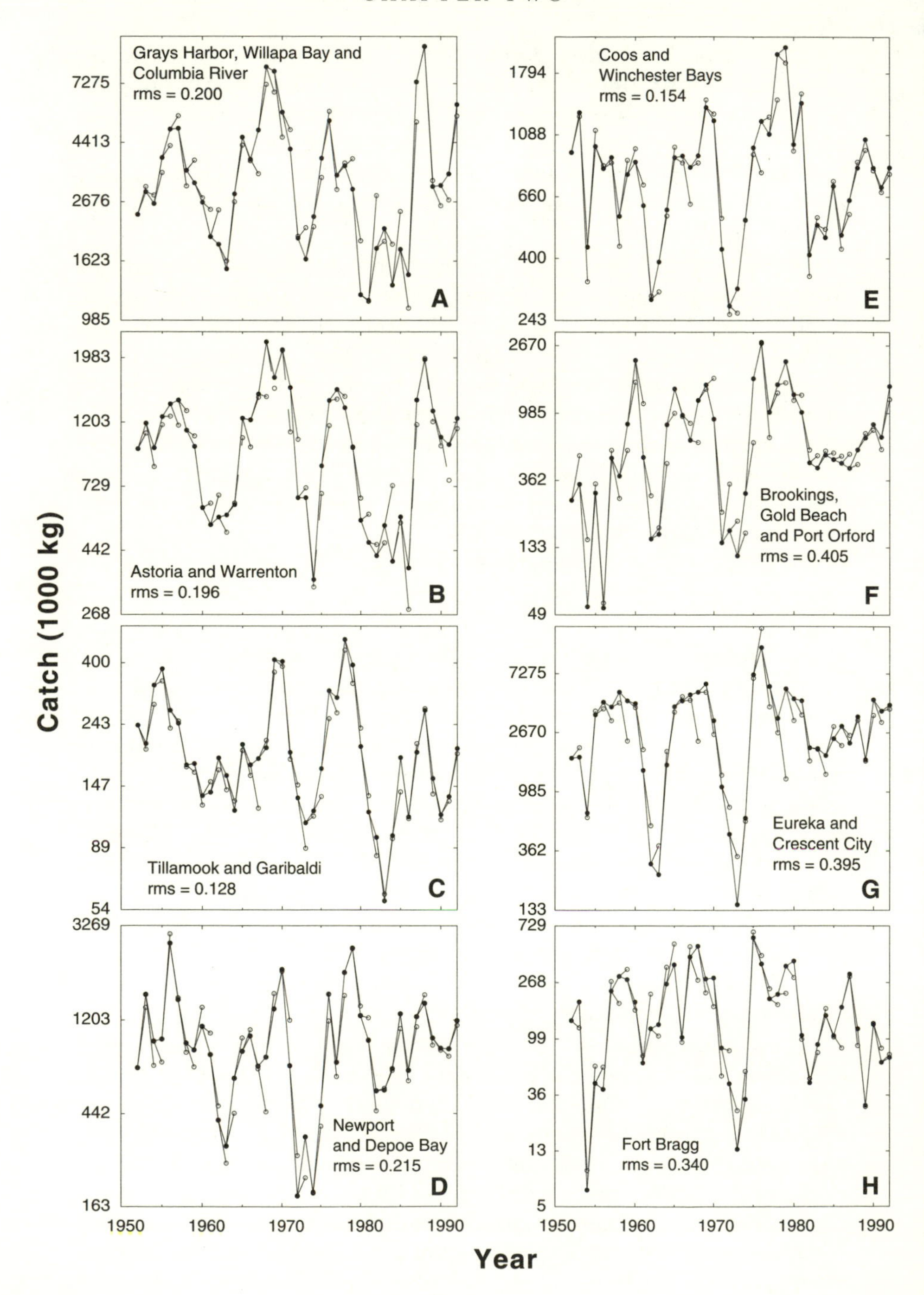

Grays Harbor, Willapa Bay and Columbia River
rms = 0.200
A
7275
4413
2676
1623
985
Coos and Winchester Bays
rms = 0.154
E
1794
1088
660
400
243
Astoria and Warrenton
rms = 0.196
B
1983
1203
729
442
268
Brookings, Gold Beach and Port Orford
rms = 0.405
F
2670
985
362
133
49
Tillamook and Garibaldi
rms = 0.128
C
400
243
147
89
54
Eureka and Crescent City
rms = 0.395
G
7275
2670
985
362
133
Newport and Depoe Bay
rms = 0.215
D
3269
1203
442
163
Fort Bragg
rms = 0.340
H
729
268
99
36
13
5
Catch (1000 kg)
1950
1960
1970
1980
1990
Year

COMPLEX DYNAMICS: THE INTERPLAY BETWEEN NOISE AND NONLINEARITIES

As noted throughout this chapter, we have opposed nonlinear deterministic dynamics to random fluctuations. To some extent, this responds to the fact that when chaos was discovered, people tried to see evidence for determinism in an ocean of seeming random behavior. It also responds to the fact that one tends to simplify the problem under study and focus on a single factor. Another reason for not considering simultaneously noise and deterministic chaos was that oftentimes one assumes that once we know the properties of random systems, and those of deterministic systems, we can integrate the information to predict properties of deterministic systems in variable environments. This way of thinking is linear, assuming that the combination is only a linear function of each one of its two components. As we shall see, this assumption is not valid when nonlinearities are at work. Interestingly, some examples of complex population dynamics could be explained neither by a nonlinear, deterministic model, nor by a stochastic one. The dynamics are better explained and predicted by a combination of both. We will consider these examples in this last section.

It seems reasonable that reality may be a combination of deterministic density-dependence and random fluctuations imposed, for example, by the weather. Ellner and Turchin (1995) had already pointed out that "that strict separation between chaotic and stochastic dynamics in ecological systems is unnecessary and misleading." These authors showed that the dynamic behavior exhibited by the populations they studied could not be entirely described with a one-dimensional spectrum from noise to chaos. Instead, they found that two axes are at least necessary, the first one explaining the relative strength of the endogenous/exogenous factors, and the other explaining the system's damping/amplification rate of exogenous perturbations.

A good illustration of the synergistic interaction of nonlinear, deterministic systems and random fluctuations is provided by Higgins et al. (1997). These authors studied yearly catch records of Dungeness crabs,

---

FIGURE 2.17. One-year-ahead model predictions and real data for Dungeness crab catch for different ports in Washington, Oregon, and California. The differences between predicted and observed values are estimates of environmental perturbations. The deterministic skeleton is unable to describe the time series. The estimates of environmental perturbation are introduced into the deterministic model in the next figure. Based on Higgins et al., 1997. Reprinted with permission of *Science*. Copyright (1997) AAAS. Figure courtesy K. Higgins.

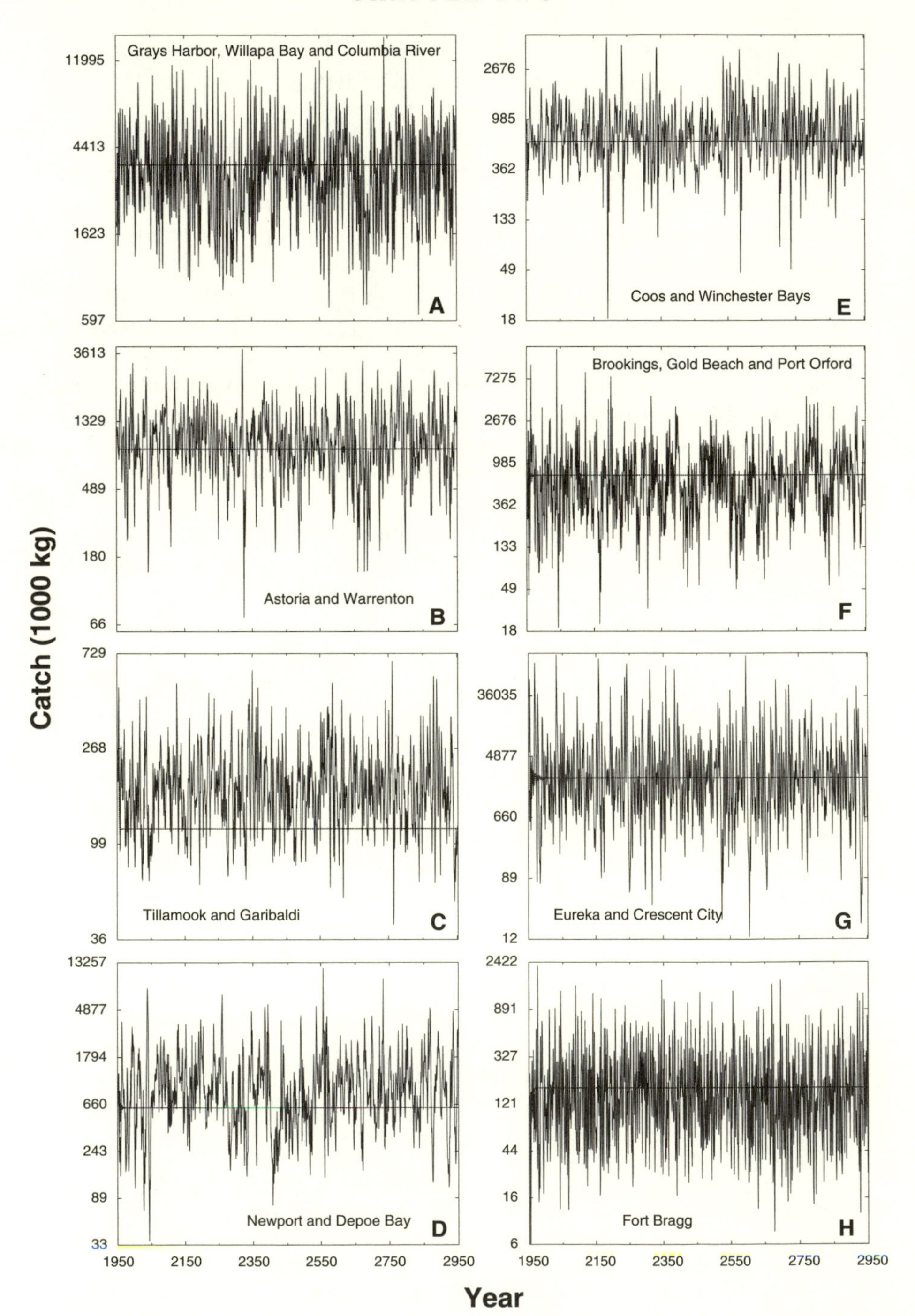
Grays Harbor, Willapa Bay and Columbia River
A
11995
4413
1623
597
Coos and Winchester Bays
E
2676
985
362
133
49
18
Astoria and Warrenton
B
3613
1329
489
180
66
Brookings, Gold Beach and Port Orford
F
7275
2676
985
362
133
49
18
Tillamook and Garibaldi
C
729
268
99
36
Eureka and Crescent City
G
36035
4877
660
89
12
Newport and Depoe Bay
D
13257
4877
1794
660
243
89
33
Fort Bragg
H
2422
891
327
121
44
16
6
1950
2150
2350
2550
2750
2950
Catch (1000 kg)
Year

Cancer magister, spanning forty-two years at eight locations in California, Oregon, and Washington. Temporal series show oscillations of a high amplitude with a cycle period of about ten years (fig. 2.17). The persistent question is, what force drives these cycles? Is it random changes in environmental factors such as currents and water temperature? Or is it nonlinear density dependence? The latter may be enhanced because of the existence of cannibalism among different age classes, the same mechanisms at work in *Tribolium* as seen above.

Higgins et al. (1997) used a nonlinear population model where all the biological information on this species is transferred. They first used this deterministic version of the model. By fitting the model to the available data, they estimated each one of the parameters. Interestingly, when the deterministic model is run with the estimated parameters, the dynamics evolves to very stable equilibrium densities. The next step was to introduce environmental fluctuations. The authors estimated the magnitude of these fluctuations by comparing the one-year-ahead catch predicted by the deterministic model to the real catch for that year. The estimated environmental fluctuations had a magnitude much smaller than that of the fluctuations in crab catch. The unexpected result is that after introducing these small doses of random fluctuations, the model generates high-amplitude fluctuations, the same kind of dynamic observed for the real-time series (fig. 2.18). As pointed out by Higgins et al. (1997), their results prove that the observed dynamics is produced by inexorably intertwined endogenous and exogenous forces. One can not tease apart each one of these components, as they interact in a complex way. The nonlinear skeleton magnifies the effects of stochastic variability. Both levels are important for understanding the observed fluctuations.

Related work has been done (Dixon et al., 1999) in relation to the problem of the episodic fluctuations in larval supply. As is well known, predicting recruitment level from reproductive stock size has been a difficult task for marine ecologists. The density-dependent relationship between stock size and subsequent recruitment accounts only for about 10% of the observed variability. Different interpretations have traditionally assumed sampling and statistical error, or high levels of

FIGURE 2.18. Simulated crab catches after parameterizing the variables with real data and incorporating the estimated environmental noise into the deterministic model. The heavy black line represents the dynamics of the deterministic skeleton. As noted, the deterministic model amplifies the environmental noise, and the resulting dynamics resembles much better the observed catches. Based on Higgins et al., 1997. Reprinted with permission of *Science*. Copyright (1997) AAAS. Figure courtesy K. Higgins.

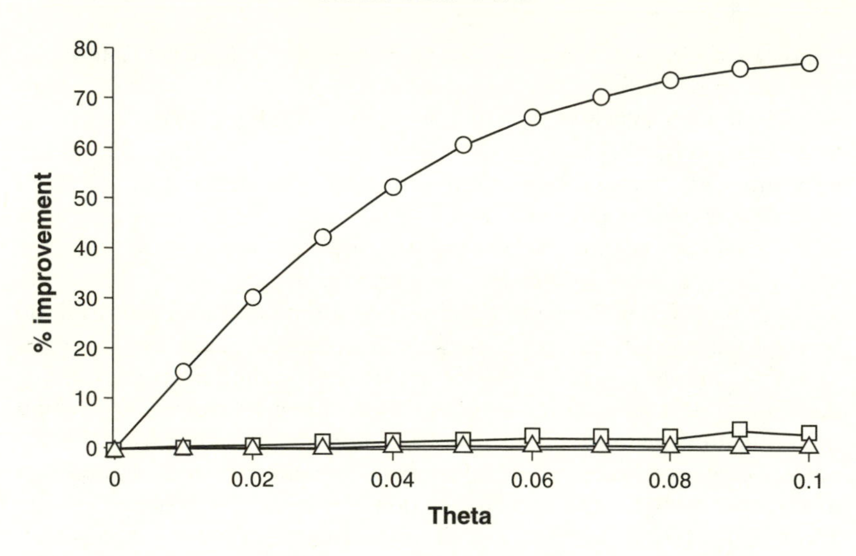

FIGURE 2.19. Percent increase in correlation coefficient between predicted and observed values for spawning, recruitment, and larval supply, as the forecasting algorithm becomes more nonlinear. As noted, predictions for larval supply become better as the nonlinearity of the prediction technique increases. Based on Dixon et al., 1999.

stochastic larval mortality. Attempts have been made to relate rates of larval supply to environmental variables. Researchers have tried linear techniques (regression analysis) with almost no success.

Dixon et al. (1999) used information on daily observations of spawning output (egg counts) and recruitment (census of recently settled juveniles) of a population of *Pomacentrus amboinensis*, a damselfish species found in the Great Barrier Reef, Australia. Their novel approach was to use techniques from nonlinear time series analysis to determine if the fluctuations in larval supply were due to random effects or to nonlinear, deterministic relationships. As noted in figure 2.19, predictability (measured as the percent increase in the correlation coefficient between predicted and observed values) for larval supply increases as the forecasting algorithms become more nonlinear. The authors concluded that these fluctuations in larval supply can be understood as nonlinear responses of larval fish to a relatively small number of environmental, stochastic variables. The reason for previous failure in predicting larval supply was not because this process is random and thus unpredictable, but because the interaction between these random variables and larval supply is multiplicative (instead of additive, as linear regression techniques assume).

CHAPTER THREE

Spatial Self-Organization

From Pattern to Process

SPACE: THE MISSING INGREDIENT

Tradition in theoretical ecology goes back to the 1920s, with the seminal work of Lotka, Volterra, Nicholson, Bailey, and others. The bulk of this work was based on continuous time equations describing two-species interactions. The only dimension was temporal; that is, space was neglected. Ecology was following the tradition of chemistry and physics by using mean field models. Mean field models are models that provide a statistical description of an average magnitude (e.g., prey density), assuming well-mixed scenarios. For example, the probability of a prey-predator interaction is made proportional to the densities of prey and predator. This means that a predator is able to reach each prey individual with the same probability. Let us illustrate this point with a couple of well-known examples.

Our first example is the so-called Lotka-Volterra two-species competition model (Lotka, 1925; Volterra, 1926). This system describes the dynamics of two competing species. If N_1 is the density of species 1, and N_2 is the density of species 2, we can write:

$$\frac{dN_1}{dt} = r_1 N_1 \left[1 - \frac{N_1 - \alpha N_2}{K_1} \right], \tag{3.1}$$

$$\frac{dN_2}{dt} = r_2 N_2 \left[1 - \frac{N_2 - \beta N_1}{K_2} \right]. \tag{3.2}$$

Here r_1 and r_2 are the growth rates, and K_1 and K_2 are the carrying capacities; α and β are the interspecific competition coefficients, a kind of equivalence from individuals of one species to individuals of the other species.

Model (3.1–3.2) has three nontrivial solutions. Two of them correspond to the extinction of one species $(K_1, 0)$, $(0, K_2)$, and the third corresponds to competitive coexistence (N_1^*, N_2^*). The value of parameters, especially α and β, determines which solution the system will present. Interspecific competition has to be lower than intraspecific competition for coexistence. That is to say, $\alpha\beta$ has to be lower than one. Otherwise, competitive exclusion will take place. Such competitive exclusion was soon confirmed by laboratory experiments, such as the ones using the flour beetle *Tribolium* or different *Paramecium* species (Gause, 1934). The integration of Lotka-Volterra theory plus these first experimental results had a great impact on ecology through the development of the ecological niche theory (Hutchinson, 1965).

One particularly simple illustrative case (to be used later) is defined by symmetric species competition, where the coefficients are the same, that is, $\alpha = \beta$, $K_1 = K_2 = K$, and $r_1 = r_2 = r$. In this case, the coexistence point is given by $N_1^* = N_2^* = (1 + \beta)K/(1 - \beta)$, and it exists and is stable provided that $\beta < \beta_c = 1$. Otherwise, the exclusion points are the only stable attractors. The bifurcation diagram for this dynamical system is shown in figure 3.1a, where we have used the parameter $\Omega = N_1^* - N_2^*$. We can see that the value of the initial condition is important in defining the future history of the dynamics. In spite of the fact that the dynamical system exhibits a strong symmetry, it makes a choice once we leave the $N_1 = N_2$ condition. This phenomenon is known as symmetry breaking and plays an important role in complex systems, particularly in biology (Hopfield, 1994). A mechanical illustration of the idea is also displayed in figure 3.1b. Here the marble defines an unstable equilibrium state. Any small perturbation will shift the marble's position toward *one* of the possible alternative minima. For the two-species competition case, this means shifting from a two-species system to a simpler, single-species situation.

The last result seems a bit ironic, since diversity is one of the main characteristics of our biosphere. On one hand theory predicts simplification and competitive exclusion. On the other hand, we want to understand self-organization and the maintenance of high diversity levels. Since most complex nonlinear systems involving nonlinear interactions display multiplicity of alternative attractors, it turns out to be the case that competitive interactions might typically lead to simplified (low diversity) systems in the long run.

It is becoming clear that competition may have been overemphasized, perhaps because of the strong influence of studies on bird communities in early ecology. Posterior analysis of field data has shown that interspecific competition may not be such a major force in community organization. For example, some *Drosophila* species rarely exclude one another in nature even if they show no traditional resource partitioning and compete strongly (Shorrocks, 1991).

Our second example is the Nicholson and Bailey (1935) model describing the dynamics of a host-parasitoid interaction. Parasitoids are arthropods that put their eggs in, on, or near the surface of their prey. It is a very specific kind of predation, very easy to model. The densities of host and parasitoid at generation $t + 1$ (H_{t+1}, P_{t+1}) are given by:

$$H_{t+1} = \lambda H_t e^{-aP_t}, \tag{3.3}$$

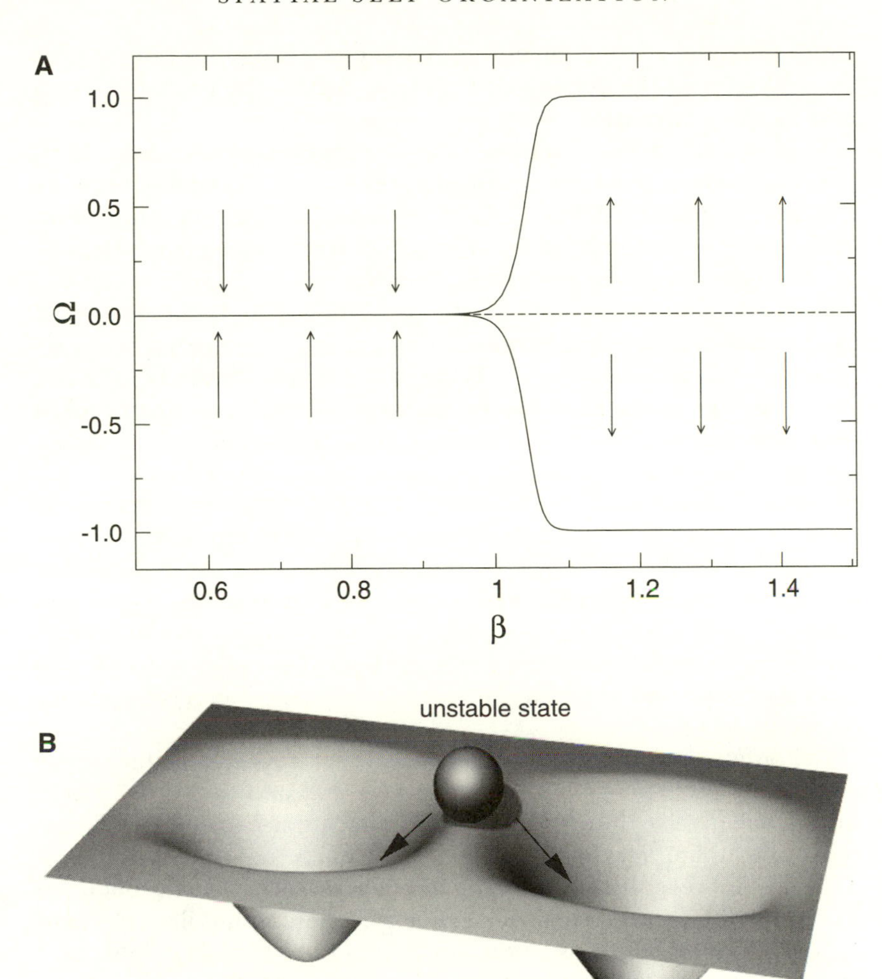

FIGURE 3.1. (A) Symmetry breaking in competitive interactions. For a totally symmetric scenario, the bifurcation diagram is shown. The parameter Ω indicates the difference in populations. After the critical point $\beta_c = 1$ is reached, two basic outcomes are possible: One of the two competitors wins (and thus either $\Omega = 1$ or $\Omega = -1$). The initial conditions totally determine the final outcome. In (B) a mechanical illustration of the symmetry-breaking process is shown (see text).

$$P_{t+1} = cH_t(1 - e^{-aP_t}), \tag{3.4}$$

where H_t and P_t are the densities at the previous generation (t), λ is the per capita host reproduction rate, a is the parasitoid per capita attack rate, and c is the parasitoid per capita emergence rate; e^{-aP_t} is the null term of the Poisson distribution, that is, the probability of a host escaping parasitism.

Interestingly, while Nicholson was intrigued by the way host-parasitoid interactions persist through time, their model leads to very unstable, nondamped fluctuations, with increasing amplitude until both species go extinct. At the same time, field data suggests that interactions lead to cyclic dynamics, though persistence is enhanced.

In both examples, we see that the prediction from theory is toward exclusion and instability. However, observation of the real world suggests that this is not the case. What is the missing ingredient? The reason for the divergence between model predictions and nature could be a combination of several factors including environmental variability, perturbations, or biotic factors not taken into account by these simplistic two-species models. However, even more important than these factors is the omission of space. Space can play very important roles in population stability, as shown by the classic experiments by Huffaker (1958). However, theory has neglected space until recently. As pointed out by Ramon Margalef (1986), this kind of ecological model accepts that "everything happens at a point, without space, a curious view to hold after centuries of making fun about how many angels can dance on the tip of a pin."

In the next sections we will illustrate different ways to introduce space into simple ecological models such as (3.1–3.2) and (3.3–3.4). These models will involve the introduction of explicit space through different types of coupling mechanisms between spatially segregated domains. Then we will review the surprising, often unexpected results obtained by introducing space in theoretical ecology. These results illustrate how spatial self-organization is key to understand population stability and species diversity. They also give new insight into traditional questions such as those regarding spatial synchrony in population dynamics. Finally, once we prove the importance of space, we will discuss the level of realism we need to incorporate since a trade-off between realism and simplicity will often be at work.

TURING INSTABILITIES

The problem of how heterogeneous, ordered spatial structures can emerge from an initially homogeneous system (without any environmental cue) was early studied by Alan Turing (1952). Although Turing's work was presented within the context of morphogenesis, its applicability and implications are widespread (Murray, 1989; Meinhardt, 1982;

Cross and Hohenberg, 1993; Camazine et al., 2001). Spatial structures displaying characteristic length scales are easy to identify in mammal coats (Murray, 1989), segmentation patterns and butterfly spots (Nijhout, 1991), cortical neural structures (Miller et al., 1989), or insect nests (Camazine et al., 2001; Theraulaz et al., 2002) to cite just a few. In ecological systems, regular structures of different types can also be identified. In general, the spatial distribution of individuals tends to be highly heterogeneous. In some cases (such as in semiarid ecosystems, see below) it is clearly regular. The spatial structures that arise from local interactions in spatially extended ecological models are dissipative structures: They are maintained away from equilibrium by a flow of energy, matter, and information. Energy is dissipated in the process (thereby generating entropy). If the external input stops, the pattern goes away. It is important to note that in this type of structure the pattern features exhibited at the macroscopic scale bear no relation to the size of its constituents. In our context, this means that the characteristic scale of population distribution patterns is much larger than the size of its constituents (interacting individuals), and this scale is robust in the face of perturbations (Ball, 1999). Since such a decoupling between spatial pattern and individual features is a key property, two important consequences immediately emerge. First, there is no need to include a detailed description of individual units. And second, emergent phenomena might arise at the level of macroscopic structures, thus introducing levels of selection and adaptation that go beyond the individual level.

Two basic ingredients are considered that are essential within our context: interactions among "species" (morphogens in Turing's work) and diffusion. Since interactions have in many cases a clearly local character, it is not obvious how large-scale structures can emerge under the effects of diffusion. Diffusion is a rather destructive force: It tends to change things toward uniformity, instead of helping create nonuniform patterns. This is easily illustrated by the following two-compartment example in figure 3.2. Let us consider a given gas confined in a box that is connected to another, empty box of the same size, through some tube with a valve. Let us call C the initial concentration of molecules in the first compartment. Once the gas starts flowing after opening the valve, the concentrations in each box (indicated as x_1 and x_2) will change in time, according to:

$$\frac{dx_1}{dt} = D(x_2 - x_1), \tag{3.5}$$

$$\frac{dx_2}{dt} = D(x_1 - x_2). \tag{3.6}$$

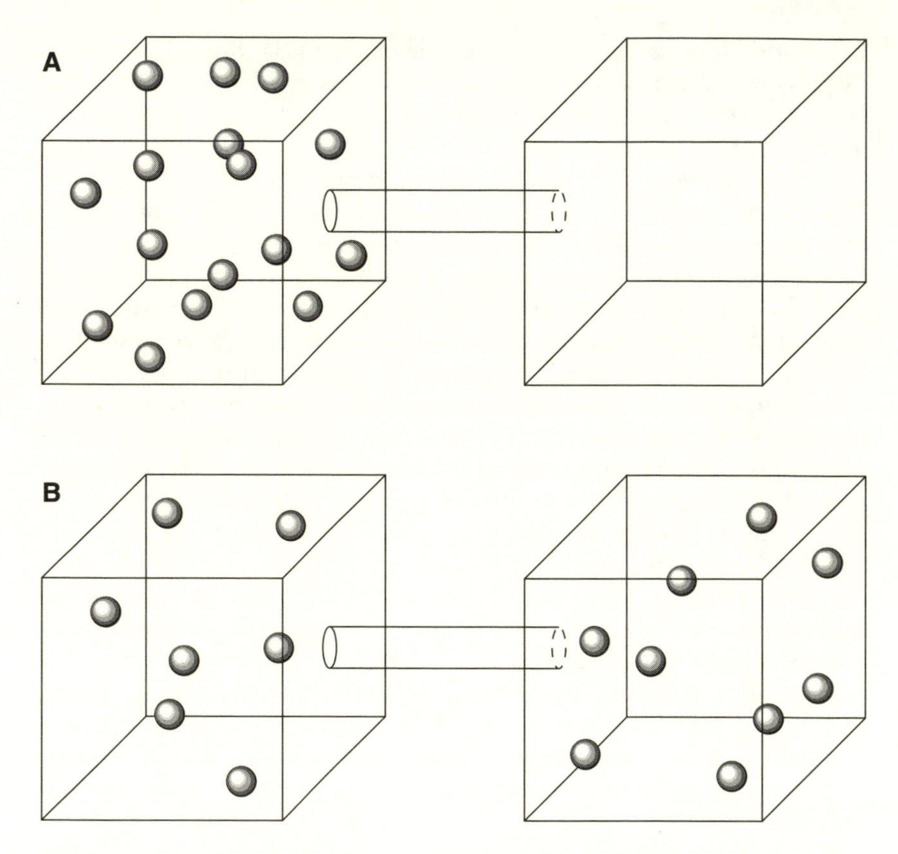

FIGURE 3.2. Role of dispersal in systems located at equilibrium: The figure illustrates a gas in a system composed of two compartments connected by a valve. At the beginning, the gas is located in the left-hand side compartment and the valve is closed (A). After the valve is opened, diffusion will distribute the gas homogeneously through both compartments (B).

Since we assume a closed system, $x_1 + x_2 = C$, we thus have:

$$\frac{dx_2}{dt} = D(C - 2x_2), \tag{3.7}$$

which gives an evolution equation:

$$x_2(t) = \frac{C}{2}\left[1 - e^{-2Dt}\right]. \tag{3.8}$$

As expected from our intuition, the final (attractor) state is simply $x_1^* = x_2^* = C/2$: The gas uniformly distributes over space. Any initial condition will eventually end in this equilibrium state.

The previous result is not surprising, given the linear nature of the diffusion term. It tells us that when interactions occur in a linear fashion, no pattern can emerge. But a key point makes diffusion necessary in our context: It actually propagates information, and in this sense, it can serve as an appropriate channel in order to create structure. The other component, interaction, is crucial in generating and maintaining the pattern. The surprising result arrived at by Turing was that, under well-defined conditions, nonlinear local interactions together with diffusion can generate complex spatial structures with well-defined characteristic scales. Although the introduction of a continuous space implies the presence of infinite degrees of freedom, a suitable perturbative procedure is able to reduce the problem to a finite number of variables (Nicolis and Prigogine, 1990). This small set of variables effectively describes the system and determines the dynamics of the process. This is ultimately a consequence of the dissipative character of far-from-equilibrium systems (Akhromeyeva et al., 1989). In this section we will summarize the main formal steps in Turing's derivation (for a detailed account, see Murray, 1989). Most of the analysis involves deterministic (and thus noise-free) models. It is worth mentioning, however, that some Lotka-Volterra models that cannot display patterns under deterministic conditions can show spatial structures under the presence of noise (Vilar et al., 2003).

The starting point in Turing's analysis is a set of partial differential equations, also known as *reaction-diffusion* (RD) equations. In one spatial dimension, they have the general form:

$$\frac{\partial C_i}{\partial t} = f_\mu^{(i)}(C_1, \ldots, C_n) + D_i \nabla^2 C_i, \qquad i = 1, 2, \ldots, n \tag{3.9}$$

where $C_i = C_i(x,t)$ indicates the local concentration of a given chemical i (population abundance in our context) at a spatial point $x \in \Gamma$ at the time step t. Here Γ is the spatial domain of interest. The simplest case considered by Turing is the two-species system, that is,

$$\frac{\partial C_1}{\partial t} = f_\mu^{(1)}(C_1, C_2) + D_1 \frac{\partial^2 C_1}{\partial x^2}, \tag{3.10}$$

$$\frac{\partial C_2}{\partial t} = f_\mu^{(2)}(C_1, C_2) + D_2 \frac{\partial^2 C_2}{\partial x^2}, \tag{3.11}$$

where the functions $f_\mu^{(i)}(C_1, C_2) (i = 1, 2)$ introduce the explicit form of the interactions, and D_1, D_2 are the diffusion constants associated to each species.

The first step here is to consider the fixed points of the nonspatial counterpart (as we did in chapter 2). If $D_1 = D_2 = 0$, the equilibrium points are $\mathbf{C} = (C_1^*, C_2^*)$ such that $f_\mu^{(1)}(C_1^*, C_2^*) = f_\mu^{(2)}(C_1^*, C_2^*) = 0$. It is not difficult to show that these points are actually solutions of the full RD system provided that the system is spatially homogeneous, that is, $C_1(x,t) = C_1^*$ and $C_2(x,t) = C_2^*$ for all $x \in \Gamma$. Following the same spirit of the linear stability analysis (chapter 2), we consider here small fluctuations around the spatially homogeneous equilibrium state, for instance,

$$C_1(r,0) = C_1^* + \delta C_1,$$

$$C_2(r,0) = C_2^* + \delta C_2,$$

where perturbations are assumed to be very small, such as, $|\delta C_i| << C_i^*$, for $i = 1, 2$. Using a simpler notation, $c_i \equiv \delta C_i$, the stability analysis is performed on the linearized equations:

$$\frac{\partial c_1}{\partial t} = L_{11}c_1 + L_{12}c_2 + D_1 \frac{\partial^2 c_1}{\partial x^2}, \tag{3.12}$$

$$\frac{\partial c_2}{\partial t} = L_{21}c_1 + L_{22}c_2 + D_2 \frac{\partial^2 c_2}{\partial x^2}, \tag{3.13}$$

where $L_{ij} = \partial f_\mu^i / \partial C_j$ are the components of the Jacobi matrix $\mathbf{L}_\mu$.

Next, an explicit form of the perturbations needs to be introduced. The standard choice is (Turing, 1952; Okubo, 1980; Murray, 1989)

$$\begin{pmatrix} c_1 \\ c_2 \end{pmatrix} = \begin{pmatrix} A_1 \\ A_2 \end{pmatrix} \cos(kx) e^{\sigma t}, \tag{3.14}$$

where the perturbation is composed of an exponential, time-dependent term ($e^{\sigma t}$) and a periodic, spatial term (here $\cos(kx)$). The reason for this choice is simple. Any spatial perturbation can be decomposed in terms of an infinite superposition of periodic functions (the so-called Fourier decomposition). Here

$$k = \frac{n\pi}{L} \tag{3.15}$$

is the so-called wavenumber. The length of the spatial domain Γ is L and n is an integer. The wavenumber is inversely proportional to the wavelength λ:

$$\lambda = \frac{2\pi}{k} \tag{3.16}$$

and thus introduces the specific length scale. The sign of σ will decide whether or not a perturbation of length λ will decay (so that no structure of this length will be observed) or get amplified (and thus be observable).

Here the spatially homogeneous solution $C_1(x,t) = C_1^*$ and $C_2(x,t) = C_2^*$ for all $x \in \Gamma$ will be assumed stable. This simply requires stability of the nonspatial model, that is, $L_{11} + L_{22} < 0$ and $L_{11}L_{22} - L_{21}L_{12} > 0$ (see chapter 2). Now, using the explicit form of the spatial perturbations and replacing them in the previous linear equations, we obtain:

$$A_1\left[L_{11} + D_1 k^2 - \sigma\right] + A_2 L_{12} = 0 \quad (3.17)$$

$$A_1 L_{21} + A_2\left[L_{22} + D_2 k^2 - \sigma\right] = 0. \quad (3.18)$$

A nontrivial solution will exist provided that

$$\begin{vmatrix} L_{11} + D_1 k^2 - \sigma & L_{12} \\ L_{21} & L_{22} + D_2 k^2 - \sigma \end{vmatrix} = 0, \quad (3.19)$$

which gives a characteristic equation for the eigenvalues:

$$\sigma^2 + \left[L_{11} + L_{22} + (D_1 + D_2)k^2\right]\sigma + \left(L_{11} + D_1 k^2\right)\left(L_{22} + D_2 k^2\right)L_{12}L_{21} = 0. \quad (3.20)$$

The spatially homogeneous state will be unstable provided that at least one eigenvalue is positive. This will occur when at least one of the two following inequalities does not hold:

$$L_{11} + L_{22} - (D_1 + D_2)k^2 < 0 \quad (3.21)$$

$$\mathcal{H}(k) = (L_{11} - D_1 k^2)(L_{22} - D_2 k^2) - L_{21}L_{12} > 0. \quad (3.22)$$

Since D_1, D_2, and k^2 are all positive, the first inequality always holds if $L_{11} + L_{22} < 0$. So we only need the last inequality to be reversed, that is, $\mathcal{H}(k) < 0$. By writing this function as follows:

$$\mathcal{H}(k) = D_1 D_2 (k^2)^2 - (D_1 L_{22} + D_2 L_{11})k^2 + Det(L_\mu) < 0, \quad (3.23)$$

we can see that it defines a parabola as a function of k^2. This defines a minimum at

$$k_c^2 = \frac{1}{2}\left(\frac{L_{22}}{D_2} + \frac{L_{11}}{D_1}\right), \quad (3.24)$$

the lower boundary at which instability will be obtained, given by:

$$L_{11} + L_{22} > 2\sqrt{d}\sqrt{det(\mathbf{L}_\mu)}, \quad (3.25)$$

where $d = D_2/D_1$. It is not difficult to show that single-species models are unable to develop spatial instabilities of this type.

Close to the critical mode k_c, the first bifurcation occurs toward a spatially unhomogeneous system. Given the fact that the original spatial symmetry (invariance) has been broken, Turing instabilities are actually

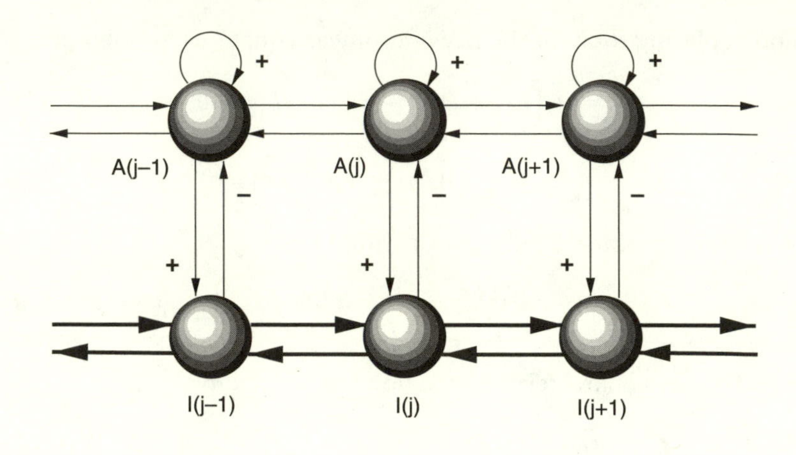
+
+
+
A(j–1)
A(j)
A(j+1)
–
–
–
+
+
+
I(j–1)
I(j)
I(j+1)

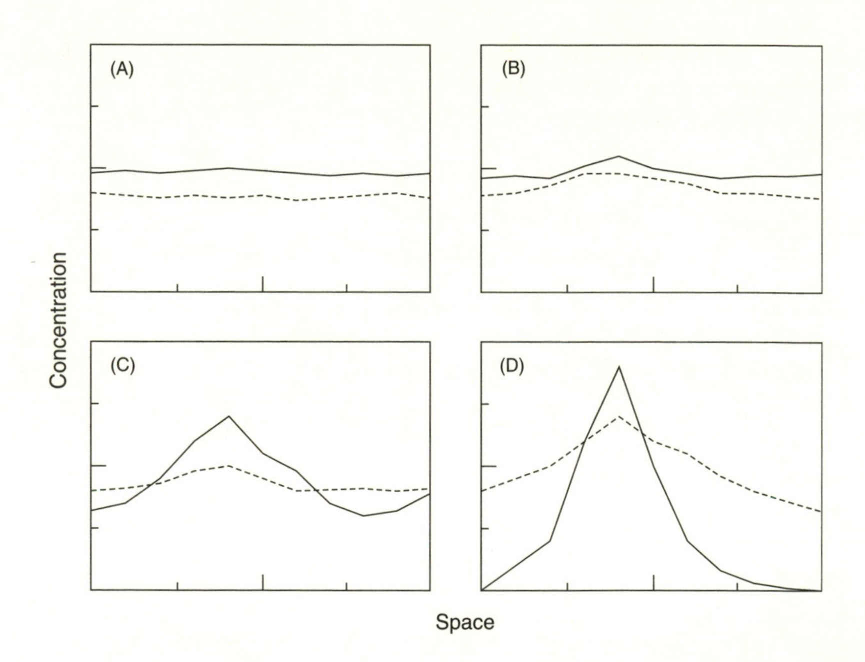
(A)
(B)
(C)
(D)
Concentration
Space

an example of *space symmetry breaking*. Intuitively, the origin of spatial structures is the cooperation between nonlinear reaction and diffusion (Nicolis and Prigogine, 1990). The activator grows autocatalitically, so those fluctuations from the steady state with the appropriate wavelength will be amplified. The mechanism is as follows. A higher concentration of activator triggers its own production (and that of its inhibitor). The inhibitor diffuses faster, and thus the rate activator/inhibitor is higher in the center of the fluctuation but decreases toward the periphery. The excess of inhibitor in the periphery produces long-range inhibition, and a peak of activator surrounded by an area of lower concentration is formed (fig. 3.3).

Turing-like instabilities are likely to occur in ecological systems (Levin and Segel, 1976; Segel and Jackson, 1972). Actually, the most common conditions under which Turing patterns arise map those expected from predator-prey models. Standard models of pattern formation require the presence of an *activator* and an *inhibitor*. The first activates its own synthesis and the growth of the second. The second inhibits the growth of the first. Additionally, it is required that the diffusion rate of the second be much larger than that of the first. Under these conditions, Turing patterns are easily observed (Meinhardt, 1982; Murray, 1989). It is not difficult to see that these conditions are fulfilled by a wide range of predator-prey or vegetation-grazer systems, and field observations support this view (Bascompte and Solé, 1998b; see below).

Turing Structures in Predator-Prey Systems

Not all models of predator-prey interactions display Turing patterns. As an example, prey-dependent models where the predator functional response does not depend on predator abundance are unable to display

---

FIGURE 3.3. Pattern formation through a diffusive-driven instability. The upper drawing schematically summarizes the basic interactions between the activator (A) and the inhibitor (I) in a multicompartment system (here each compartment is indicated by $j = 1, \ldots, N$). Horizontal arrows indicate diffusion among compartments. Below, each panel represents the concentration of two morphogens: activator (solid line) and inhibitor (broken line) through a spatial dimension, at a given time. (A) represents an equilibrium distribution, from which there are small fluctuations. Some fluctuations will be amplified due to the interaction between local autocatalysis of the activator and long-range inhibition by the inhibitor. At the end, a time-invariant inhomogeneity (Turing structure) is obtained: There is a peak in the activator concentration at some particular location. Graph reprinted from *Encyclopedia of Evolution*, edited by Mark Pagel. Copyright 2002, used by permission of Oxford University Press, Inc.

spatial patterns (Segel, 1972). For this particular system, we have

$$\frac{\partial N}{\partial t} = f(N)N - g(N)P + D_N \frac{\partial^2 N}{\partial x^2}, \tag{3.26}$$

$$\frac{\partial P}{\partial t} = eg(N)P - \mu P + D_P \frac{\partial^2 P}{\partial x^2}, \tag{3.27}$$

and it can be shown that the components of the Jacobi matrix are:

$$L_{11} = f(N^*) + N^* \left.\frac{df}{dN}\right|_{N^*} - \left.\frac{dg}{dN}\right|_{N^*} P^*, \tag{3.28}$$

$$L_{12} = -g(N^*) = -\frac{\mu}{e}, \tag{3.29}$$

$$L_{21} = e \left.\frac{dg}{dN}\right|_{N^*} P^*, \tag{3.30}$$

$$L_{22} = eg(N^*) - \mu. \tag{3.31}$$

It can be seen that L_{22} must cancel. Thus, the stability condition (3.21) will become $L_{11} < 0$, while condition (3.25) will become

$$L_{11} > \frac{2\sqrt{d}}{d}\sqrt{det(L_\mu)} > 0, \tag{3.32}$$

and therefore, the two conditions cannot be fulfilled simultaneously.

But other models are able to show diffusive instabilities. A classical example is provided by Segel and Jackson (1972), where predator density-dependent mortality rates plus an autocatalytic effect on prey growth rates are assumed. More recently, other interesting examples are provided by ratio-dependent predator-prey models (Bartumeus et al., 2001) and predator-dependent (Alonso et al., 2002) predator-prey models. Assuming predator density-dependent mortality rates, patterns can emerge (Segel, 1972). The predator functional response is the essential link describing a predator-prey model. Beddington (1975) was the first to call attention to the effect of mutual interference between predators on searching efficiency, and proposes a simple formulation for the rate of prey consumption for an average predator:

$$g(N, P) = \frac{bN}{B + kP + N}, \tag{3.33}$$

where b is a maximum consumption rate, k is a (positive) predator interference parameter, and B is a saturation constant. This expression actually defines a quite general functional response excluding the case where predators benefit from co-feeding.

In terms of dimensionless variables, the model is described by the following reaction-diffusion equations:

$$\frac{\partial n}{\partial t} = (1-n)n - \frac{\beta np}{p+n} + \frac{\partial^2 n}{\partial x^2}, \tag{3.34}$$

$$\frac{\partial p}{\partial t} = \epsilon\beta\frac{np}{p+n} - \eta p + d\frac{\partial^2 p}{\partial x^2}, \tag{3.35}$$

where n and p are prey and predator densities, respectively.

Defining a new dimensionless parameter, $\Delta = \epsilon\beta/\eta$, the system has a homogeneous coexistence point, (n^*, p^*), where $n^* = 1 - (\eta/\epsilon)(\Delta - 1)$ and $p^* = (\Delta - 1)n^*$. So, in order to have a feasible coexistence point, two conditions are needed $\eta > \epsilon(\beta - 1)$ and $\eta < \epsilon\beta$. Using the standard definition of the elements of the community matrix and defining four auxiliary functions:

$$g(\eta) \equiv (\epsilon - 1)\,\eta^2 - \epsilon^2\beta\,\eta + \beta\epsilon^2(\beta - 1), \tag{3.36}$$

$$f(\eta) \equiv \eta^2\,(\epsilon\beta - \eta)\,\epsilon^2\beta\,(\eta - \epsilon\,(\beta - 1)), \tag{3.37}$$

$$F(\eta) \equiv \sqrt{f(\eta)}, \tag{3.38}$$

$$G(\eta, d) \equiv \frac{d\,(\beta - 1)\,\epsilon^2\beta}{2\sqrt{d}} - \frac{\epsilon^2\beta}{2\sqrt{d}}\,\eta + \frac{(\epsilon - d)}{2\sqrt{d}}\,\eta^2. \tag{3.39}$$

The set of three conditions for Turing instability can be rewritten respectively as follows:

$$g(\eta) < 0, \tag{3.40}$$

$$f(\eta) > 0, \tag{3.41}$$

$$G(\eta, d) > F(\eta). \tag{3.42}$$

When the previous inequalities hold, spatially uniform steady-state predator-prey coexistence is no longer stable. Small random fluctuations will be strongly amplified by diffusion leading to nonuniform population distributions. An example of the patterns is shown in figure 3.4.

The generation of these structures can be understood in intuitive terms as explained before for the general case of an activator and an inhibitor (fig. 3.3). We just have to think in terms of prey as activator and predator as inhibitor. Increased local predator density has a negative effect in per capita growth rates. The functional form of the prey-predator terms ensures a higher prey consumption at higher

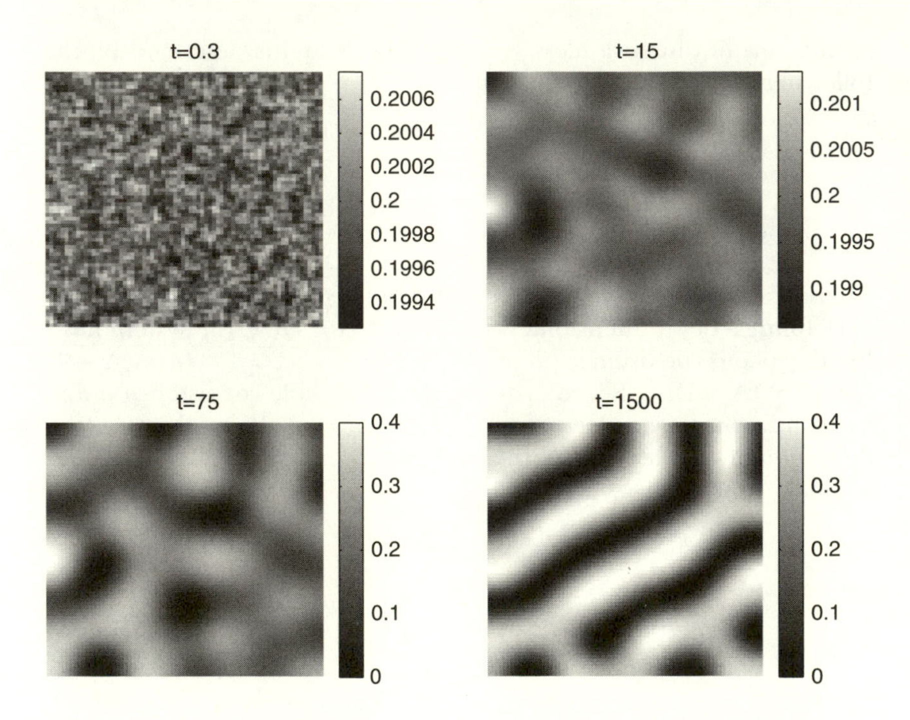

FIGURE 3.4. Pattern formation in a ratio-dependent model with diffusion. Here the two-dimensional patterns are shown for a two-dimensional domain at four different times.

predator densities. The inhibitor mechanism on prey densities is then guaranteed. Predator interference in the functional response is responsible for the self-inhibition effect of increased predator densities on the growth of predator population. Thus, as random fluctuations increase local prey populations beyond their equilibrium value, prey populations experience accelerated growth. Simultaneously, predator population increases, but as predators diffuse faster than prey, they disperse away from the center of prey outbreaks, and spatially uniform population distributions break down, and two different spatial domains arise. At the outbreak center, prey growth rate remains positive. As a consequence, preys increase out of predator control. Henceforth, the proportion of prey to predator in those central areas increases. Nevertheless, on the border of prey outbreaks, the proportion of prey to predator becomes lower and lower. If relative diffusion d is large enough, prey growth rate will locally reach negative values and prey will be driven by predators to very low levels. The final result is the formation of patches of high prey density surrounded by areas of low prey densities.

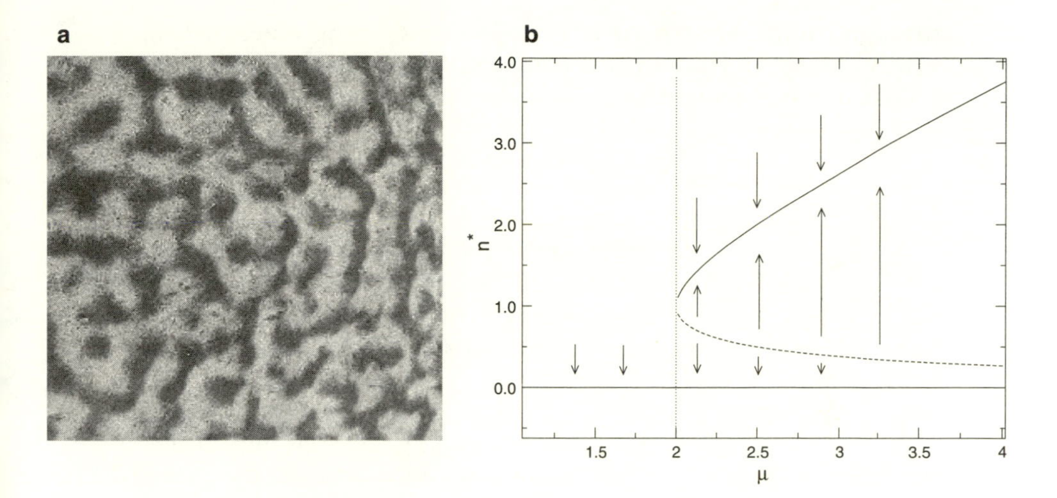

FIGURE 3.5. Pattern formation in semiarid ecosystems. In (a) an example of the regular spatial patterns is shown, illustrating the presence of a labyrinth structure with a characteristic wavelength. In (b) the bifurcation diagram for the nonspatial Klausmeier model is shown. For $\mu < \mu_c = 2$, the only available equilibrium point is the desert state ($n^* = 0$). After the bifurcation point μ_c is crossed, two additional states are available. We can appreciate a shift between the desert and vegetated states at the bifurcation point.

The presence of spatial derivatives in partial differential equation models is not restricted to linear diffusion terms. Other types of spatial dependence might also appear and give rise to spatial patterns. The following two sections consider a particularly important example.

Regular Patterns in Semiarid Vegetation

A remarkable example of regular spatial patterns in ecology is provided by vegetation distribution in semiarid regions. In figure 3.5a, an example of these patterns (here the perennial grass *Paspalum vaginatum*) is shown: Darker and lighter areas correspond to vegetated and nonvegetated parts, respectively. There are actually different types of patterns, and a surprising range of structures is displayed by the spatial distribution of plants, including bands, stripes, spots, holes, and labyrinths (von Hardenberg et al., 2001). Mathematical models involving Turing instabilities have been proposed to explain these patterns (Lefever and Lejeune, 1997; Klausmeier, 1999; von Hardenberg et al., 2001; Meron et al., 2004; Shnerb et al., 2003).

The driving forces that dominate vegetation distribution in arid lands are water availability, competitive interactions, and water redistribution through runoff. One of the first models considering water dynamics, which accounts for the striped patterns, was proposed by Klausmeier (2001), and it is defined by the following pair of equations:

$$\frac{\partial W}{\partial t} = A - LW - RWN^2 + V\frac{\partial W}{\partial r}, \tag{3.43}$$

$$\frac{\partial N}{\partial t} = RJWN^2 - MN + D\left(\frac{\partial^2 N}{\partial x^2} + \frac{\partial^2 N}{\partial y^2}\right), \tag{3.44}$$

where water (W) and plant biomass (N) interact on a two-dimensional domain. Here water is uniformly supplied at a rate A and lost through evaporation at a rate LW. Vegetation grows by taking in water at a rate $RJWN^2$ where J is the yield of plant biomass and R is a constant. Plant mortality (both due to senescence and grazing) is introduced by the linear term MN. An important difference between the previous RD models is the nature of the spatially dependent term of water dynamics, which introduces water flow downhill (here in the negative x-direction). Here $V\partial W/\partial r$ is called the *advection* term and V is the (constant) downhill runoff flow velocity. This model is further simplified to its adimensional form (Klausmeier, 1999):

$$\frac{\partial w}{\partial t} = a - w - wn^2 + v\frac{\partial w}{\partial r}, \tag{3.45}$$

$$\frac{\partial n}{\partial t} = wn^2 - mn + \left(\frac{\partial^2 n}{\partial x^2} + \frac{\partial^2 n}{\partial y^2}\right), \tag{3.46}$$

and thus only three relevant parameters must be considered, associated to: water input (a), plant losses (m), and downhill flow (w). By analyzing the nonspatial counterpart, it can be shown that multiple states are present: one vegetated and the other bare. The bare state, $P_0 = (w^* = a, n^* = 0)$ always exists and is stable. This can be shown by computing the eigenvalues of the Jacobi matrix,

$$L_\mu(P^*) = \begin{pmatrix} -1 - n^2 & -2wn \\ n^2 & 2wn - m \end{pmatrix}_{P^*}, \tag{3.47}$$

which for P_0^* is:

$$L_\mu(P_0^*) = \begin{pmatrix} -1 & 0 \\ 0 & -m \end{pmatrix}, \tag{3.48}$$

and thus the eigenvalues are $\sigma_1 = -1$ and $\sigma_2 = -m$. Two additional equilibrium points are present, given by $P_\pm^* = (w^* = m/n_\pm^*, n*_\pm)$ with

$$n_\pm^* = \frac{1}{2}\left[\frac{a}{m} \pm \sqrt{\frac{a^2}{m^2} - 4}\right]. \tag{3.49}$$

These nontrivial states exist provided that $a > 2m$ (i.e., water input is at least twice as large as plant loss). One of these points is always unstable and can be ignored. The second is linearly stable for ecologically reasonable parameter values (Klausmeier, 1999). The bifurcation diagram for this system is shown in figure 3.5b, using $\mu = a/m$ as a bifurcation parameter.

When considering the full model with diffusion, it is shown to present instabilities similar to those displayed by Turing patterns. The result of this analysis shows that, for a fixed water velocity, the domain of stripe patterns appears between the bare soil phase and the spatially homogeneous phase. In the stripe phase, the wavelength of the patterns decreases as water input rate grows, consistent with field observations. Similarly, increased levels of grazing (for a constant a) lead to bare soil after a threshold is reached, and decreasing levels lead to a replacement of stripes by homogeneous vegetation. Both transitions have been actually reported.

Hysteresis and Catastrophic Shifts

A more sophisticated model is able to explain not only the stripe patterns, but also the presence of spots and holes (von Hardenberg et al., 2001). Again, the model considers biomass density n and ground water density w as the key variables. The model (in nondimensional form) reads:

$$\frac{\partial w}{\partial t} = p - (1 - \rho n)w - w^2 n + \delta \nabla^2 (w - \beta n) - v \frac{\partial (w - \alpha n)}{\partial r}, \tag{3.50}$$

$$\frac{\partial n}{\partial t} = \phi(w)n - n^2 - \mu n + \nabla^2 n. \tag{3.51}$$

Some of the previously introduced terms in Klausmeier's model are easily identified. Here p stands for precipitation, $\nabla^2 n$ for plant dispersal, and $-\mu n$ for vegetation decay. The rest of the terms require some attention. Vegetation now grows if water is available, but the growth term, here $\phi(w) = \gamma n/(1 \delta w)$, is a saturating function. We can see that the first equation can actually be written in a familiar, logistic form,

$$\frac{\partial n}{\partial t} = n \left[\gamma_w - n \right] + \nabla^2 n, \tag{3.52}$$

where $\gamma_w = \phi(w) - \mu$. Water dynamics also includes interesting features. First, the loss term representing evaporation is now $-(1 - \rho n)w$: Vegetation reduces evaporation due to shading and other mechanisms. The local uptake of water by plants, $-w^2 n$, is based on real transpiration curves (Hillel, 1998). Similarly, biological data support the use of the term $\nabla^2(w - \beta n)$, where suction of water by roots is considered. Finally,

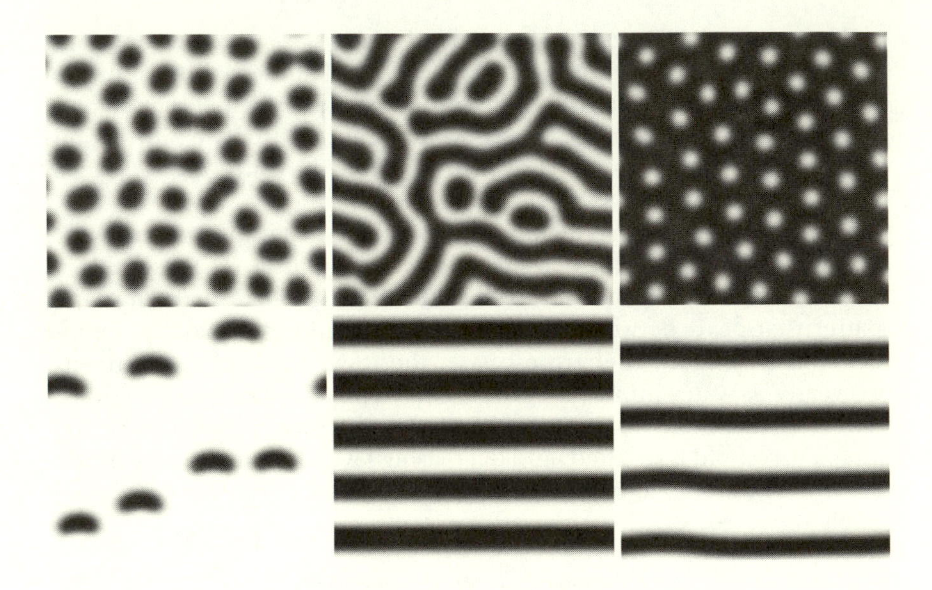

FIGURE 3.6. Spatial patterns generated by the von Hardenberg et al. (2001) model. The upper row shows three different snapshots of the stable structures obtained for plain landscape ($v = 0$) and different precipitation regimes (increasing p from left to right). The lower sequence shows the banded and dashed snapshots for the same domain of parameter values but this time including water runoff (figures kindly provided by E. Meron).

the runoff term is slightly modified to a new form $-v(\partial h/\partial x)$ with $h \equiv w - \alpha n$ defining the runoff height, which in the absence of vegetation reduces to the same term used in Klausmeier's model. Here the effects of vegetation density are introduced as a linear drop in water runoff.

Using a realistic range of parameters, von Hardenberg et al. numerically explored the dynamics exhibited by their model, using the precipitation rate p as a bifurcation parameter. When runoff was not considered (i.e., flat landscape) the sequence spot-stripe-hole was recovered by tuning p from small to high values (fig. 3.6, upper row). When runoff is included ($v \neq 0$), bands of vegetation are the common structures (fig. 3.6, lower row). Additionally, transients toward each of these stationary states involve other types of forms, such as rings, which have also been observed in the field. The numerical analysis allows definition of several phases:

1. Dry subhumid: Only uniform vegetation or vegetation patterns are allowed.

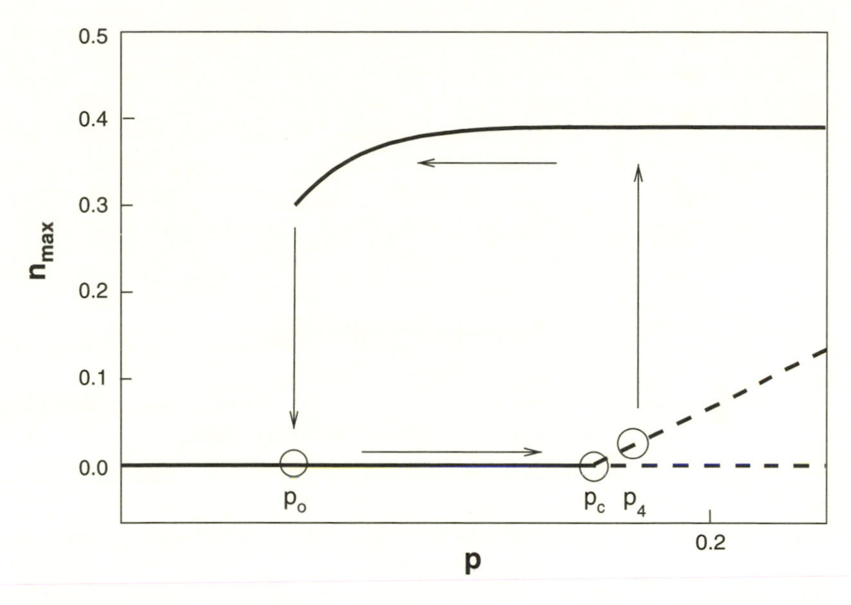

FIGURE 3.7. Biomass amplitude n_{max} in the low precipitation regime. The figure illustrates the view of desertification as a hysteresis loop (redrawn from von Hardenberg et al., 2001).

2. Semiarid: The system does not support uniform vegetation or bare soil. Patterns are the only allowed structures.
3. Arid: Either bare soil or vegetation patterns are possible attractors, with n^* being of low amplitude (low biomass density).
4. Hyper-arid: The only stable state is bare soil.

Each of these phases is determined by well-defined precipitation thresholds and the transitions are sharp.

An important consequence of these models, made by dedicated field observations, is the presence of hysteresis (see chapter 2) already predicted by nonspatial models of soil degradation in terrestrial systems (van de Koppel et al., 1997). In figure 3.7, the hysteretic loop associated to climatic changes (different precipitation regimes) is shown. Here the biomass amplitude n_{max} is plotted for different precipitation ranges. We can see that, starting from the upper branch (vegetation pattern), reduced water availability eventually shifts the system toward the bare soil (desert) state (lower branch). Once this state is reached, perhaps as a consequence of persistent drought, the system cannot return to its previous state just by a return to the previous precipitation level. In fact, it might happen that the hysteresis loop can be closed at much higher levels of precipitation. A similar scenario can be described in terms of overgrazing (Von Hardenberg et al., 2001). In both cases, catastrophic shifts from vegetated to desert scenarios are possible. It is worth mentioning that the fact that the spatial distribution of vegetation changes with p might allow one to predict whether such a shift is approaching so that appropriate (and economically feasible) measures can be adopted.

COUPLED MAP LATTICE MODELS

In the previous sections we considered continuous space models, space and time being treated as continuous variables. In other situations, space is discretized (as time and number of individuals, see previous chapter). Discretization of space is a useful shortcut when computational requirements are an important constraint to modeling approaches, and it is a common strategy in statistical physics (Marro and Dickman, 1999). But the use of discretized variables is not only a good qualitative practical solution. In many cases, toy models of dynamical systems defined in a continuous space are actually very well represented by some discretized counterpart. Discretization can be introduced at different levels: Time can be discrete (as with models based in discrete maps), space can be discrete (as it occurs with the lattice models we will introduce in this section), and the available range of states adopted by system variables can also be discrete. The extreme approach is well represented by cellular automata (see chapter 4) where space, time, and states are all discrete (Farmer et al., 1984; Wolfram, 1984, 1986; Langton, 1989; Wuensche and Lesser, 1992; Ermentrout and Edelstein-Keshet, 1993; Ilachinski, 2001). Although different approaches will be more appropriate for different goals, it is often possible to go from one approach to another under some suitable conditions. As an example, cellular automata can be modeled by means of partial differential equations when a system's responses are highly nonlinear, saturating functions (Omohundro, 1984; Nijhout, 1992). More important, discretization is often successful because it captures the essential features of the interactions that ultimately are responsible for macroscopic patterns, whose properties are not dependent upon the interaction (lattice) scale. The success of modeling complex fluid dynamics by means of so-called lattice gases (Wolf-Gladrow, 2000) is a good example of its power. In other situations, discrete space is a better description of reality than are continuous space models, where populations move around a set of discrete patches of available habitat. In the last scenario, the best strategy is to model space as a discrete lattice of points.

One easy way to introduce space into discrete time models such as (3.3–3.4) is by means of the coupled map lattice formalism. Coupled map lattices (CMLs) were first defined in 1981 by the Japanese physicist Kunihiko Kaneko as a simple model for studying spatiotemporal chaos (Kaneko 1984, 1987, 1992, 1993, 1998; Crutchfield and Kaneko, 1988). A CML can be defined as a dynamical system with discrete time, discrete space, and continuous local populations. As its name indicates, the idea is to couple a set of n maps or discrete time models by means of dispersal. In this way, space and time are defined on a lattice substrate, but dynamics occur within a continuous range.

This type of theoretical approach has been successfully used in many different contexts. It has been used as a powerful model of pattern formation in biology, physics, and chemistry (Kaneko and Tsuda, 2000). Coupled map lattices were first introduced into ecology in the early 1990s (Solé and Valls 1991; Solé et al., 1992a, b, c; Bascompte and Solé, 1995). Similar modeling approaches had independently been used by other ecologists (Hastings, 1993; Hassell et al., 1991, 1994; Comins et al., 1992). The number of papers using CMLs as a spatiotemporal modeling approach in ecology has since then risen quickly (Rohani and Ruxton, 1999, White et al., 1996; Bascompte and Solé, 1995, 1998a).

Different ways of coupling can be used. For example, one can couple one lattice site with its nearest neighbors or with the entire lattice (global mixing). Imagine a map of the form $x_{t+1} = F_\mu(x_t)$, which introduces the particular features of local dynamics in the absence of coupling. As a case study, we might consider the logistic map (chapter 2) that is, $F_\mu = \mu x(1 - x)$. A simple example of a spatially extended CML is obtained by considering a one-dimensional lattice Γ (for example, vegetation patches following a shoreline). The simplest type of coupling is global. Globally coupled maps (GCMs) have been extensively analyzed and serve as a mean field approach to nonlocal systems. Global mixing is introduced as follows:

$$x_{t+1}(i) = (1 - D)F(x_t(i)) + \frac{D}{N-1} \sum_{j=1, j\neq i}^{N-1} F(x_t(j)), \qquad (3.53)$$

(with $i = 1, \ldots, N$) denoting the spatial position. Here D is the fraction of individuals leaving its patch and moving to any of the $N - 1$ other patches. If $D = 0$, then we have a set of uncoupled maps. The above model has a remarkably rich behavior, in spite of its simplicity. One particularly key feature is that it exhibits clustering of elements that oscillate synchronously, and thus heterogeneous groups of elements emerge in spite of the global connection (Kaneko, 1984). Such richness only increases when local interactions are considered.

A different situation is that in which dispersal is reduced to the nearest neighboring patches. Following the one-dimensional system, dispersal can only take place to one of the nearest neighbors (right or left sites). We can write the population density in site i at generation $t+1$ ($x_{t+1}(i)$) as follows:

$$x_{t+1}(i) = (1-D)F(x_t(i)) + \frac{D}{2}\Big[F(x_t(i-1)) + F(x_t(i+1))\Big], \qquad (3.54)$$

for $i = 2, \ldots, N-1$. Patches located at the boundary $\partial\Gamma$ have to be updated in a particular way (here $\delta\Gamma = \{1, N\}$). Three main boundary conditions (BC) are usually considered: periodic, reflecting, and absorbing. In periodic BC, each end of the lattice is linked to its opposite one as in a torus, so individuals moving north of the first row are reappearing at the last row. In reflecting BC, individuals that would disperse out are maintained in the same cell, that is, they are sent back to the same patch. In absorbing BC, the boundary is surrounded by a ring of permanently empty patches where dispersers are lost. Some quantitative results may depend on the kind of BC, but all the qualitative results here reviewed are the same. A diffusion-like approach can also be used (Waller and Kapral, 1984; Kapral, 1985; Alstrom and Stassinopoulos, 1992) where the standard Laplacian operator is introduced as a discrete term. However, there is a risk of obtaining unbounded solutions when diffusion-like terms are translated into the CML formalism (Jackson, 1991; Hassell et al., 1995). This is why it is more reasonable and safe to employ the previous implementation, in which the diffusive processes are separated from the f_μ-interaction term.

Equation (3.54) is a simple dynamical system with a big potential to describe a wide range of situations in biology where we have a set of coupled oscillators. Note that CMLs are very general; we could be considering a phenotype space (instead of a physical space, see last section in this chapter), where D would have the meaning of mutation. In the following sections, equation (3.54) or its two-dimensional counterpart describes the density of a population in a patch as a combination of the dynamics in that patch (e.g., density dependence), and dispersal to nearest patches.

A wide range of new results on systems such as (3.54) have been obtained. These are related to the properties of spatiotemporal systems where dispersal or diffusion couples the dynamics of local oscillators. Some of these results include the existence of supertransients, pattern formation, multiple attractors, spatiotemporal chaos, and so forth (see Kaneko 1992, 1993, 1998 for a review). As stated before, these concepts were rooted in the physics literature. We will not develop in detail these concepts. Rather, we will explore in the next sections the implications of such concepts for ecosystem dynamics.

Self-Organizing Spatial Patterns and Competitive Coexistence

Let us consider again the problem of species coexistence. A discrete time version of the Lotka-Volterra model (3.1–3.2) with populations normalized ($x = N_1/K_1, y = N_2/K_2$) can be written as:

$$x_{t+1} = \Phi_x(x_t, y_t, t) = x_t \exp\left[r_x(1 - x_t - \alpha y_t)\right] \tag{3.55}$$

$$y_{t+1} = \Phi_y(x_t, y_t, t) = y_t \exp\left[r_y(1 - y_t - \beta x_t)\right] \tag{3.56}$$

where parameters are as in equations (3.1–3.2). The properties of this system are also similar to those of system (3.1–3.2), with the difference that since now we are dealing with discrete time equations, cycles and chaos are possible (see chapter 2). However, we will focus on the domain of stability, that is, the domain of steady states. The main properties of the above system are as before. Coexistence is possible only when interspecific competition coefficients are lower than intraspecific ones.

What happens when space is introduced? One possibility is the same result as for the nonspatial model with more replicates (one for each spatial patch). If this was the case, then space would not introduce any new qualitative result. It would be negligible, and we could omit it from our models. But if by introducing space we obtain qualitatively different results, and these results are in agreement with what we observe in nature, then space would be an important factor. To answer this, let us consider a two-species competition CML based on model 3.55–3.56):

$$x_{t+1}(k) = (1 - D_x)\Phi_x[x_t(k), y_t(k)] + \frac{D_x}{4}\sum_{j=1}^{4} \Phi_x[x_t(j), y_t(j)], \tag{3.57}$$

$$y_{t+1}(k) = (1 - D_y)\Phi_y[x_t(k), y_t(k)] + \frac{D_y}{4}\sum_{j=1}^{4} \Phi_y[x_t(j), y_t(j)], \tag{3.58}$$

where the vector $j = (i, j)$ denotes the spatial coordinates of the four nearest neighbor patches from point k, and D_x, D_y are the dispersal rates of species x and y.

As a case example consider the symmetric scenario studied at the beginning of the chapter with high interspecific competition rates, that is, both species having the same parameter values ($r_x = r_y, D_x = D_y$, and $\alpha = \beta > 1$). Under these conditions, the nonspatial model (3.55–3.56) predicts competitive exclusion, as discussed in the first section (this is

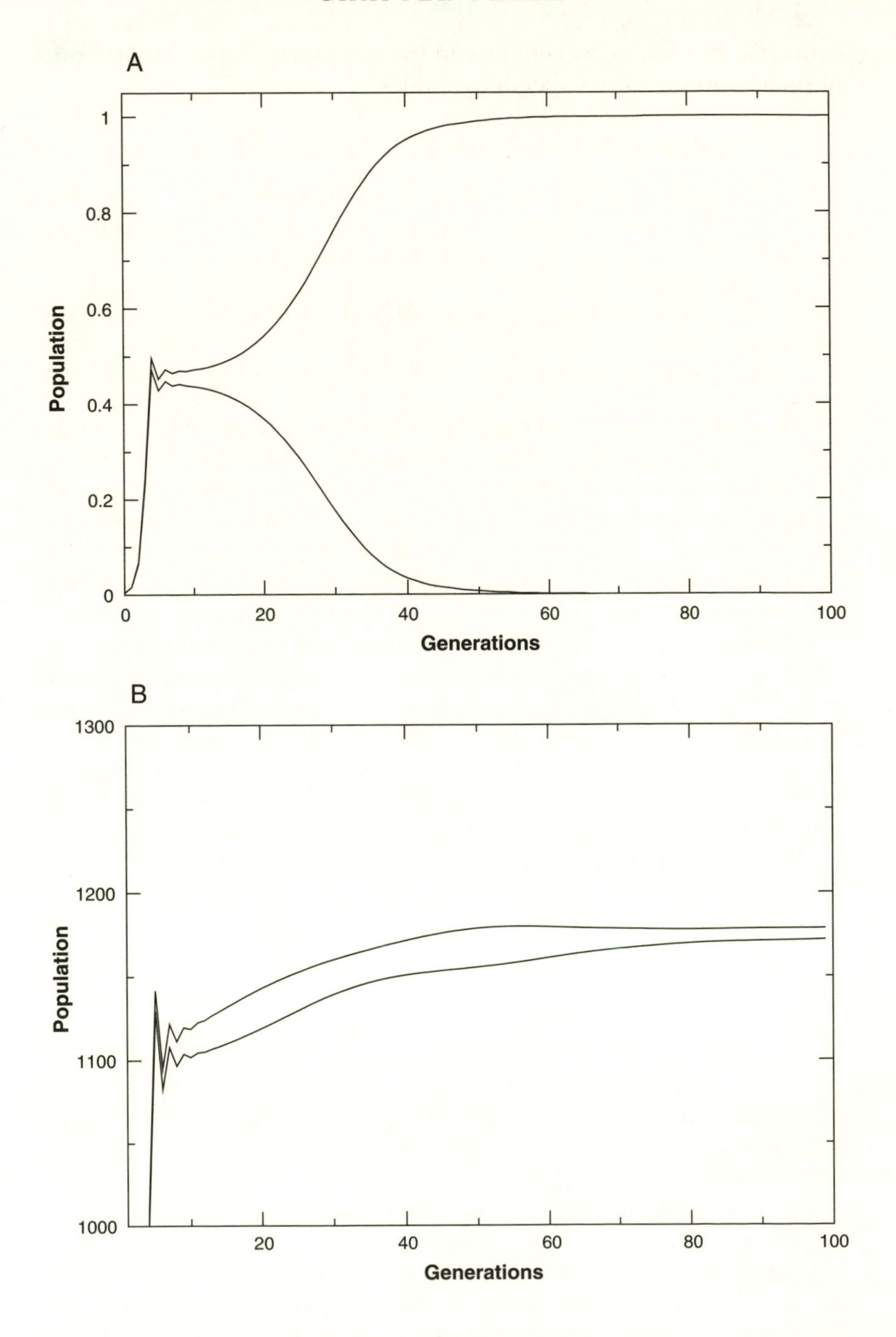
A
Population
1
0.8
0.6
0.4
0.2
0
0
20
40
60
80
100
Generations
B
Population
1300
1200
1100
1000
20
40
60
80
100
Generations

observed at a specific lattice point in fig. 3.8a). One of the two species will go extinct and the second will reach its carrying capacity. Which species goes extinct depends on the initial conditions. Now, what happens with the spatially extended system (equations 3.57–3.58)? Let us consider an almost homogeneous initial distribution, where initial densities at each point are given by $x_0(k) = y_0(k) = \phi + \zeta(k)$, where $\zeta(k)$ is a small noise term ($\zeta(k) << \phi$) different at each patch and for each species. If we plot the initial distribution, we can see a homogeneous pattern, qualitatively similar for both species.

Figure 3.8b shows the global dynamics (the sum of the abundance at each lattice point) of both species. As noted, both species reach a steady state and persist through time. Coexistence is now enhanced. What is the mechanism leading to such a coexistence? Competitive exclusion takes place at each point. That is to say, there is only one species at each patch. However, the outcome is dependent on initial conditions, and since initial distributions are slightly different everywhere, the competitive outcome is also different at different patches as seen for the nonspatial model before. The small initial differences are amplified through time. So, it is not that competitive exclusion does not happen, but that it happens at the local scale. Globally, there is coexistence because the output of competition is different at different patches (Solé et al., 1992a). What is the consequent spatial distribution of abundances? Figure 3.9 shows the stationary distribution through the lattice for both species. As can be observed, the distribution for each species is highly heterogeneous, with clusters where the species is abundant surrounded by clusters where the species is absent. These patterns are time invariant, and reminiscent of the Turing patterns described above for reaction-diffusion models. The take-home message here is that dynamics evolves from a homogeneous distribution to a more heterogeneous one. Thus, coexistence is enhanced by self-organizing spatial pattern

FIGURE 3.8. Dynamics of a spatially extended two-species competition model (the coupled map lattice version of model for symmetric conditions). (A) plots the time series in one point in space. Since interspecific competition coefficients are very high, one species goes extinct as predicted by the nonspatial theory. (B) Time series of the global abundances for each species (sum of local abundances throughout the lattice). Global coexistence is compatible with local exclusion because local exclusion is dependent on initial conditions, and the outcome is different at each location. Lattice size is 50×50, $r_x = r_y = 1.5$, $\alpha = \beta = 1.2$, and $D_x = D_y = 0.05$. Dispersal is constrained to the eight nearest neighbors, and boundary conditions are absorbing.

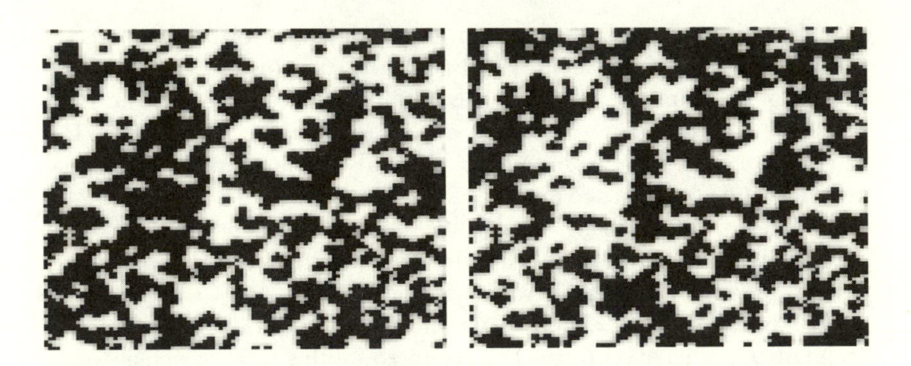

FIGURE 3.9. Spatial distribution of abundances for the two-species competition model corresponding to the previous figure. Each plot represents the abundance of one species at equilibrium. As noted, there is spatial segregation through self-organization, the mechanism that allows local exclusion but global coexistence.

formation (Solé et al., 1992a; Hassell et al., 1994). As earlier mentioned, some studies (Levin, 1974; Slatkin, 1974) indicate that space enhances species coexistence. The explicit consideration of spatially extended areas confirms that spatial self-organization is linked to species coexistence. The mechanisms involved are symmetry-breaking operating at a local scale, and the formation of patterns at a global scale.

Hassell et al. (1994) extended this problem to consider the case of three species, a set of two competing species and a common predator. They also used a coupled map lattice, and the results are qualitatively similar: Species coexistence is possible due to self-organizing spatial patterning. In their case, however, the pattern was not Turing-like, but traveling waves, a spatial pattern that is reviewed in the next section.

Traveling Waves and the Stability of Host-Parasitoid Interactions

Our second example, in the Introduction, was the Nicholson-Bailey's (1935) model. We had seen that the dynamic is unstable, leading to extinction. Hassell et al. (1991) and Comins et al. (1992) extended the Nicholson-Bailey model to incorporate space, by using a coupled map lattice.

The dynamic at each generation consists again of two phases—dispersal and interaction. In the first phase, a fraction μ_h of adult hosts and a fraction μ_p of adult parasitoids leave the patch where they were born and distribute themselves into the eight surrounding patches (considering four nearest neighbors as before does not change any of the qualitative results). The equations for the dispersal stage can be written as:

$$H'_{i,t} = (1 - \mu_h)H_{i,t} + \mu_h \bar{H}_{i,t} \tag{3.59}$$

$$P'_{i,t} = (1 - \mu_p)P_{i,t} + \mu_p \bar{P}_{i,t}, \tag{3.60}$$

where $H_{i,t}$ and $P_{i,t}$ denote the densities of adult hosts and parasitoids at patch i and generation t prior to dispersal, and $H'_{i,t}$ and $P'_{i,t}$ denote postdispersal densities. $\bar{H}$ and $\bar{P}$ indicate the average host and parasitoid densities over the eight nearest neighboring patches. Hassell et al. (1991) used absorbing boundaries, but again similar results are obtained by using different boundary conditions. The second part of the dynamic is host and parasitoid reproduction, which takes place according to the Nicholson-Bailey model (equations 3.3–3.4).

Depending on the fraction of hosts and parasitoids leaving their birth patch (μ_h and μ_p), three kinds of self-organizing spatial patterns are defined by Hassell et al. (1991): spiral waves, crystal lattice, and spatial chaos. These three patterns are shown in figure 3.10. With high host dispersal rates, population densities form spiral waves which rotate around the lattice from a focus (fig. 3.10a). The corresponding local dynamic can be either periodic, quasiperiodic, or chaotic (Solé et al., 1992b).

Crystalline structures are stable patterns characterized by patches with high densities surrounded by lower density patches (fig. 3.10b). These are qualitatively similar to the Turing patterns described in the last section. Finally, spatial chaos consists of spatially and temporally aperiodic patterns that change continuously (fig. 3.10c). The last pattern resembles spatial randomness, densities vary in space and time without any clear pattern—although the model is absolutely deterministic. Figure 3.10d depicts the phase space defined by the dispersal rate of hosts (μ_h) and parasitoids (μ_p) and the regions giving place to each one of the above spatial patterns. A similar range of spatiotemporal patterns had been found by Solé and Valls (1991) for a predator-prey CML and by Solé et al. (1992b) for a similar host-parasitoid model with host density dependence. For additional details see Hassell et al. (1991), Solé and Valls (1991), Comins et al. (1992), and Solé et al. (1992b).

The important point as stressed by Hassell et al. (1991) is that by considering only space and local dispersal among patches, an otherwise unstable interaction is able to persist. Even if local populations go extinct, they are colonized from empty patches. It is this spatial heterogeneity that makes this possible, since different patches are out of phase. This simple example proves the stabilizing role of space. Once more, adding space does not only bring "more of the same," but

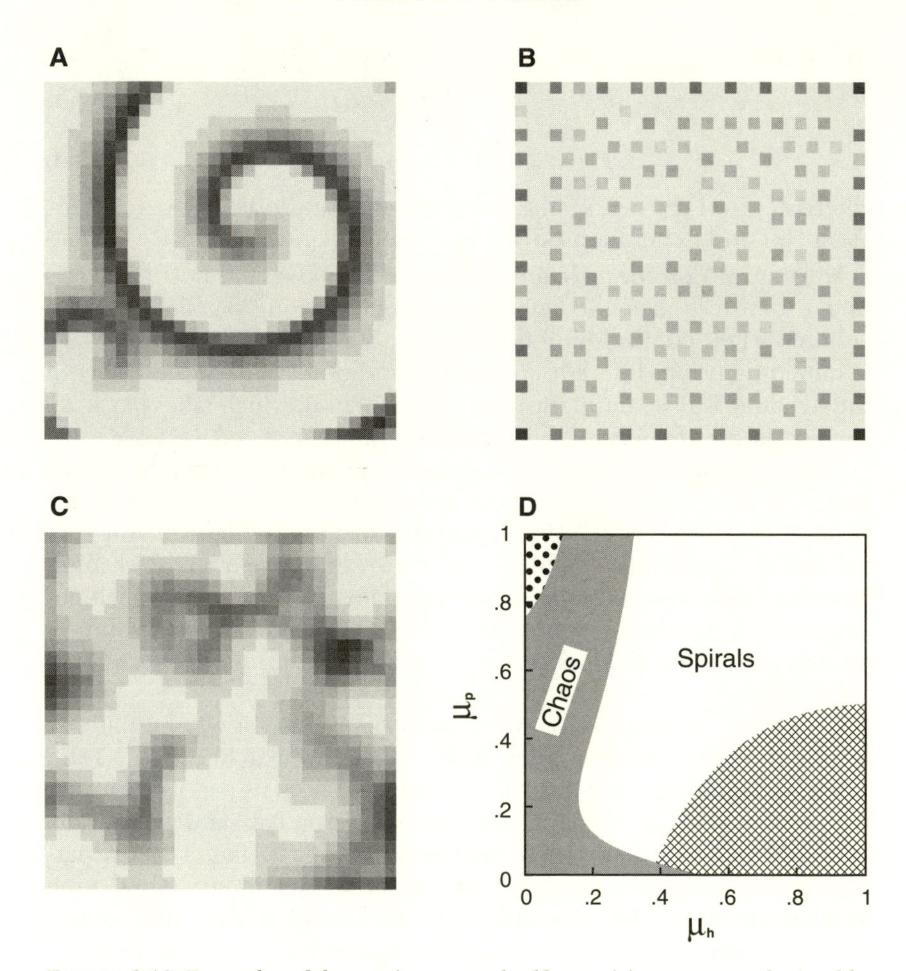

FIGURE 3.10. Examples of the spatiotemporal self-organizing patterns obtained by the host-parasitoid coupled map lattice. Each pattern is obtained for a different combination of host and parasitoid dispersal rates. The lattice size is 30 × 30. At each site, the density of the host is plotted with a degree of shading proportional to its abundance. (A) Spiral waves. (B) Crystal lattice patterns. (C) Spatial chaos. (D) shows the phase space given by dispersal rates of both hosts μ_h and parasitoids μ_p indicating the domains for each of the spatial patterns. The dotted area indicates the region for crystal lattice, and the hatched area indicates "hard to start" spirals. Modified from Hassell et al., 1991. Reprinted with permission of Blackwell Publishing.

in fact introduces qualitative new results not observed when working with nonspatial models. Competitor coexistence and host-parasitoid persistence are just two examples.

FIGURE 3.11. Spiral wave in the spatial distribution of host for the host-parasitoid coupled map lattice. Parameters are $\lambda = 2$, $a = 1$, $c = 1$, $\mu_h = 0.5$, $\mu_p = 0.3$.

The formation and maintenance of spiral waves (fig. 3.11) is a common process in spatially extended, far-from-equilibrium dynamical systems. Such spiral waves are well known in chemical reactions (such as the Belousov-Zabotinski reaction), in the patterns of activity of the brain, in the aggregation of the slime mold *Dictyostelium discoideum*, et cetera. (Murray, 1989; Winfree, 1972). These systems are known as excitable media. Small perturbations from the homogeneous state are damped down, but if these perturbations go beyond a critical threshold, they are amplified. If the medium contains a so-called pacemaker, that is, a small domain able to generate critical perturbations, targetlike patterns and spiral waves can be generated and maintained. The first evidence for this kind of behavior came at the beginning of the 1950s, when the Russian scientist Boris Belousov found a complex chemical reaction in a system similar to the one involved in the Krebs cycle. The whole reaction seems to be alive, since it alternates between two different states (colors) with clocklike regularity. This seemed strange at the time, a violation of the concept of thermodynamic equilibrium, by means of which reactions lead to a final chemical species and stop there.

If there is a spatial component in these excitable media (for example the reactions are set in a petri dish), one observes the formation of traveling waves. Belousov's discovery was so unexpected that his work was systematically rejected, and only after his death was it accepted as a common behavior of far-from-equilibrium chemical reactions. The interesting point is that the basic mechanism of excitable media is found in a wide range of systems, despite the very different nature of the elements (molecules, electric activity, populations, etc.). Once again, complex patterns show qualitatively the same patterns despite the different nature of the elements. Thus, once the key ingredient of their dynamic is introduced, further details can be omitted as irrelevant.

The discovery of the emergence of spiral waves in spatially explicit ecological models (Solé and Valls, 1991; Hassell et al., 1991; and Solé et al., 1992b) inspired extensive research on the implications of such patterns for ecosystem dynamics (White et al., 1996; Bascompte et al., 1997; Rohani et al., 1997). The stability induced by spiral waves has also played an important role in the context of pre-biotic evolution (Boerlijst and Hogeweg, 1991; Boerlijst et al., 1993).

The apparent ubiquity of spiral waves in models of ecological systems has led some scientists to ask about their robustness. Since the models introduced until now are deterministic, one could ask whether the spiral waves shown by these deterministic models, such as the host-parasitoid CML (fig. 3.11) are robust in relation to the unavoidable noise characteristic in real systems. If such patterns are destroyed by small amounts of noise, then they are not that relevant for real systems. Rohani et al. (1997) explored this question and found that spiral waves are indeed robust in relation to moderate amounts of noise. Similarly, Bascompte et al. (1997) studied an individual-oriented simulation of the snowshoe hare–vegetation interaction and found the same kind of spatial patterns predicted by the simpler, deterministic CMLs. This is an interesting result, because individual-based simulations (IBS), as mentioned in the previous chapter, are a very different modeling approach. In conventional ecological models, we describe a macroscopic magnitude, like average density. The level of description is the population, so we cannot account for any individual difference. On the other hand, IBS define a series of individuals, with unique properties (age, size, position, ...). The nature of these traveling waves can be understood easily with this particular example. Hares eat vegetation and deplete it from a point. There is a refractory period that consists of the time necessary for the vegetation to regrow in such a patch.

Hares disperse toward the periphery, following areas rich in vegetation. One interesting point is that the time delay necessary for vegetation regrowth is key for the maintenance of the observed population cycles. Interestingly, snowshoe hares *Lepus americanus* show well-defined cycles with a period of ten years. Bascompte et al. (1997) showed how self-organizing spatial patterns and temporal cycles are intrinsically related. They arise from the same nonlinear mechanisms. These large-scale, macroscopic patterns imply that local population densities will be correlated over large spatial scales, an issue we will consider in forthcoming sections.

LOOKING FOR SELF-ORGANIZING SPATIAL PATTERNS IN NATURE

The previous section suggests that basic biological rules are able to create large-scale macroscopic patterns with far-reaching implications for ecosystem dynamics. If this is in fact the rule, the suggestion is that some of the complexity we observe in nature may have relatively simple causes. One first step toward resolving this question has been to test whether such patterns are robust in relation to noise. The next and most important step is, of course, to try to detect these patterns in nature.

A few attempts have been made to find the hallmark of self-organizing spatial patterns in nature. For example, Maron and Harrison (1997) studied a host parasitoid system, the western tussock moth (*Orgya vetusta*) and its natural enemies. Tussock moths are characterized by a highly patchy distribution despite the continuous distribution of their resource. Several characteristics of this system seem to suggest that patchiness may be a result of spatial self-organization, a kind of Turing structure. Among these characteristics it is worth noticing the low mobility of tussock moths, the high mobility of their parasitoids, and the high parasitism rates observed.

We have already seen the characteristics of Turing patterns. In this kind of pattern the ratio of inhibitor abundance/activator abundance increases as one moves away from the peaks. In agreement with this prediction, Maron and Harrison (1997) observed the following two facts: first, there was always a minimum distance between the outbreaks of the tussock moths (as there is a minimum distance between the Turing peaks due to lateral inhibition). Second, parasitism rates (which provide an index of the ratio between tussock moths and their parasitoids) increased from the center of the outbreak until some distance was reached, decreasing thereafter.

To further confirm these observations, Maron and Harrison (1997) created artificial outbreaks by introducing larvae from previously collected eggs in bushes at different distances from the outbreak of the previous year. Thus, they were making use of perturbation analysis to prove the existence of the long-range lateral inhibition characteristic of Turing diffusive instabilities. As observed by the authors, population growth was suppressed at short distances from the natural outbreaks and increased (via a reduction in parasitism rate) at farther distances. This provides clear evidence for self-organized pattern formation. A similar case study comes from Turchin et al. (1998). The authors studied the spatial patterning of the system composed by the southern pine beetle, *Dendroctonus frontalis*, its host plant *Pinus taeda*, and one of its major predators, the clerid beetle *Thanasimus dubius*. Different factors make of this system a good candidate to apply a reaction-diffusion framework (Turchin et al., 1998).

A suggestion for the existence of a traveling wave in population dynamics came from two studies by Ranta et al. (1997a) and Ranta and Kaitala (1997). The first studied synchrony in population fluctuations of the Canadian lynx (*Lynx canadensis*). Below we will consider the issue of spatial synchronization in more detail. But it can be advanced here that lynx (as well as other species of mammals, birds, and insects) show a surprising amount of synchronization over long distances. It can be observed that synchrony first decreases with distance to increase again after a certain distance. This U-shape pattern in the synchrony versus distance plot could be compatible with several spatial structures such as traveling waves. We further discuss this possibility below. Using similar techniques, Ranta and Kaitala (1997) suggested evidence of traveling waves in the dynamics of vole populations in northern Europe.

But the best evidence of a traveling wave in population dynamics comes from a study by Bjørnstad et al (2002). These authors studied data on the outbreaks of the larch budmoth (*Zeiraphera diniana*), a species that, as seen in the previous chapter, shows well-defined cycles. The peaks of these cycles correspond to outbreaks that have been recorded in the European Alps every eight to nine years. Since these outbreaks have caused extensive defoliation of larch forests, spatiotemporal data have been recorded for four decades, an effort initiated by Baltensweiler. Bjørnstad et al. (2002) analyzed 135 time series expanding these four decades of records. They applied a lagged cross-correlation function, a technique that describes how the cross-correlation between a spatial variable (density) in successive years changes with distance. They detected a traveling wave that moves at an average speed of 219.8 km a year. Interestingly, these authors also

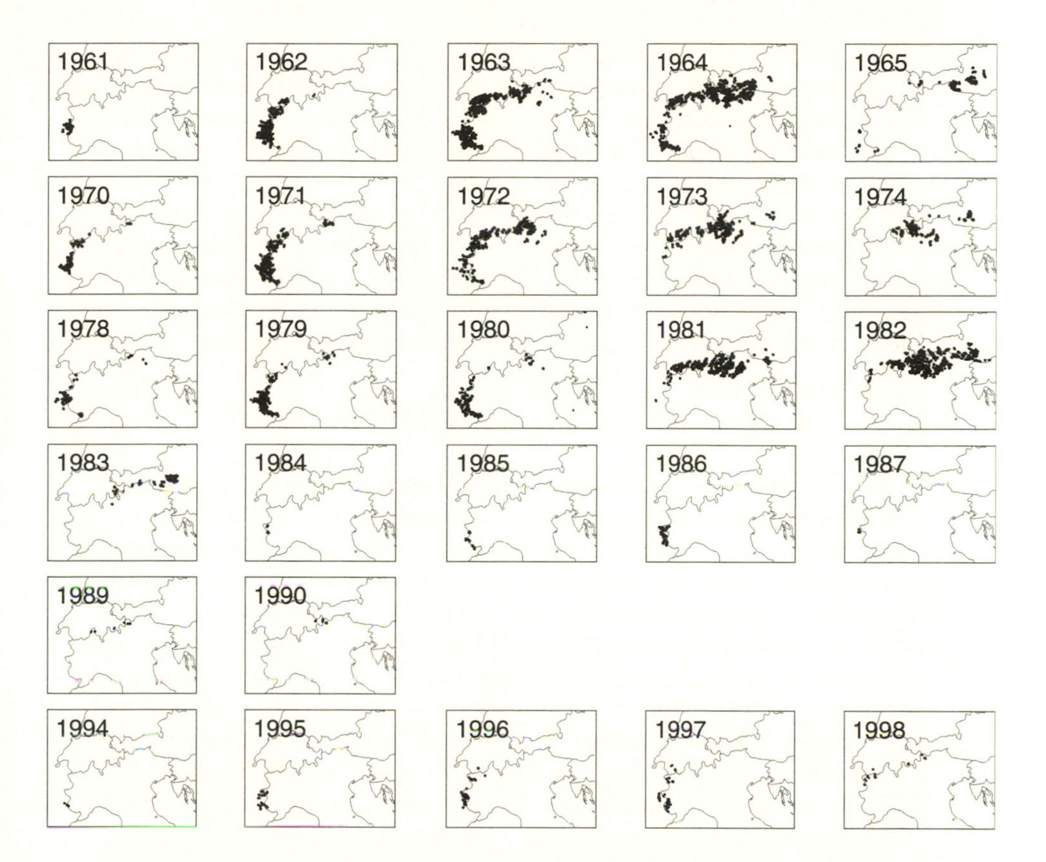

FIGURE 3.12. Spatiotemporal dynamics in the outbreaks of larch bud moths *Zeiraphera diniana* in the European Alps. These outbreaks occur every eight to ten years in a given locality. Only years for which defoliation was observed are plotted. Dots indicate areas defoliated. As noted, the moths generate self-organized directional waves of distribution. From Bjørnstad et al., 2002. Data kindly provided by S. Liebhold.

detected a directionality of the wave: It moves toward a direction northeast by east (65–80 degrees from the north). An example of these waves is shown in figure 3.12. In the previous section, the spiral waves generated by CML models had no direction, in contrast to the directional waves of the larch budmoth. As seen before, coupled map lattices assume generally homogeneous dispersal, that is, a lack of a preferential direction of dispersal. To check whether directional dispersal can produce the shape of the traveling wave observed by data, Bjørnstad et al. (2002) used several variants of the host-parasitoid CML (model [3.3–

3.4] with dispersal rules given by [3.59–3.60]). This encapsulates the belief that cycles in the larch budmoth likely arise from its nonlinear interactions with parasitoids. The authors showed that directional waves are generated if dispersal is directionally biased or there is a gradient in habitat productivity across the landscape.

In summary, the previous empirical papers confirm the predictions made by simple spatially extended models. There are examples in real nature of spatial self-organization due to the interaction between highly nonlinear interactions and short-range dispersal. This interaction is able to create large-scale, heterogeneous patterns of abundance.

DISPERSAL AND COMPLEX DYNAMICS

We have already seen that space can change the stability properties of ecological interactions. It can also modify the nature of the dynamic. Different work has shown that dispersal can induce instabilities in otherwise stable maps. Examples of dispersal-induced chaos are abundant in continuous time models (Pascual, 1993). This is not a new result. It has been well known in the realm of physics that the coupling of periodic oscillators can trigger complex instabilities and deterministic chaos (Kuramoto, 1984). The implications for ecology are important. Part of the original criticism of the role of chaos in real ecosystems, as reviewed in the previous chapter, was based on two points (Berryman and Milstein, 1989a). On one hand, deterministic chaos is only observed for high values of the growth rate. These values can be too high indeed to be realistic. On the other hand, deterministic chaos implies high oscillations. These high oscillations lead populations to very low numbers. Under such circumstances, small fluctuations could drive the population extinct. Thus, chaos would not be adaptive (Berryman and Millstein, 1989a; Lomnicki, 1989; Mani, 1989; Nisbet et al., 1989).

Once again, the persistence of deterministic chaos can only be understood within a spatial context. Diffusion-induced chaos is another route to chaos that does not require high growth rates. It is the interplay between nonlinear dynamics and dispersal, once more, that leads to chaotic oscillations. But, as we have seen in the previous section, despite local unstable dynamics, the spatially extended system can coexist because of recolonizations from other patches. In fact, the scaling relationship $\xi \sim \lambda^{-1}$ between coherence length (ξ) and the Lyapunov exponent (λ, see previous chapter) has been found close to the onset of chaos (Rasmussen and Bohr, 1987). Interestingly, as noted by Rasmussen and Bohr (1987), the latter relationship implies a kind of symmetry between time and space. The inverse Lyapunov exponent is a correlation time, the time horizon beyond which two originally initial trajectories fluctuate independent of each other (see previous chapter). Thus, correlation time and correlation space remain close. The argument is that two points cannot remain correlated if their distance is large enough that a local perturbation in one point grows sizable due to the existence of a positive Lyapunov exponent (see chapter 2) before the perturbation reaches the other point (Rasmussen and Bohr, 1987).

One can further observe by numerical analysis that the more chaotic the system is, the faster the spatial correlation decreases with distance. That is, distant patches oscillate independently of one another. Thus, "ups" and "downs" are compensated and if one plots global dynamics, a kind of global stability can be observed. Ironically, this global stability is induced by the same mechanism generating local instability. Solé et al. (1992c) coined the term *chaotic stability* to describe this phenomenon. This chaotic stability has been found to play a major role in a model of a dominant perennial grass (Bascompte and Rodríguez, 2000).

From Local Instability to Global Stability

Building on a model by Tilman and Wedin (1991), Bascompte and Rodríguez (2000) used the following spatial model of biomass-litter interaction:

$$B_{t+1}(r) = cN\frac{e^{(a-bL_t(r))}}{1+e^{(a-bL_t(r))}} \tag{3.61}$$

$$L_{t+1}(r) = pL_t(r) + (1-\epsilon)ckN\frac{e^{(a-bL_t(r))}}{1+e^{(a-bL_t(r))}} + \frac{\epsilon}{4}\sum_{j=1}^{4} ckN\frac{e^{(a-bL_t(j))}}{1+e^{(a-bL_t(j))}} \tag{3.62}$$

where $B_t(r)$ and $L_t(r)$ are the litter and living biomass at year t (measured in gm^2). N is the total soil nitrogen (mg N per kg soil); k is the rate of conversion from biomass to litter; and p is the litter persistence, that is, the fraction of litter persisting form one year to the next. Quantities a, b, and c are positive constants, and ϵ is the fraction of the litter produced in a site from the nearest neighboring sites affecting growth there.

As shown in figure 3.13, the amplitude of biomass and litter fluctuations (averaged over a spatial scale) decreases quickly in the area sampled. This is in agreement with an early suggestion by Tilman and Wedin (1991) regarding the difficulty of detecting chaos at larger spatial scales even when present at local scales. However, the presence of nonlinear dynamics has an important role at a spatial level. The very same mechanism generating cycles and chaos, that is, the inhibitory effect of litter on the biomass of the dominant grass, can also cause spatial heterogeneity. When litter is very high, its inhibition may eliminate

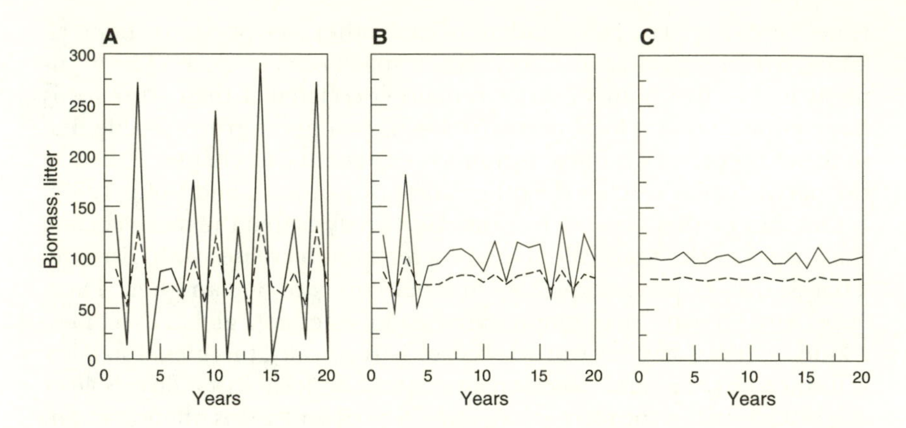

FIGURE 3.13. Fluctuations in biomass (solid line) and litter (broken line) for the model of a dominant perennial grass. The three plots represent data averaged at different spatial scales. Thus, (A) represents fluctuations at a site, while biomass and litter is averaged at clusters of 9 (B) and 100 (C) sites. While chaos is present at a local scale, it may be difficult to detect at larger scales. Parameters are $N = 1300$, $a = 5$, $b = 0.1$, $c = 0.5$, $k = 0.4$, $p = 0.5$, and $\epsilon = 0.1$. Lattice size is 50×50.

biomass after a certain time lag. A gap is then created that can be colonized by other species that are competitively inferior. Litter favors the propagation of fire, which is beneficial for the dominant species. But in the absence of fire, the effect of litter is similar to the effect of small-scale disturbances that are spatially uncorrelated. These disturbances may increase the number of coexisting species (Bascompte and Rodríguez, 2000). Figure 3.14 shows the correlation between nonlinear dynamics and spatial heterogeneity (gap formation) for model (3.61–3.62).

The previous mechanism may explain the well-known observation that the highest diversity in tallgrass prairies is observed when they are left undisturbed, in a departure from the predictions of the intermediate disturbance hypothesis (Connell, 1978) stating that the highest diversity levels will occur at intermediate values of frequency and intensity of perturbations such as fire (Collins, 1992; Gamarra and Solé, 2002). The difference is due to the combination of two factors. First, the major perturbation favors the dominant species. Second, the inhibitory effect of litter production acts as a *self-disturbance* (Bascompte and Rodríguez, 2000).

Similarly, Allen et al. (1993) showed that chaos reduces species extinction by amplifying local population noise. Ruxton (1994), building on this last result, showed that even in the absence of noise, the high dependence on initial conditions typical from chaotic systems (see previous chapter) could induce asynchrony, and so, prevent global ex-

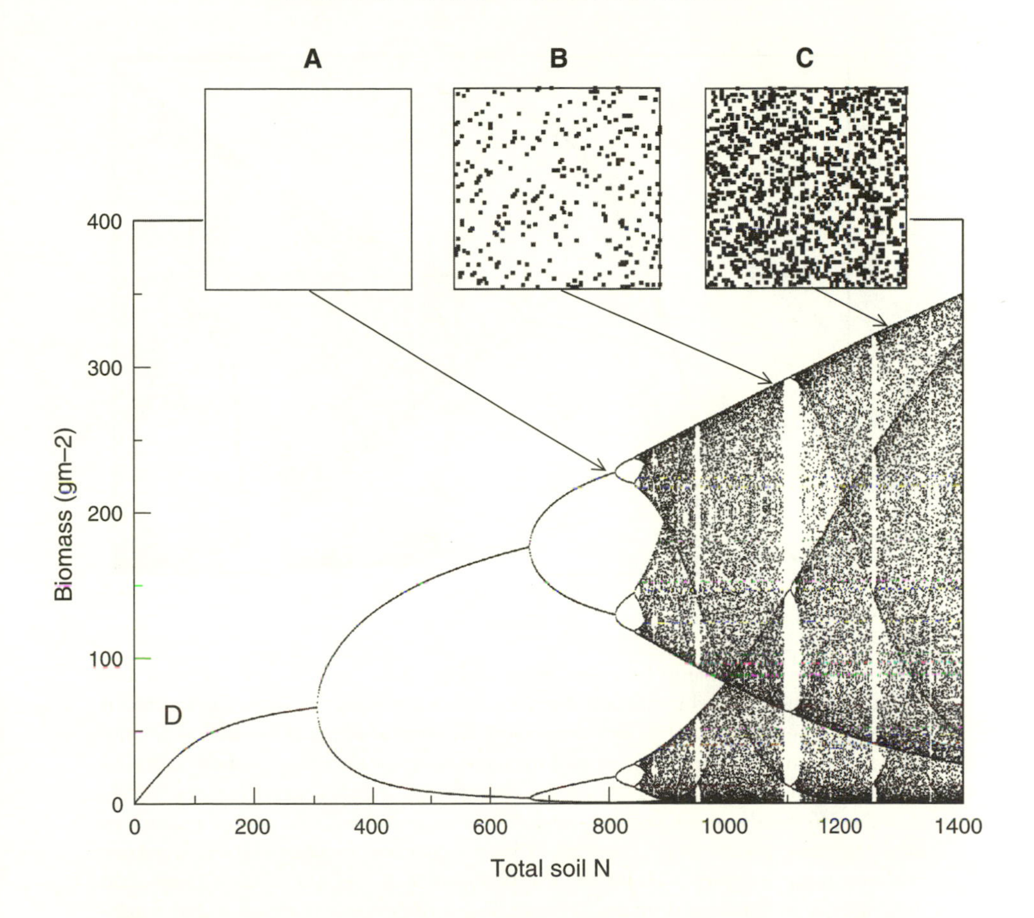

FIGURE 3.14. Relationship between nonlinear dynamics and spatial heterogeneity for the model of a dominant perennial grass. (D) represents the bifurcation diagram for the nonspatial model, generating a period-doubling route to chaos as total soil nitrogen is increased. (A–C) represent a snapshot of the spatial system for the values of nitrogen indicated by the arrows. Black dots indicate gaps where the local biomass is depressed by litter to a value lower than 1 gm^{-2}. These sites could be colonized by other species. For low vales of N, biomass is present over the entire lattice (white plot in A).

tinctions. Figure 3.15, based on Allen et al. (1993) illustrates this point. The figure plots two extinction probabilities: that of local extinction for a Ricker map with added noise, and that of global extinction for a metapopulation of ten Ricker models globally coupled. Noise is both local and global, a simple additive random perturbation to the one-dimensional map. Superimposed is the bifurcation diagram corresponding to a single map without noise. As noticed, local population

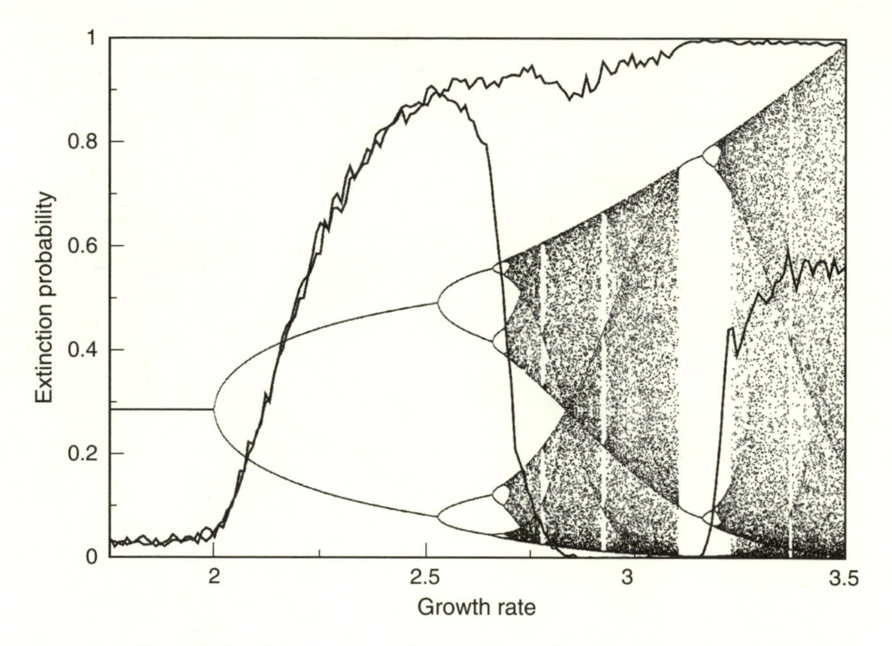

FIGURE 3.15. Relationship between deterministic chaos and the risk of local and global extinction. The bifurcation diagram corresponds to the Ricker map, and it is used as a benchmark. The lines represent the probabilities of extinction for the local map with noise (upper line), and for a metapopulation of ten globally coupled Ricker maps with local and global noise (lower line). For low growth rates (corresponding to steady states and low-period cycles), both local and global extinction have the same probability. The risk of local extinction increases as the dynamic becomes more complex. However, the probability of metapopulation extinction decreases in the chaotic domain. The reason is the dependence on initial conditions of deterministic chaos. Trajectories in nearby local populations oscillate out of phase, and so local extinctions can be recolonized from other patches. Extinction probability is calculated as the fraction of 1,000 replicates failing to persist for at least 1,000 generations. Dispersal rate is $D = 0.01$, local noise is distributed as $U(-0.0001, 0.0001)$; global noise takes place with probability $p = 0.025$ and it is distributed as $N(-1, 1)$. Based on Allen et al., 1993.

extinction always increases with the reproductive rate, that is, as the dynamics goes from stationary to periodic and chaotic. However, the probability of global extinction (all local populations going extinct simultaneously) clearly decreases when the chaotic domain is entered. Once more, the reason is the dependence on initial conditions of deterministic chaos and how it is translated into a set of uncorrelated local populations. Thus, inferences about risk of extinction and chaos change radically when we consider the spatial dimension.

Supertransients

In relation to the effect of dispersal on local dynamics, controversial results have been obtained for CMLs. Dispersal-induced chaos has been proven in two patch models with age structure (Hastings, 1992). Bascompte and Solé (1994) presented wide evidence of dispersal-induced instabilities in a simple spatially extended version of Hassell's (1975) model. Bascompte and Solé's formulation for dispersal, however, was a discretization of a continuous time model, and the resulting dispersal form was somewhat unrealistic. Building on that result, Hassell et al. (1995) showed that local dynamics are not affected by dispersal when a more realistic dispersal formulation (the one used throughout this chapter) is used. In reality, dispersal does not shift the bifurcations to cycles and chaos, but, as we will see below, it still has an important role because it destroys the periodic windows. This does not necessarily mean, however, that dispersal-induced instabilities are not present in discrete time models when other considerations such as age structure are considered, as shown by various authors (e.g., Hastings, 1992). Other studies had proven the opposite, that is, that dispersal in two-patch systems can cause chaotic dynamics to be replaced by periodic dynamics (Hastings, 1993). Thus, one can conclude from these results that the interplay between local dynamics and dispersal may be more complex than expected.

One good example of how dispersal can make local dynamics more complex is the existence of supertansients as seen by Hastings and Higgins (1994). A transient, a concept introduced in the previous chapter, is the number of time steps necessary for a deterministic dynamical system to reach its attractor, that is, its time-invariant solution. Normally transients are short and we always focus on long-term dynamics. This is true for nonspatial systems, but as we will see, the scenario changes suddenly when space is considered.

In their influential study, Hastings and Higgins (1994) used a one-dimensional CML to describe the dynamics of marine organisms whose pelagic larvae are distributed along the coast each generation. The model assumes a finite length L of favorable habitat, x ($0 < x < L$) denoting position along the coast. The number of larvae produced at position x during year t is provided by the Ricker model, that is,

$$l(t,x) = N(t,x)e^{r[1-N(t,x)]}. \qquad (3.63)$$

Dispersal takes place after reproduction, and the number of individuals at location x comes from the larvae released in a neighborhood around x:

$$N(t+1,x) = \int_0^L l(t,y)g(y,x)dy, \tag{3.64}$$

where the probability that larvae released at y settles at x is given by the Gaussian distribution:

$$g(y,x) = \frac{e^{-D(y-x)^2}}{\sqrt{\pi/D}}, \tag{3.65}$$

where D measures the dispersal distance.

Analysis of the previous model reveals the existence of extremely large transients. Normally one would expect this kind of model to quickly settle into its attractor, but instead we may observe that dynamics do not settle down for as long as 15,000 generations. This is shown in figure 3.16 for the coupled map lattice version of the previous model with dispersal restricted to the nearest neighbors. Also, this dynamics is dependent on initial conditions and can shift through time. That is to say, the dynamics observed within a time frame of, let's say, 1,000 to 2,000 years may be very different from the dynamics observed within a temporal scale of 10,000 to 12,000 years (Hastings and Higgins, 1994; Hastings, 1998). It is the interplay between complex nonlinear dynamics and spatial structure that induces such long transients.

The last result has far-reaching implications in ecology and evolutionary biology. As stated before, emphasis has been put on long-term dynamics, but if these supertransients are in fact a property of spatially extended systems, the take-home message is that transient dynamics may be much more relevant than long-term dynamics for understanding ecological systems (Hastings and Higgins, 1994).

As suggested by Hastings and Higgins, supertransients could play a major role in several ecological phenomena such as the paradox of the plankton, or insect outbreaks. Interestingly enough, supertransients seem to be more relevant as the strength of density dependence increases. Ironically then, we may find it more difficult to detect density dependence in spatially extended systems when this density dependence is stronger. Posterior papers have explored the importance of transients in spatially extended models, and the distinction between supertransients and mesotransinets (transients at shorter time scales) has been made (Saravia et al., 2000). While supertransients are not so common (they are observed only for a narrow range of parameter combinations), mesotransients seem to be more widespread (Saravia et al., 2000).

The problem of supertransients that Hastings and Higgins brought to the ecological context is not a new concept. Kunihiko Kaneko studied this kind of dynamical property in coupled map lattices. Kaneko (1990,

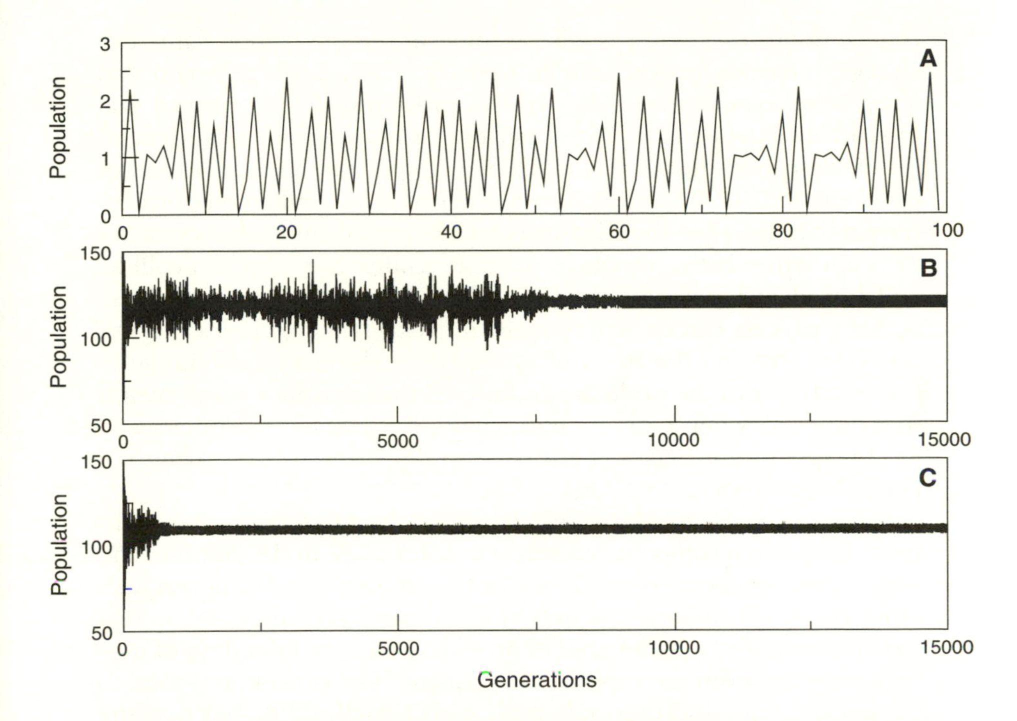

FIGURE 3.16. Long transients in spatially extended models. (A) shows the dynamics of the Ricker map for $r = 3$ corresponding to chaos. (B) and (C) represent the spatially extended version of the Ricker map, a coupled map lattice in one dimension (e.g., a coastline), with individuals dispersing to the two neighboring patches. The same growth rate is used as in (A). Note the large time scale in (B) and (C). Spatially extended systems may show incredibly large transient dynamics. Lattice size is 100, and dispersal rate is $D = 0.2$ (B) and $D = 0.5$ (C). As shown, both transient time and shape of dynamics depend on D.

1998) did intensive work describing the existence of supertransients and how they are related to several parameters such as lattice size and dispersal rate. He defined two kinds of supertransients, the so-called type I and type II. In type I supertransients, the length of the transient scales algebraically with lattice size (with the form N^{σ}, where σ is fitted to 2). In type II supertransients, the length of the transient scales exponentially with lattice size. Interestingly, there is a threshold in the value of the dispersal rate (degree of coupling) separating both types of supertransients.

Kaneko's (1990) work has another important implication in relation to the role of chaos. Another criticism against chaos reviewed in the last chapter is that there is a deep nest of periodic windows within the chaotic domain. Thus, a small variation in the bifurcation value can change dynamics from chaotic to periodic. Technically, it is said that chaos is structurally unstable. However, Kaneko (1990) proved that in spatially extended maps, the periodic windows from local dynamics are destroyed due to the interplay of supertransients and another phenomenon called spatiotemporal intermittency. Spatiotemporal intermittency describes the situation in which chaotic motion propagates to neighboring lattice points as a wave. Different waves can collide, and turbulent bursts are created. These bursts propagate in turn, and the process can be maintained forever. This precludes the system from entering the basin of attraction of the homogeneous solution associated to the periodic windows. This important result means that dispersal, when high enough, changes the stability properties of local maps since it makes chaos structurally stable by destroying the periodic windows.

Supertransients may have important implications for species richness. Even when competitive exclusion takes place in the equilibrium, long transients may prevent this equilibrium from ever being reached. Once more, the interaction between complex dynamics and spatial structure can play a major role in providing an understanding of how high diversity levels are supported in nature. This concept is similar to the intermediate disturbance hypothesis (Connell, 1978), but now the source of "perturbation" is not external, but it is part of the intrinsic dynamic as shown above for the model of a perennial grass (3.61–3.62).

Here we have dealt with the effects of dispersal on local dynamics. In previous sections, we focused on the effects of space on global dynamics, and we proved that space and dispersal stabilize otherwise unstable systems. This is always true when dispersal is constrained to a set of nearest neighboring patches. If dispersal is global, however, space becomes less relevant and global extinctions are more likely.

Multiple Attractors

Another relevant phenomenon of spatiotemporal systems is the existence of multiple attractors. In some spatially extended systems, the observed dynamic is dependent on the initial conditions. For the same parameter combination, we can observe one kind of dynamic or another as a function of tiny differences in the initial conditions (Solé et al., 1992b; Hastings, 1993; Briggs et al., 1999). For example, consider the two-patch model studied by Hastings (1993). At each patch the dynamic is described by the logistic map, and a fraction of individuals disperses to the other patch after the density-dependent stage. Figure 3.17 shows the dynamics of the entire metapopulation (the sum of

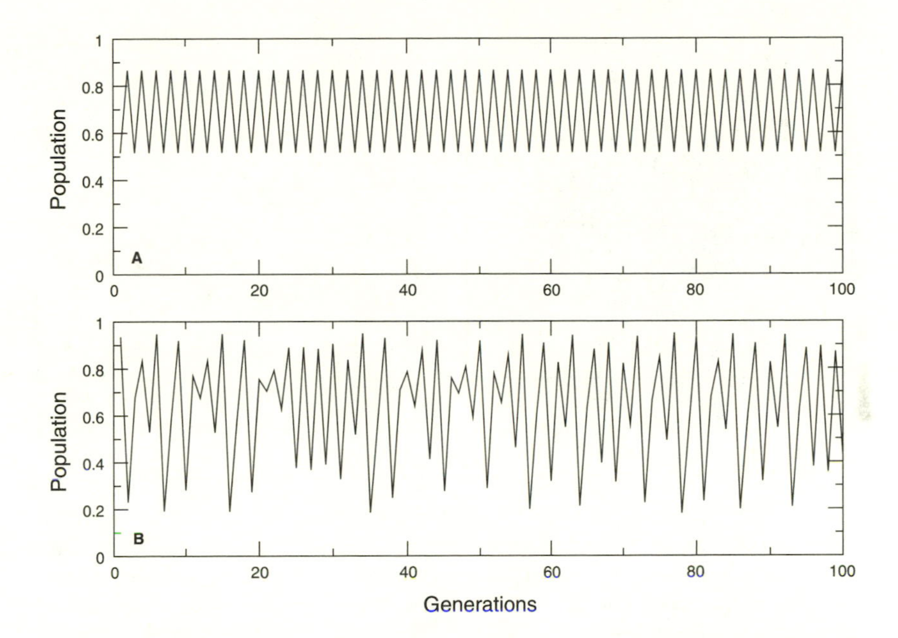

FIGURE 3.17. Multiple attractors in a two-patch model. At each generation, density-dependent dynamics described by the logistic map take place at each patch, and then a fraction D of the population migrates to the other patch. Global population (sum of populations at each patch) is plotted versus time. Both series correspond to the same parameter combinations but different initial conditions. $\mu = 3.8, D = 0.15$, and initial conditions are $x_0 = 0.2, y_0 = 0.4$ (A) generating cycles, and $x_0 = 0.2, y_0 = 0.2$ (B) generating chaos.

individuals at each patch). Both figures correspond to the same parameter combination, and the only difference is that the initial number of individuals in one of the patches is different. As we can see, the system evolves toward different attractors: a period-two cycle in figure 3.17a and deterministic chaos in figure 3.17b. This result is important because traditionally the thinking has been that similar conditions lead to similar dynamics, and that different dynamics correspond to different biological conditions. If deterministic chaos puts a limit on our capacity to predict, multiple attractors go a step further into such a limitation. Since the boundary separating the domain of attraction of different attractors is extremely complex (fractal), we may never have enough information to predict from an initial condition what kind of dynamic the system will show, even in the case where we completely understand our system (Solé et al., 1992b; Hastings, 1993; Bascompte and Solé, 1995).

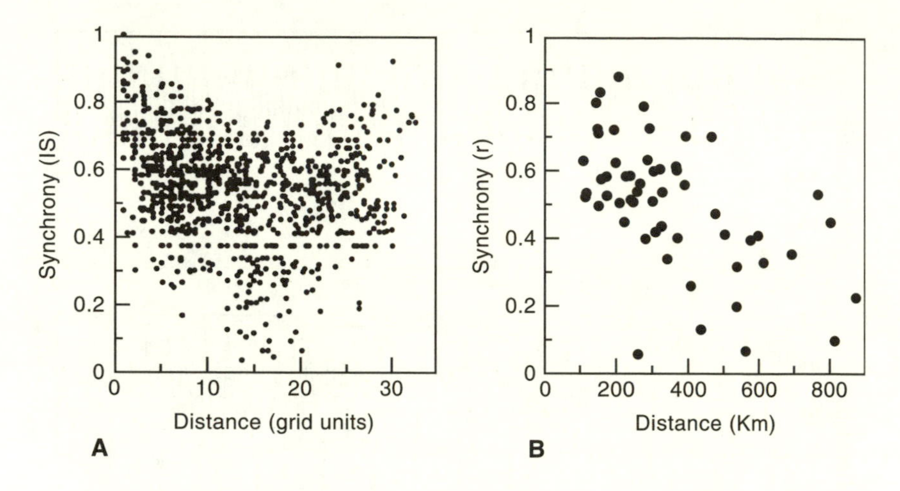

FIGURE 3.18. Spatial synchrony in population fluctuations. Synchrony is plotted versus distance for the Canadian snowshoe hare (A) and the Finnish mountain hare (B). In (A) spatial distance is measured as grid distance, where each grid corresponds roughly to 900 square miles. Synchrony is measured by index (3.66) in (A) and by the correlation coefficient in (B). See text for details. Modified from Ranta et al., 1997a, data kindly provided by E. Ranta.

SPATIAL SYNCHRONY IN POPULATION CYCLES

We had anticipated that not only do some animal species fluctuate in regular cycles, but that they do it coherently over large spatial geographic areas. The problem of describing and understanding the mechanism driving spatial synchrony has become one of the most interesting areas in spatial ecology during the last few years (see Ranta et al., 1998, and Bjørnstad et al., 1999a, for reviews). Regional synchrony in the dynamics of local populations is common in animal populations ranging from parasites (Bolker and Grenfell, 1996) to insects (Pollard, 1991; Hanski and Woiwod, 1993; Myers and Rothman, 1995; Williams and Liebhold, 1995), fish (Myers et al., 1995, 1997), birds (Ranta et al., 1995, 1997b), and mammals (Moran, 1953; Steen et al., 1990; Royama, 1992; Sinclair et al., 1993). In summary, all these studies seem to prove that (1) populations of a variety of species from different taxa fluctuate in phase over very large spatial scales, (2) this synchrony tends to decrease with distance (fig. 3.18), and (3) for some species there is a clear *U*-shape pattern in the synchrony versus distance plot. That is, synchrony first decreases, and then increases again with further distance (Ranta et al., 1998, fig. 3.18a).

Despite the abundance of papers describing synchrony in population abundance, almost no studies have addressed the theoretical question of the meaning of the covariance-distance function. In a theoretical paper, Bjørnstad and Bolker (2000) study the expected relationship between synchrony and distance in populations when dispersal is the synchronizing agent. This is the simplest scenario since oftentimes not just dispersal but correlated environments and other factors (see below) also influence synchrony. These authors obtain two characteristic canonical functions depending on the relative frequency of dispersal in relation to birth and death. If dispersal is rare relative to birth and death, then covariance at short distances will follow the dispersal distribution, although for large distances covariance is found to tail off according to an exponential or Bessel function. On the other hand, if dispersal events are common in relation to birth and death processes, covariance will follow a Gaussian distribution with exponential or Bessel tails far from the origin (Bjørnstad and Bolker, 2000).

Patterns of Synchrony versus Distance

Different statistical techniques can be used to measure spatial covariance. The most common is to calculate the zero-lag cross-correlation between the temporal series of log abundances (or alternatively, growth rates). One averages cross-correlations among different patches for a specific spatial scale. This would be the first step. The second would be to calculate the rate at which spatial covariance drops with distance. Parametric covariance functions have been used extensively, although some authors have claimed that a better approach is the nonparametric covariance function because it avoids the problem of using a discontinuous step function when approximating the relationship between synchrony and distance (see Bjørnstad et al., 1999a, for a review of the statistical tools). The main point here is that by means of these statistical tools, one can plot how spatial covariance decreases with distance, and test for statistical significance.

The first step in studies of synchronization is to detect a pattern. For example, Ranta et al. (1997a) studied the synchrony patterns of the Canadian snowshoe hare (*Lepus americanus*), a problem that had already been studied by P.A.P. Moran (Moran, 1953). Ranta et al. (1997a) used Smith's (1983) analysis. The data was provided in grids, each covering an area of about 900 square miles. Ranta and colleagues labeled each grid site (i) according to whether the population value had been the same during the previous year ($S(i) = 0$), had increased ($S(i) = 1$), or had decreased ($S(i) = -1$). One can then compare pairs of grid points (i,j), these two points being in phase if $S(i) = S(j)$ for a large number of time steps. An index of local synchrony between two locations i,j (IS_{ij})

can then be used as follows:

$$IS_{ij} = \frac{1}{2}\left(2 - \frac{1}{t}\sum_{t=1}^{n} |S_i(t) - S_j(t)|\right) \tag{3.66}$$

where n is the number of years. If synchrony between two points is perfect, the value of IS will be 1, while the value will be 0 if there is no synchrony at all between these two points (Ranta et al., 1998). Figure 3.18a plots the values of synchrony versus distance for the snowshoe hare. As can bee seen, synchrony is positive for large spatial distances, slowly decreasing with distance and then increasing again. For comparison, the spatial synchrony for the Finnish mountain hare (*Lepus timidus*) is plotted in figure 3.18b. Historical data were available in this case for the eleven Finnish provinces, and synchrony between any two sites is calculated by means of the Pearson correlation (Ranta et al., 1995). As noted, we again find high values of synchrony for long distances, and synchrony decreases with distance. In this case we do not find the U-shaped pattern (Ranta et al., 1995).

Disentangling the Mechanisms Affecting Synchrony

Once the pattern of synchrony has been detected, the next question, of course, involves the mechanism behind such a pattern. Three different hypothesis have traditionally being discussed: The first was proposed by Moran (1953) in relation to his study of the snowshoe hare and lynx dynamic. The so-called Moran effect is based on the idea that density-independent fluctuations that are correlated over large spatial scales may serve to synchronize population dynamics. The second hypothesis is based on dispersal. By dispersing from one place to another, animals can couple different regions and so induce synchrony among them. Finally, the third hypothesis deals with nomadic predators. Predators can be highly mobile in search of preys, and so they can induce some synchrony by coupling the abundance of different regions in space (see Bjørnstad et al., 1999a, for a review of these hypotheses).

Despite the existence of these hypotheses for spatial covariance, it is just recently that people have tried to disentangle the contributions of these mechanisms. Three papers have explored the contributions of dispersal and correlated environmental synchrony in spatial covariance (Lande et al., 1999; Kendall et al., 2000; Ripa, 2000). These papers provide analytical insight into the problem of synchrony. Since analytical work is only possible for a single species, they focus on two of the hypotheses while leaving out the hypothesis of nomadic predators (which we shall focus on later in this section).

In order to disentangle the relative contribution of dispersal and environmental synchrony, Kendall et al. (2000) studied a simple first-order autoregressive process. It is a two-patch model with density-independent dispersal and environmental stochasticity. It is one of the simplest models we can think of; its simplicity will help us in finding simple and general analytical results.

Kendall et al. (2000) assume that the local population density fluctuates around a stable equilibrium that is defined by

$$N(t+1) - N^* = b(N(t) - N^*) + \varepsilon(t), \tag{3.67}$$

where $N(t)$ is the current population value, $N(t+1)$ is the population value at next time step, N^* is the equilibrium density, b (between -1 and 1) is the rate of return to the equilibrium (i.e., the strength of density-dependent regulation), and $\varepsilon(t)$ is a white noise process.

The previous equation can be rearranged into the following expression:

$$N(t+1) = bN(t) + (1-b)N^* + \varepsilon(t). \tag{3.68}$$

We can couple two local populations by means of dispersal. The coupled model will read:

$$N_1(t+1) = (1-D)\left[bN_1(t) + (1-b)N^* + \varepsilon_1(t)\right] + D\left[bN_2(t) + (1-b)N^* + \varepsilon_2(t)\right], \tag{3.69}$$

$$N_2(t+1) = (1-D)\left[bN_2(t) + (1-b)N^* + \varepsilon_2(t)\right] + D\left[bN_1(t) + (1-b)N^* + \varepsilon_1(t)\right], \tag{3.70}$$

where D, as usual, represents the dispersal rate, the fraction of individuals moving from one patch to the other.

The dynamic of the total population density, $M(t) = N_1(t) + N_2(t)$, is described by:

$$M(t+1) = bM(t) + 2(1-b)N^* + [\varepsilon_1(t) + \varepsilon_2(t)], \tag{3.71}$$

which is also a first-order autoregressive process.

We are now ready to calculate the covariance between $N_1(t+1)$ and $N_2(t+1)$. Recall that the covariance of two sums is the sum of the covariances of all of the cross terms: $cov(a+b, c+d) = cov(a,c) + cov(a,d) + cov(b,c) + cov(b,d)$. Let us assume that the noise is density-independent: $cov[N_i(t), \varepsilon_i(t)] = 0$. The noise variance, $var(\varepsilon_i)$, is defined to be σ^2, and the noise covariance, $cov(\varepsilon_1, \varepsilon_2)$, is assumed to be ρ.

Because $cov[N_1(t), N_2(t)]$ is time-independent, $cov[N_1(t+1), N_2(t+1)] = cov[N_1(t), N_2(t)]$. Kendall et al. (2000) calculated $cov[N_1(t+1),$

$N_2(t+1)]$ by calculating the covariance of both sides of (3.69–3.70):

$$cov[N_1(t+1), N_2(t+1)] = 2D(1-D)\sigma^2 + [D^2 + (1-D)^2]\rho + b^2 D(1-D)\{var[N_1(t)] + var[N_2(t)]\} + [D^2 + (1-D)^2] b^2 cov\,[N_1(t), N_2(t)]. \qquad (3.72)$$

By using the well-known relationship $var(N_1) + var(N_2) = var(M) - 2cov(N_1, N_2)$, we can calculate the variance terms in equation (3.72). The dynamics of the total population size (M) is also a first-order autoregressive process, and so its variance can be written as follows:

$$var(M) = \frac{2\sigma^2 + 2\rho}{1 - b^2}, \qquad (3.73)$$

and so

$$var(N_1) + var(N_2) = \frac{2\sigma^2 + 2\rho}{1 - b^2} - 2cov(N_1, N_2). \qquad (3.74)$$

Now we substitute (3.74) into (3.72), and apply the identity $cov(N_1(t+1), N_2(t+1)) = cov(N_1(t), N_2(t)) \equiv cov(N_1, N_2)$. After some rearrangements, one can obtain:

$$cov(N_1, N_2) = \frac{2\sigma^2 D(1-D) + \rho\left[1 - b^2 - 2(1 - 2b^2)D(1-D)\right]}{(1 - b^2)\left[1 - b^2(1 - 2D)^2\right]}. \qquad (3.75)$$

The covariance increases linearly with ρ (the environmental covariance) and σ^2 (the environmental variance). Although more difficult to observe from the previous equation, covariance also increases with D. Interestingly, the covariance also increases with the strength of population regulation—$cov(N_1, N_2)$ diverges to infinity as $|b|$ approaches one. This is an unexpected result that suggests how different factors may interact in complex ways even in this simple model. Due to the divergence of the covariance as b is increased, it may be more useful to consider the correlation instead of the covariance, since its value is bounded. Figure 3.19 illustrates how the correlation in population fluctuations depends on D, b^2, and r.

Now that we have an expression for the total covariance, we can partition the contribution of both dispersal and environmental correlation. The dispersal-induced covariance is the fraction of the total covariance that does not involve environmental covariance. We can find this component by setting ρ equal to zero in equation (3.75). By doing this, we obtain:

$$cov_d = \frac{2\sigma^2 D(1-D)}{(1 - b^2)\left[1 - b^2(1 - 2D)^2\right]}. \qquad (3.76)$$

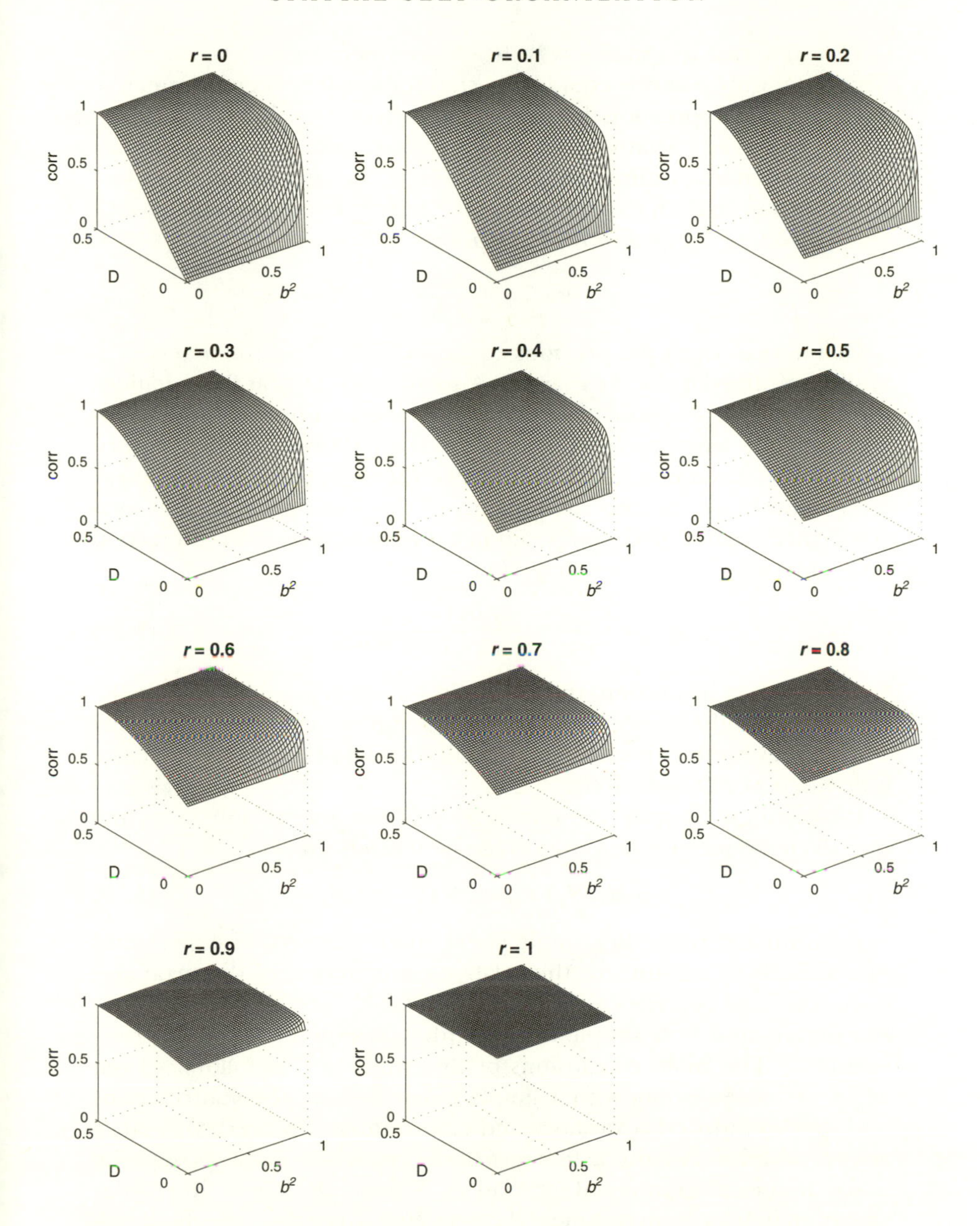

FIGURE 3.19. The correlation between the fluctuations of two populations in a two-patch model is plotted as a function of dispersal rate (D), the strength of population regulation (how fast the population returns to equilibrium after perturbed, b), and correlation in the environmental noise (r). From Kendall et al., 2000. Reprinted with permission of the University of Chicago Press.

We can define in a similar way the environment-induced covariance. This is not only ρ as one could suspect, but the effect of environmental covariance on population density, which is affected by the particular dynamics observed. That is to say, environmental perturbations interact with local dynamics, and its effect may be more complex than expected. Then we can find an expression for the environment-induced covariance by setting D equal to zero in equation (3.75):

$$cov_e = \frac{\rho}{1 - b^2} \tag{3.77}$$

One notices from the previous equation that environmental covariance is amplified by local dynamics. Also, from looking at the equations for the dispersal-induced covariance and the environment-induced covariance, one sees that they do not account for 100% of the total covariance (3.75). Even in a simple system like this one, there is a nonlinear interaction between both components of the covariance, so that total covariance cannot be decomposed in these two terms. An interaction term emerges that is represented by:

$$cov_i = -\frac{2\rho D(1 - D)}{(1 - b^2)[1 - b^2(1 - 2D)^2]}. \tag{3.78}$$

Upon closer inspection, this is simply $-rcov_d$, where $r = \rho/\sigma^2$ is the correlation in the noise. Thus the interaction covariance is small only when either r or cov_d is small; it is a small fraction of the total covariance only when r or D is small (fig. 3.20).

Interestingly, despite the importance of the interaction term, the overall covariance decomposition can be written simply:

$$cov(N_1, N_2) = cov_e + (1 - r)cov_d. \tag{3.79}$$

In summary, from this simple linear model, analytical expressions can be built to account for the relative contribution of dispersal and environmental correlation on population synchrony. By doing so, we can understand how simple demographic parameters alter this relationship. The main conclusions of this study are as follows. Even when we were dealing with a simple linear system, the contributions of dispersal-induced covariance and environment-induced covariance are not additive. An interaction term emerges, which is always opposite in sign to the environmental correlation. This means that in the regular situation, in which environmental correlation is positive, the dispersal-induced covariance and environment-induced covariance are subadditive. Second, local density-dependence lowers spatial covariance. The stronger the local regulation, the more uncorrelated the local populations are going to be (Kendall et al., 2000).

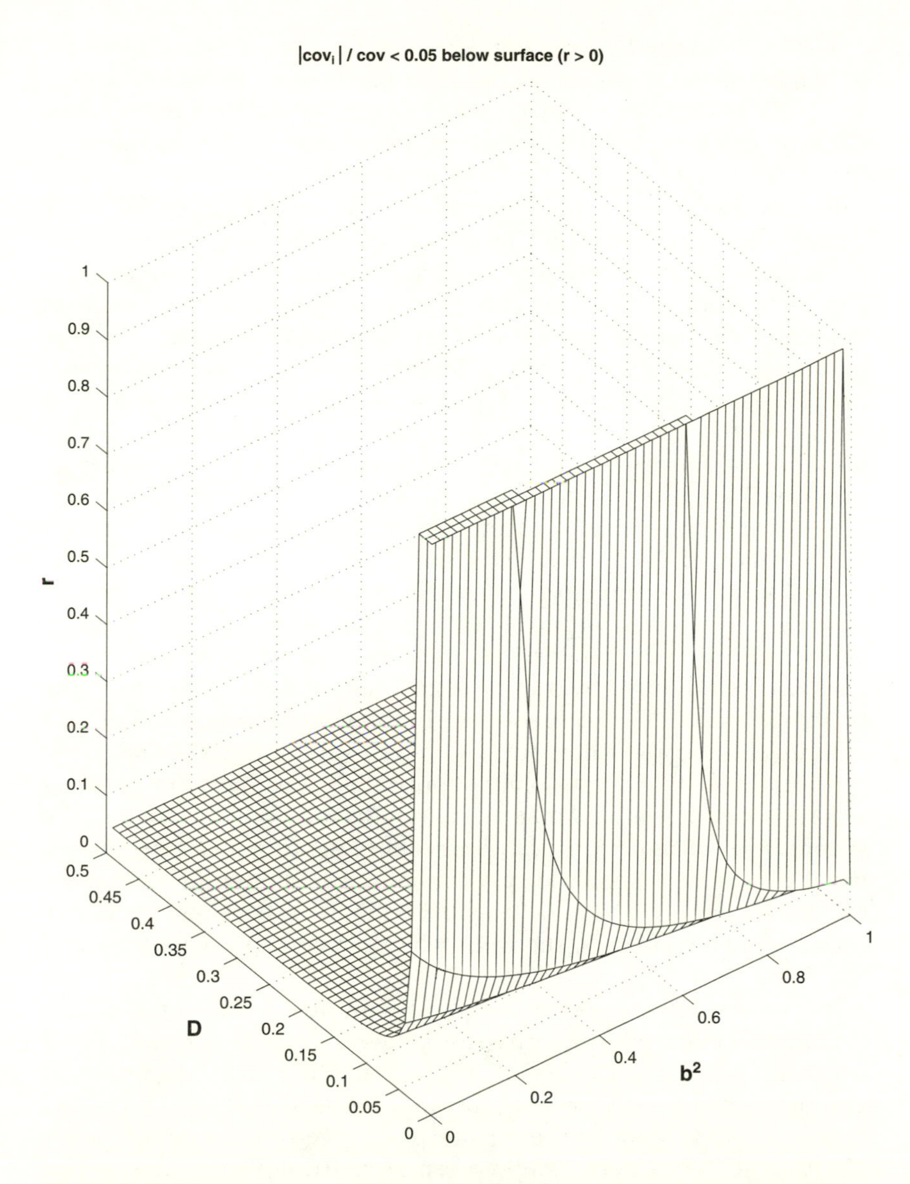

FIGURE 3.20. The components of dispersal and environmental noise are nonadditive in the total covariance. The surface captures the parameter combinations for which the component of population covariance due to the interaction between dispersal (D) and environmental noise (r) is 5% of the total covariance. For the parameter combination lying below the surface, the relative magnitude of the interaction is less than 5%. b is the strength of population regulation. From Kendall et al., 2000. Reprinted with permission of the University of Chicago Press.

The previous results are not only interesting, but they can provide insight into conservation biology or pest control. In conservation biology our goal is to decrease spatial synchronization, so that local populations fluctuate out of phase, thus reducing the chances of global exclusion. In the case of biological control, the situation is reversed.

Results similar to the ones arrived at by Kendall et al. (2000) were independently provided by Lande et al. (1999) and Ripa (2000). The models are slightly different in these latter (Ripa used an n-patch discrete time model, and Lande et al. used a continuous time, continuous space model), but the main results are surprisingly similar. All three papers emphasize the role of local density-dependence in lowering spatial covariance. Two of them (Ripa and Kendall et al.) also point out the importance of the timing in the population censuses for the characterization of synchrony. One difference between these three papers is that Ripa finds dispersal-induced covariance and environment-induced covariance additive (that is, no interaction term arises). Despite this difference, most of the results are the same.

Until now, these three papers have been able to disentangle the relative contributions of dispersal and environmental correlation. What about the effect of predators? Is there any particular spatial covariance function induced by trophic interactions and distinctive from the patterns observed for single species? In this case we can no longer provide analytical approaches, but have to rely on numerical simulations. Bjørnstad and Bascompte (2001) approached this question by using a host-parasitoid coupled map lattice (model 3.59–3.60) as a reference system. Such an approximation is also aimed at bridging a gap between the predicted self-organized spatial patterns generated by CMLs and the statistical tools used in field spatial ecology. Is there any unequivocal one-to-one relationship between self-organizing spatial patterns and the shape of the SCF?

As we have already pointed out, host-parasitoid CMLs can generate three distinctive spatial patterns (spiral waves, crystal lattice, and spatial chaos, see fig. 3.10). Figure 3.21 shows the SCF for the host (top row) and the spatial cross-corelation function between host and parasitoid obtained for each one of these patterns. Three results emerge. First, all three SCFs are distinctive and representative of the characteristic spatial pattern. Second, two out of three SCFs (the ones corresponding to spiral waves and crystal lattice, fig. 3.21a and b) are different than the SCF patterns obtained for single species. Third, spiral waves show a characteristic second-order SCF (fig. 3.21a). By this we mean that synchrony first decreases with distance, becomes negative, and then grows again with further distance. This *U*-shape pattern is similar to the pattern observed by Ranta et al. (1997a) for the snowshoe hare.

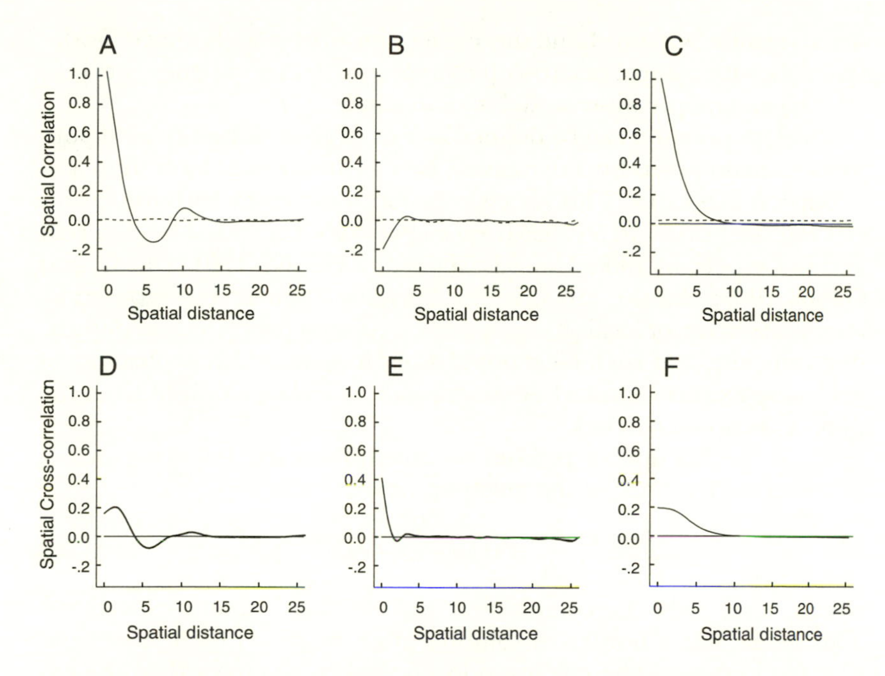

FIGURE 3.21. Top row: Spatial correlation functions (SCF) as a function of distance for hosts in a host-parasitoid CML. Bottom row: Spatial cross-correlation functions between host and parasitoid (SCCF) plotted as a function of distance. Each SCF and SCCF corresponds to one of the spatial patterns depicted in figure 3.10. (A) and (D) correspond to a spiral wave; (B) and (E) correspond to a crystal lattice; (C) and (F) correspond to spatial chaos. As noted, each spatial pattern shows a distinctive pattern of SCF. From Bjørnstad and Bascompte (2001). Reprinted with permission of Blackwell Publishing.

The main difference is that synchrony never becomes negative for the snowshoe hare. Bjørnstad and Bascompte (2001) explored whether this difference could be due to global, correlated noise, but their answer was negative.

WHEN IS SPACE RELEVANT? A TRADE-OFF BETWEEN SIMPLICITY AND REALISM

Throughout this chapter we have emphasized the role of space in population dynamics. However, there are different levels of detail we may consider. Since explicit space is computationally expensive and oftentimes precludes analytical solutions, one can wonder what level of detail we want. On one hand there are mean-field models, that consider space but with a single macroscopic variable such as occupancy. Levins's (1969a) metapopulation model is an example.

A metapopulation can be defined as a set of geographically local populations maintained by a dynamical balance between colonization and extinction processes. This idea was first introduced by Richard Levins (1969a), and since then it has been successfully expanded by different authors, especially Ilkka Hanski (Gilpin and Hanski, 1991; Hanski and Gilpin, 1997; Hanski, 1998, 1999). Imagine a universe composed by an infinite array of local sites or patches. At each patch, a local population can exist, and such local populations have a certain probability of becoming extinct. Similarly, extinct patches can be colonized from the pool of occupied patches.

If x is the fraction of patches occupied, then the following model captures the dynamic of the metapopulation:

$$\frac{dx}{dt} = cx(1 - x) - ex, \tag{3.80}$$

where c and e are the colonization and extinction rates, respectively. This model has a nontrivial solution given by $x^* = 1 - e/c$ if $e \geq$ and $x^* = 0$ otherwise. The colonization rate has to be larger than the extinction rate for the metapopulation to persist.

The last model is a good example of a mean-field model (similar examples are abundant in the ecological and epidemiological literature). We describe the system macroscopically, without keeping track of spatial details. This model is simple and one can solve it analytically. However, one assumes spatial homogeneity.

Mean field models have constrained our understanding of dynamical processes that occur in spatially heterogeneous environments. Only recently has there been a surge of interest in the role of space in community dynamics (see Hanski and Gilpin, 1997; Tilman and Kareiva, 1997; and Bascompte and Solé, 1998a, for reviews). The alternative approach is to use spatially explicit models. In a metapopulation context, for example, a spatially explicit simulation describes each patch as occupied or empty, and describes local colonization events and neighborhood effects. One example is the coupled map lattices described through this chapter. Another way to explicitly consider space would be by using cellular automata if our description of a patch is either occupied or empty (finite states). Introducing all these spatial details requires additional degrees of freedom.

Spatially explicit systems can provide much more detail than a mean [illegible]d model, since we deal only with one variable (the average occupancy) in the latter, while for the explicit system we have as many variables as sites. The relevant question is: Can we get a good enough description of spatial dynamics by incorporating only one additional variable into the homogeneous model? Trying to find an optimal point between maximal accuracy and lower complexity is at the core of the strategy of model building in population biology (Levins, 1966). The process of simplification is based on a relationship between processes and patterns through different spatial scales, on what detail is essential and what superfluous (Levin, 1992; Hiebeler, 1997; Levin and Pacala, 1998).

Pair Approximation

One intermediate alternative between the simplicity of mean-field models and the complexity of spatially explicit simulations is pair approximation. Pair approximation is a technique that traces the nearest neighbor correlation of states (occupied, empty) by constructing ordinary differential equations. These equations do not form a closed system as they contain higher order terms, and need to be closed by neglecting such higher order correlations (Dickman, 1986; Matsuda et al., 1992; Harada and Iwasa, 1994; Sato et al., 1994; Harada et al., 1995; Levin and Durret, 1996; Levin and Pacala, 1998; Pacala and Levin, 1998; Ives et al., 1998).

As an example, let us consider the pair approximation version of the Levins model. We have a set of infinitely numerous local patches, each one of which can be either occupied (1) or empty (0), and colonization takes place to the z nearest neighbors. We indicate the fraction of sites occupied by ρ_1, and the fraction of empty sites is denoted by $\rho_0 = 1 - \rho_1$. These variables are called singlet densities. Let us denote by ρ_{ij} the fraction of ij pairs of patches, where $i, j = 0$ or 1. These variables, called doublet densities, represent the probability that two randomly chosen nearest neighbor sites are in the configuration ij. Finally, $q_{i/j}$ is the conditional probability of a site being in state i given that one of its nearest neighbors is in state j. That is to say, it is a measure of nearest neighbor correlation. If we normalize the colonization rate by setting the extinction rate equal to one ($a = c/e$), we can write the time derivatives for singlet densities as follows:

$$\frac{d\rho_1}{dt} = a\rho_1 q_{0/1} - \rho_1, \tag{3.81}$$

$$\frac{d\rho_0}{dt} = \rho_1 - a\rho_1 q_{0/1}. \tag{3.82}$$

For doublet densities, we can write:

$$\frac{d\rho_{00}}{dt} = 2\rho_{01} - 2a\left(1 - \frac{1}{z}\right)q_{1/00}\rho_{00}, \tag{3.83}$$

$$\frac{d\rho_{11}}{dt} = 2a\left[\frac{1}{z} + \left(1 - \frac{1}{z}\right)q_{1/01}\right]\rho_{01} - 2\rho_{11}, \tag{3.84}$$

$$\frac{d\rho_{01}}{dt} = \rho_{11} + a\left(1 - \frac{1}{z}\right)q_{1/00}\rho_{00} - a\left[\frac{1}{z} + \left(1 - \frac{1}{z}\right)q_{1/01}\right]\rho_{01} - \rho_{01}. \tag{3.85}$$

As noticed by looking at the equations, the dynamics of doublets contain higher order terms. Pair approximation or doublet decoupling approximation consists in neglecting such higher order terms by assuming that $q_{i/jz} \approx q_{i/j}$. Once the previous system of equations is closed, we have only three independent variables. We can then solve the system and find the steady state. More details are provided in chapter 5, where pair-approximation is applied to the problem of spatially correlated habitat loss.

Pair approximation significantly contributes to introducing spatial heterogeneity (and local colonization) within an analytical framework where analytical solutions can be found (however, these may be very complex). One can argue that we do not need all the details of a spatially explicit simulation, but just by considering an extra moment, that is, by considering spatial correlation, we can get enough realism. Different authors have compared spatially explicit simulations with mean field and pair approximation (see, for example Matsuda et al., 1992; Harada and Iwasa, 1994; Sato et al., 1994; Harada et al., 1995; Levin and Durret, 1996; Levin and Pacala, 1998; Pacala and Levin, 1998; Ives et al., 1998; Bolker and Pacala, 1999; see also chapter 5).

Pair approximation makes a number of assumptions. Higher order correlations are neglected, that is, only pair correlations are considered. This makes sense since interactions are restricted to the nearest neighbor in this basic contact process. However, this simplification may break down in some situations such as near the extinction thresholds, which are the equivalent in physics of a critical point where a phase transition takes place. Close to such critical points, correlations scale with the system following a defined power-law. Higher-order terms may become here more relevant. The deviations between pair approximation and simulations are higher near the extinction threshold (Matsuda et al., 1992; Harada and Iwasa, 1994; Sato et al., 1994; Harada et al., 1995; Levin and Durret, 1996). Other approaches may avoid this problem by estimating spatial correlation from a spatially explicit simulation and substituting the value into the analytical expression. In this way, we do not make any assumption about the length of spatial correlation. We shall give an example of this procedure next.

Aggregate Statistical Measures

Bascompte (2001) applied techniques from population genetics to derive a metapopulation model incorporating an index of spatial correlation. He first introduced a statistical measure that describes the spatial correlation (φ) between the occupancy state of two neighboring sites. Since we are interested in how spatial structure modifies the occupancy levels predicted by mean field models, we want to know how the relative occupancy level near an occupied site departs from what is assumed in a homogeneous case (i.e., density in the neighborhood of an occupied site is the same as total density over the landscape). A measure of spatial clustering, heterogeneity, or correlation can be defined as follows:

$$\varphi = p(1|1) - p(1) = p(1|1) - x, \tag{3.86}$$

where $p(1)$ is the probability of a site being occupied, which is estimated by the fraction x of sites occupied, and $p(1|1)$ is the probability that a site is occupied conditional to one of its nearest neighbors being occupied. This last quantity is called doublet density (Matsuda et al., 1992) or local density (Harada and Iwasa, 1994). Mean field models assume "zero-correlation" or spatial homogeneity and hence $p(1|1) = p(1)$. Thus, φ captures spatial heterogeneity as an increase in local density above and beyond what we would observe in a homogeneous system. The goal is to incorporate φ into a mean field model as a way to deal with otherwise neglected spatial heterogeneities.

In the process of deriving the model, Bascompte (2001) followed a strategy commonly used in population genetics where total change is partitioned into meaningful, different components (Price, 1970; Frank, 1997a, b; Frank, 1998). In particular, we can partition the increase in the fraction of sites occupied x_+ into two components in the following way:

$$x_+ = (1 - \varphi)\theta_a + \varphi\theta_b. \tag{3.87}$$

The total contribution to x's increase has been partitioned into two parts. The first represents the noncorrelated component, the increase in x in a homogeneous situation. It can be written as $\theta_a = cx_t(1 - x_t)$, where c is the probability of an empty site being colonized when all the remaining nearest neighbor sites are occupied (Levins, 1969a; Hanski et al., 1996). The second part describes the correlated component, the increase in x in a situation with perfect correlation among the occupancy status of neighboring patches. This component describes a situation in which all occupied sites are surrounded by occupied sites. Thus, its contribution to x_+ is zero: $\theta_b = 0$. By substituting θ_a and θ_b into (3.87), we get:

$$x_+ = (1 - \varphi)cx_t(1 - x_t) + \varphi 0. \tag{3.88}$$

On the other hand, the decrease in x, denoted by x_- is due to extinction. This extinction term does not depend on the correlation state. A fraction ex_t of sites becomes extinct (e being the extinction probability) and a fraction $(1 - e)x_t$ of previously occupied sites remains occupied. The aggregate statistical metapopulation model (ASM) (Bascompte, 2001) is:

$$x_{t+1} = cx_t(1 - x_t)(1 - \varphi) + (1 - e)x_t. \tag{3.89}$$

Note that when $\varphi = 0$, space becomes homogeneous, and equation (3.89) coincides with the discrete time version of the Levins model (Hanski et al., 1996).

Here we are interested in the equilibrium patch occupancy, defined as:

$$x^* = 1 - \frac{e}{c(1 - \varphi^*)}, \tag{3.90}$$

where $e/c \leq 1 - \varphi^*$, the extinction threshold being defined by the equality; φ^* is the spatial correlation at the steady state. The equilibrium patch occupancy (3.90) is plotted in figure 3.22 as a function of the colonization rate for different φ^*-values. By comparing it with the homogeneous situation ($\varphi^* = 0$), it can be easily seen that the correlation between the occupancy status of neighboring sites reduces the equilibrium patch occupancy. Also, the effect of increasing φ^* is higher, as φ^* is already high.

The extent to which the ASM model (3.89) can encapsulate the essential features of a spatially explicit simulation has been tested (Bascompte, 2001). The results indicate that by only adding a single macroscopic parameter, the ASM model behaves very similarly to the spatially explicit simulation. Similar approaches have been used by Frank and Wissel (2002) and Snyder and Nisbet (2000).

In summary, the recent development of analytical techniques such as pair approximation allows us to bridge the gap between complex, spatially explicit simulations and simpler, analytically tractable models. By doing so we are getting valuable information on what is important and what can be simplified. Space is definitely a dimension of great importance in ecological processes. But different levels of detail can be

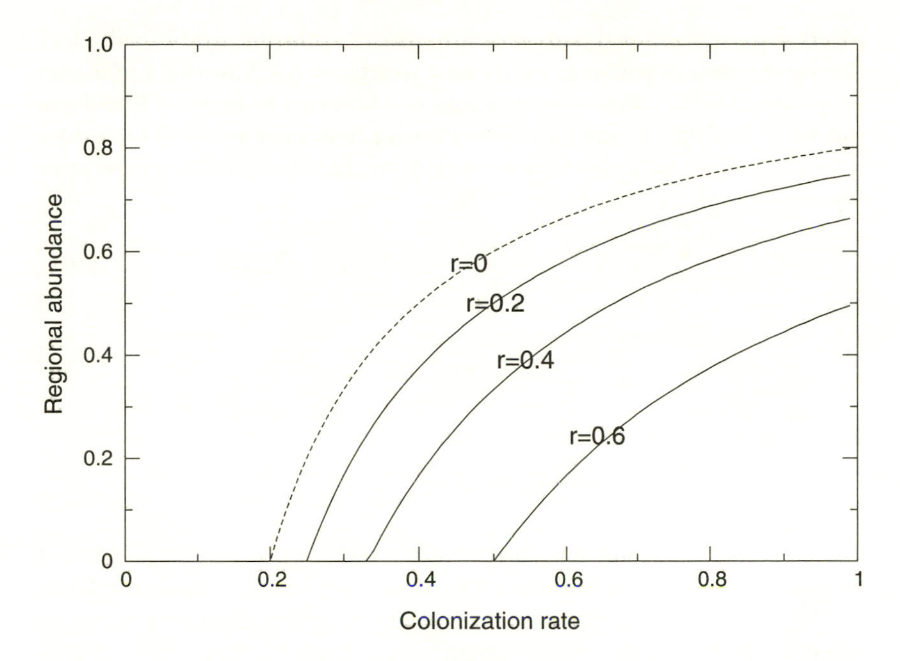

FIGURE 3.22. Effect of the colonization rate on the regional abundance for different values of spatial correlation (φ^*, r in figure).

used depending on our particular goal. A complementary approximation to the ideas described in this chapter is developed in chapter 5, where the problem of habitat loss is discussed.

COEVOLUTION AND DIFFUSION IN PHENOTYPE SPACE

The reaction-diffusion models considered in previous sections entail dispersal of organisms on a given spatial domain. Additionally, all standard models include parameters that are kept constant through time. This is a consequence of the underlying assumption of a scale separation between ecological and evolutionary processes. Models of coevolutionary dynamics incorporate both scales in different ways (Ikegami and Kaneko, 1990; Rand et al., 1996; Van der Laan and Hogeweg, 1995). In this section, models that explicitly include variations in their characteristic parameters are considered.

One possible way of introducing changes in parameters in a Lotka-Volterra model (and thus coevolutionary responses) is to consider a linear relation between genotype and phenotype. Such a linear relation is not likely to be true in many cases, but it is a good starting point. The phenotype space Γ will be a one-dimensional interval, and Γ will affect the interaction between prey (x) and predator (y) (Van der Laan and Hogeweg, 1995). Mutations (at a rate μ) allow one to explore Γ and are modeled as a simple diffusion process on phenotype space. The model is built by defining a discretization of Γ so that the following system of equations is obtained:

$$\frac{\partial x_i}{\partial t} = ax_i - bx_i \sum_{j=1}^{n} x_j - cx_i \sum_{j=1}^{n} \alpha_{ij} y_j + \mu \nabla^2 x_i \qquad (3.91)$$

$$\frac{\partial y_i}{\partial t} = -dy_i + ecy_i \sum_{j=1}^{n} \alpha_{ji} x_j + \mu \nabla^2 y_i \qquad (3.92)$$

where here the diffusion term reads:

$$\nabla^2 z_i = \frac{1}{2}[z_{i-1} + z_{i+1}] - z_i \qquad (3.93)$$

and periodic boundary conditions are used. Here x_i and y_i indicate prey and predator populations with genotype $i = 1, \ldots, n$, respectively.

The interaction matrix α_{ij} is determined from the difference between the phenotype of prey and predator:

$$\alpha_{ij} = \frac{1}{\sigma} \exp\left[-\frac{\sigma_{ij}^{min}}{2\sigma^2}\right] \qquad (3.94)$$

where the distance between the two genotypes is calculated from

$$\sigma_{ij}^{min} = min\left\{|i-j|, n - |i-j|\right\} \qquad (3.95)$$

(i.e., the shortest distance along phenotype axis). The parameter α ranges from $1/\delta$ to $1/\delta exp(-1/2\sigma^2)$. High (small) values of σ imply generalist (specialist) predators. Given the chosen form of the α_{ij} distribution, generalists will have lower total interaction strength.

The relevant parameter in this model is thus the degree of predator's specialization σ. In figure 3.23 we can see three examples of the coevolutionary dynamics under different degrees of predator's specialization. At the beginning, the two populations start identically with the same phenotype $\alpha_{ii} = 1/\sigma$. The plots display the formation of branched structures indicating speciation. Evolutionary branching is actually a common phenomenon under frequency-dependent selection (Dieckmann and Doebeli, 1999). The model, in spite of its simplicity, is rich in terms of the dynamics displayed by the coevolving populations. The following phenomena are observed:

1. Formation of a heterogeneous ensemble of species with related genotypes (a *quasi species*).

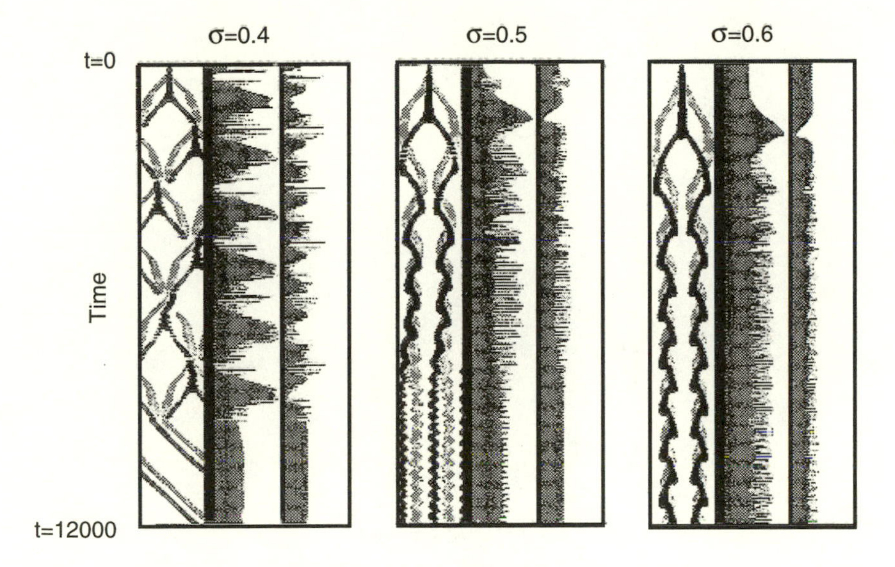

FIGURE 3.23. Examples of the time series of evolutionary and population dynamics of the predator-prey coevolution model. Here starting from a symmetric initial condition (both species located at the same point in phenotype space), the resulting patterns are shown over 12,000 steps. Here prey and predators are shown in gray and black, respectively. Three different σ values have been used.

2. Speciation takes place and is patent in phenotype pace through the presence of branching patterns. Pattern formation actually implies a segregation of the population into differentiated species.
3. The model exhibits (typically) multiple evolutionary attractors.
4. The transient behavior (and some attractors) exhibit "Red Queen" dynamics (van Valen, 1973): Predators chasing prey in Γ are in fact changing all the time without improving their adaptation (i.e., no real "progress" occurs).

The previous features seem to be widespread in other coevolutionary models involving population dynamics. One particularly interesting model by Kaneko and Ikegami (1992) explored the coevolution of a host-parasite CML model with evolving mutation rates. Using a species description coded by bit sequences, it was found that mutation rates change such that a species-rich ecology is obtained displaying weak, high-dimensional chaos (homeochaos). This type of evolving web will be studied in chapter 8.

CHAPTER FOUR

Scaling and Fractals in Ecology

SCALING AND FRACTALS

A striking, widespread feature of many complex systems is that some of their properties are reproduced at different scales in such a way that we perceive the same patterns when looking at different subparts of the same system. This property, known as scale invariance, is widespread in many systems under nonequilibrium conditions. This is the case for ecological systems, where flows of energy enter into the system and are dissipated at different, interconnected scales. The origin of such fractal patterns, named after the pioneering work by Benoit Mandelbrot, is a fundamental problem in many areas of science (Bunde and Havlin, 1994; Schroeder, 1991). Actually, empirical evidence has been mounting in support of the unexpected possibility that many different systems arising in disparate disciplines such as physics, biology, and economy may share some intriguingly similar scale invariant features (Stanley et al., 2000).

Fractals have been observed in the patterns of surface growth, river networks, erosion in landscapes (Barabási and Stanley, 1995; Rodriguez-Iturbe and Rinaldo, 1997), and other physical systems (Mandelbrot, 1977, 1983; Takayasu, 1990; Vicsek, 1992; Turcotte, 1992). But this is a widespread phenomenon and seems to be common in biological systems too (Hastings and Sugihara, 1993; Gisiger, 2001). It is known to be present, for example, in branching networks such as blood vessels and plant vascular systems. In this context, it has been shown that there are general principles constraining the possible, fractal-like architectures observable in different organisms (Brown and West, 2000). In this book we are mainly concerned with dynamical patterns of population change, and, although the general models derived for allometric relations have deep ecological consequences, we restrict ourselves to a higher level of description.

Deterministic Fractals

The simplest examples of fractals have a deterministic origin. In figure 4.1a, one of them, the so-called Sierpinski gasket, is shown. It has been generated in a recursive way by sequentially removing the open central upside-down equilateral triangle with half the side length of the starting (black) triangle. Some iterations of this process are shown at

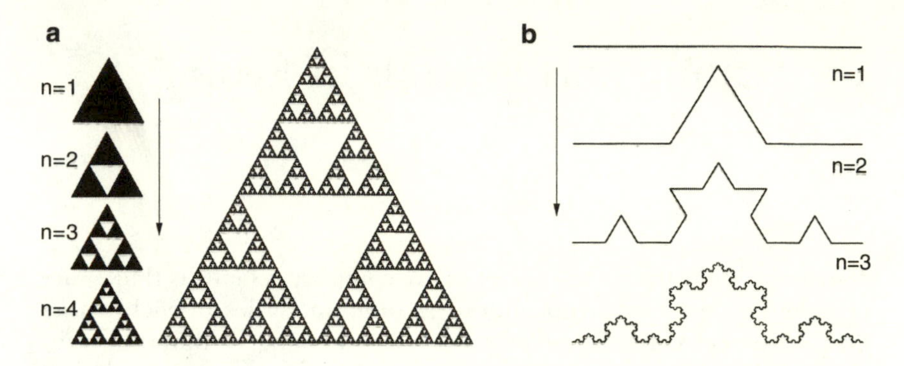

FIGURE 4.1. (a) Geometric fractal: The so-called Sierpinski gasket is generated iteratively as indicated in the left colum. The right figure is a picture of the final result, displaying the same features at all scales. (b) A fractal geometric coastline: the Koch curve. The first iterations are shown in the upper part, and the resulting fractal curve is shown at the bottom.

left. The resulting object displays the same basic pattern at all scales. Another example is shown in figure 4.1b: the Koch curve. The generation process is also displayed, starting at the top with a single line of unit length. The final pattern is again self-similar, and a simple calculation can give us an idea of the unexpected properties of fractals with nontrivial consequences. The Koch curve is clearly confined to a finite domain in the two-dimensional plane. However, if we try to calculate its length $\mathcal{L}$, we reach a strange conclusion: $\mathcal{L}$ goes to infinity. To see this, we just need to compute $\mathcal{L}_n$ at different steps (n) of the generation process ($\mathcal{L}$ is the limit set $\mathcal{L}_n$ when $n \to \infty$). At the beginning we have $\mathcal{L}_0 = 1$, the first iteration gives $\mathcal{L}_1 = 4/3$ and the second $\mathcal{L}_2 = (4/3)^2$. At an arbitrary step, we have $\mathcal{L}_n = (4/3)^n$, which goes to infinity as n grows.

Fractals can also be generated in a dynamical way, and actually the time-dependent scenarios leading to fractal patterns are the main topic of this chapter. Perhaps the simplest example is provided by cellular automata (Wolfram, 1984, 1986; Wuensche and Lesser, 1992; Langton, 1994). Cellular automata (CA) are dynamical systems with discrete space, discrete time, and a discrete number of states. They were first introduced by von Neumann around 1950 and have been widely applied in many domains of science. In biological modeling they have been used to model developmental systems, game theory, neural networks, immunology, and population dynamics (Ermentrout and Edelstein-Keshet, 1993). As an example, let us consider a one-dimensional CA defined on a linear lattice with N sites. The state of each site (which can correspond, for example, to the species number occupying it) at a given time step will be indicated as $S_i(t) \in \Sigma = \{0, 1, \ldots, S - 1\}$ (for $i = 1, \ldots, N$). The set of values of the entire lattice at a given time is called a *configuration*. The dynamic is defined by a rule table: Each time step, all the elements are updated following a deterministic rule (if the CA is deterministic):

$$S_i(t+1) = \Phi(S^i(t)) = (\Phi(S_{i-n}(t), \ldots, S_{i-1}(t), S_i(t), S_{i+1}(t), \ldots, S_{i+n}(t)) \tag{4.1}$$

This rule indicates that the state of each unit is updating depending on its own state plus the state of some nearest neighbors. The string $S^i(t)$ is the so-called *neighborhood* of $S_i(t)$, which includes $2n + 1$ variables. The transition rule $\Phi(S^i(t))$ specifies the transition $S_i(t) \rightarrow S_i(t+1)$ for each given $S^i(t)$.

The simplest CA are two-state, one-dimensional systems. For these CA, the previous rule reduces to:

$$S_i(t+1) = \Phi(S_{i-1}(t), S_i(t), S_{i+1}(t)) \tag{4.2}$$

For this case, the set of possible neighborhoods is reduced to a 2^3 list:

$$\{000, 001, \ldots, 111\} \tag{4.3}$$

and the transition table is defined by means of a string $\mathcal{R} = (\alpha_0, \alpha_1, \ldots, \alpha_7)$ corresponding to the following list of transitions: $\Phi(000) = \alpha_0$, $\Phi(001) = \alpha_1, \ldots, \Phi(111) = \alpha_7$. Clearly, each rule is specified by means of a given string $\mathcal{R}$ and for this particular case we have 256 different choices. A number $N(\Phi)$ can be associated to each rule,

$$N(\Phi) = \sum_{i=0}^{7} 2^i \alpha_i \tag{4.4}$$

The dynamics for rule $N(\Phi) = 22$ is shown in figure 4.2, together with the complete set of transitions. In spite of the simplicity of the CA rules, we can see that a very complex pattern emerges. In (B) the resulting space-time diagram for an $N = 100$ system has been generated from a random initial condition. A fractal pattern of triangles of many different sizes emerges from the initial randomness, clearly suggesting the presence of positive feedback processes. This pattern becomes more obvious if we start with a nonrandom initial condition: The central site is $S_1 = 1$ and all the other sites are set to zero. Using this single-seed start, we obtain a Sierpinski gasket, this time under the dynamical evolution of a CA. It is worth noting that these models are able to reproduce, for example, the intrincate shell patterns displayed by a wide variety of mollusc species (Meinhardt, 1995). The intrinsic complexity of CA goes far beyond any expectation, and some of the simpler models have been shown to display a high unpredictability: The long-term behavior can only be known by simulating it.

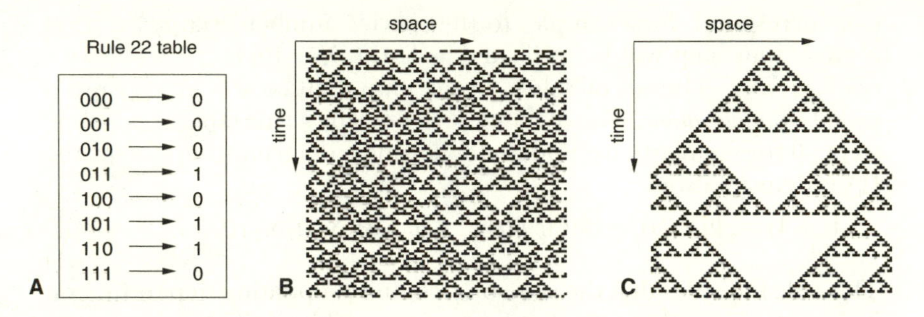

FIGURE 4.2. Complex patterns in cellular automata models (rule 22). Here each square represents a given automaton. Black and white squares indicate the states $S_i = 1$ and $S_i = 0$, respectively. The rules defining the dynamics in elementary CA (A) involve three states $(S_{i-1}(t), S_i(t), S_{i+1}(t))$ for each unit. The new state $S_i(t+1)$ is determined from a simple rule (see text). Starting from a given random initial condition (B) a complex pattern of triangles of different sizes is formed, displaying large correlations. In (C) the same rule is used but the initial condition corresponds to a single black seed at the center of the line. A Sierpinski pattern is now obtained.

The pattern shown in figure 4.2(c) is a fractal in the statistical sense. Statistical self-similarity means that a given feature, measured on some part of an object at high resolution is proportional to the same feature measured over the whole system at coarser resolution (Bassingthwaighte et al., 1994). Consider a given quantitative property $L(r)$ (which can be, say, length, mass or population abundance) measured at some given scale of resolution r. Now consider the same quatity measured at a different scale $r' = \alpha r$. If $\alpha < 1$ then we are looking at a finer resolution. Statistical self-similarity means that $L(r)$ is proportional to $L(\alpha r)$. We can write this as follows:

$$L(\alpha r) = kL(r) \tag{4.5}$$

where k is a constant (which may depend on α). This definition implies that the statistical features of a fractal set are the same when measured at different scales.

Scaling Laws

Self-similar patterns can be viewed under a statistical point of view by means of the so-called power laws (or scaling laws). These laws are again widespread, and in many cases underly the presence of fractal structures. One of the best known is Zipf's law, which states that the fraction of cities $N(n)$ with n inhabitants shows a power-law dependence ($N(n) \propto n^{-r}$ with $r \approx 2$) that does not depend on cultural, social, or historical factors or on the short or long-term economic or political plans

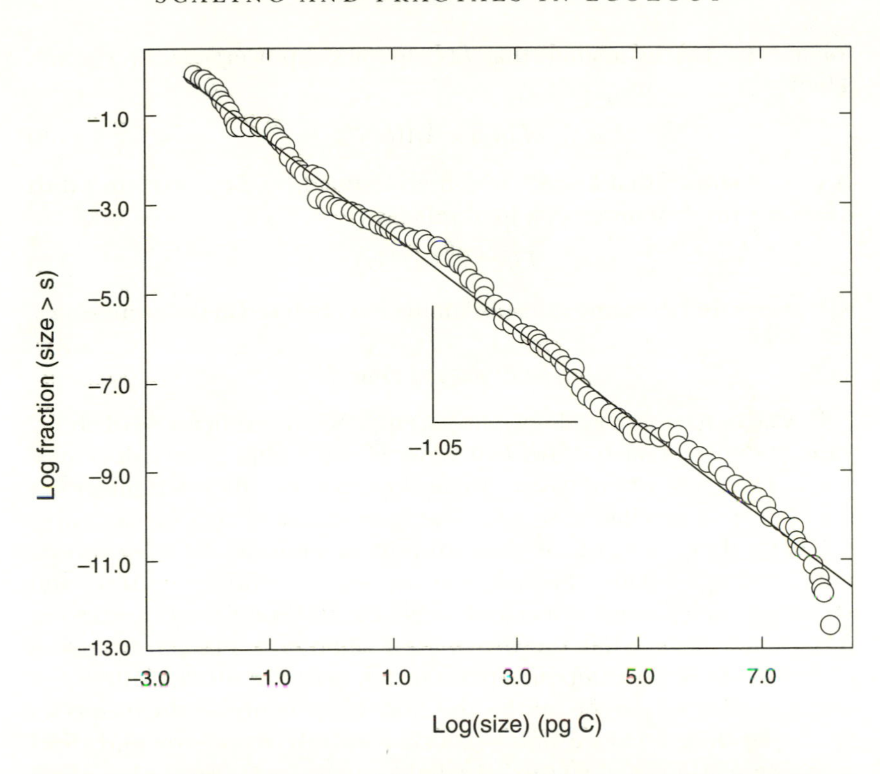

FIGURE 4.3. Scaling in the biomass distribution of all organisms in Lake Konstanz (Gaedke, 1992). Here the log-log plot shows the cumulative distribution $N(> n)$ against biomass.

of different countries. This, Zipf's law, seems to be present in other systems such as the distribution of words in human language (Schroeder, 1991). An impressive example of an ecological scaling law is provided by the frequency distribution of biomass, the scaling behavior spanning up to ten decades (Gaedke, 1992). This is shown in figure 4.3a, where the distribution

$$N(> n) = \int_1^n N(n')dn' \tag{4.6}$$

is displayed. Here for a scaling law $N(n) \approx n^{-\alpha}$, we get $N(> n) \approx n^{-\alpha+1}$.

To show that power laws are scale-invariant, we only need to see the effect of a scale transformation. Self-similarity implies that

$$\frac{L(r)}{L(\alpha r)} = k \tag{4.7}$$

for $\alpha < 1$. Let us assume that $L(r)$ follows a power law, $L(r) = Ar^{\eta}$. Then

$$kL(\alpha r) = kA(\alpha r)^{\eta}. \tag{4.8}$$

It can be shown that $k = \alpha^{\eta}$, which gives us $L(\alpha r)/L(r) = \alpha^{\eta}$ and thus $L(\alpha r) = A(\alpha r)^{\eta}$. This gives a final relation:

$$L(\alpha r) = k\alpha^{\eta}L(r), \tag{4.9}$$

which says that the same statistical pattern is obtained at different scales.

The Ecology of Fractals

Fractal behavior in ecology is apparent from different levels of observation. Figure 4.4a–b shows two views of the fractal nature of ecosystems. The large weevil (*Gymnopholus lichenifer*) shown at left illustrates the idea in a qualitative sense: The individuals of this species carry lichen on their back, which gives support to a microhabitat for several mites and springtails. Such a microecology (including bacteria and other microorganisms associated to lichen and mites) is organized at this smaller scale, while the weevil itself belongs to a larger ecological network. This is a common situation that arises in other contexts: A single species of parrot can be the host of as many as thirty species of feather mites. Since mites (which feed oily secretions and dead cells) are territorial, they can actually occupy specific parts of feathers (Wilson, 1992).

But fractal patterns can show up in more subtle ways. The right plot in figure 4.4 is an example. It shows part of a rainforest plot in Panama (Welden et al., 1991). Specifically, this is a 500×500 meter square (each square is a 5×5 meter area) where the points with a canopy height h less than a given value h_c are indicated in black (white otherwise). As we will discuss, this pattern is the outcome of a dynamical process involving tree growth, competition, and tree fall. When looking at this picture, it is possible to perceive the presence of a common pattern at different scales. Many canopy gaps are observed, some quite large. And their spatial distribution, which can be characterized by appropriate measures, is fractal (Solé and Manrubia, 1995a, 1995b).

The simplest characterization of fractal behavior is provided by the so-called fractal dimension, D (Mandelbrot, 1977, 1983). In order to define D let us consider a given object Ω embedded in a given d-dimensional domain Γ. Let us first consider regular systems, such as a line Ω_1 of length L, a square Ω_2 of area L^2, and a cube Ω_3 with a volume L^3. Imagine that we want to cover these systems using a set of identical, nonoverlapping segments, squares, or cubes of side lenght ϵL (with $\epsilon < 1$), respectively. The number of segments required to cover

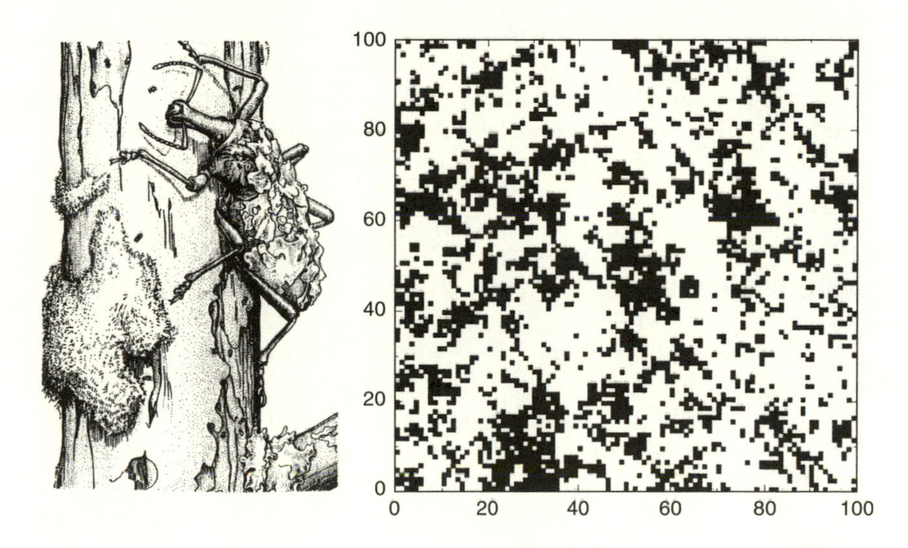

FIGURE 4.4. Ecological fractals. Left: An individual of the beetle species *Gymnopholus lichenifer* walks on the surface of a trunk carrying lichens on its back, thus offering opportunities for the formation of a microecology (drawing by A. Bartlett Wright). Right: A plot of the spatial distribution of low-canopy points in a rain forest in Barro Colorado (Panama). Black squares indicate places where the canopy height is less that $h = 15$ m. Clusters of many different sizes are observable.

Ω_1 will be $N(\epsilon) = L/(\epsilon L) = \epsilon^{-1}$. For the square, $L^2/(\epsilon L)^2 = \epsilon^{-2}$ and in general

$$N(\epsilon) = \epsilon^{-d} \tag{4.10}$$

where $d = dim(\Omega_d)$. The last is an example of a power law and allows us to define the dimension of Ω_d as:

$$d = -\frac{log\, N(\epsilon)}{log\, \epsilon}. \tag{4.11}$$

Here ϵ is in fact the scale of observation (a ruler) used in the covering process. The last definition is, however, not very satisfactory. An arbitrary object with rough boundaries (such as a coastline) will be properly covered only by using small objects. This implies that the rigorous definition needs a limit:

$$d = \lim_{\epsilon \to 0} \frac{log\, N(\epsilon)}{log\, \epsilon}. \tag{4.12}$$

This definition can be easily applied to a deterministic fractal, using the recurrence rule to define the scale of measurement. Consider for example the Sierpinski gasket Ω_s (fig. 4.1). It is the result of an iterative process that we can indicate as $\Omega_s^0 \to \Omega_s^1 \to \cdots \to \Omega_s^n \to \cdots \to \Omega_s$. If $L = 1$ (the initial side length of the triangle), we need one triangle of size $\epsilon_0 = 1$ to entirely cover Ω_s^0; Ω_s^1 needs $N_1(\epsilon) = 3$ triangles of side length $\epsilon_1 = 1/2$ and in general Ω_s^n will need $N_n(\epsilon) = 3^n$ triangles of size $\epsilon = (1/2)^n$. The (fractal) dimension will read:

$$d(\Omega_s) = -\lim_{n \to \infty} \frac{log\, N_n(\epsilon)}{log\, \epsilon_n} \tag{4.13}$$

or, in other words,

$$d(\Omega_s) = -\lim_{n \to \infty} \frac{log(3^n)}{log(2^{-n})} = \frac{log\, 3}{log\, 2} = 1.5849, \tag{4.14}$$

which is a non-integer value between one (a linear object) and two (a surface). Such a non-integer dimension is a typical feature of fractal objects, for which $d(\Omega) < d(\Gamma)$, that is, their dimension is below that of their embedding space. A similar calculation for the Koch curve gives $D_{Koch} = log\, 4/log\, 3 \approx 1.26$.

Since natural objects display statistical (nondeterministic) self-similarity, it is not obvious how to compute $d(\Omega)$ in a systematic way. A simple method for solving this problem in two-dimensional space is the so-called box-counting algorithm. The procedure is simple: We cover Ω with square, non-overlapping boxes of size ϵ^2 and repeat the procedure using a range of ϵ values. This range will be limited to the resolution scale ϵ_m (the pixels of our system) and by the system size ϵ_M. For each $\epsilon \in (\epsilon_m, \epsilon_M)$, the number of boxes $N_b(\epsilon)$ containing at least one point of Ω will be counted. It is not difficult to see that $N_b(\epsilon)$ will scale as $N_b(\epsilon) \sim \epsilon^d$ where d is the fractal dimension, here obtained in practice by estimating the slope of the scaling relation on a log-log plot (Schroeder, 1991). Similarly, the same method gives the length $L(\epsilon)$ of the perimeter of Ω, which now scales as $L(\epsilon) \sim \epsilon^{1-d}$.

Random Walks

One particularly important example of statistical fractal behavior is provided by a particle performing a random walk (RW). An RW is described as a stochastic process in which a given object moves on a d-dimensional space Γ by performing random jumps. At each step it can jump into some near point with some probability. In figure 4.5.a we plot the result of a computer simulation of an RW on a two-dimensional lattice, where available sites are restricted to the points $(i, j) \in Z^2$. For a one-dimensional system (fig. 4.5b–c), we can see that an enlarged view of a small part of the trajectory confirms that the path is fractal. Strictly speaking, the pattern displayed by the one-dimensional RW is not self-similar, but *self-affine* (Mandelbrot, 1983; Turcotte, 1992),

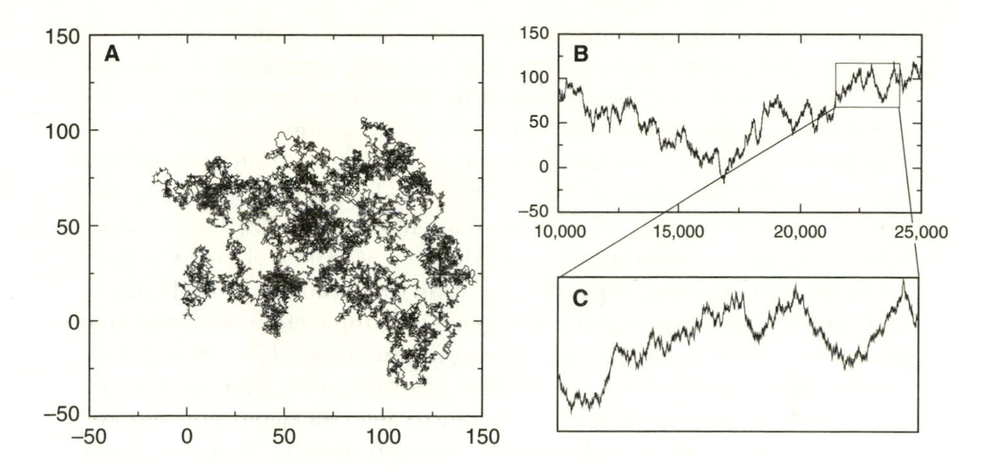

FIGURE 4.5. Random walks: In (A) an example of the path exhibited by a two-dimensional random walker is shown. The corresponding behavior for a one-dimensional walk is shown in (B) and the self-affine character of this fluctuation is apparent by looking at finer scales (C).

since the time and space dimensions do not scale in the same way. But when looking at the two-dimensional RW, given that the jumps are made in an isotropous way, the previous concepts apply and a box-counting calculation gives $d(RW) = 2$. This result actually means that an RW will eventually be plane-filling. But in fact there is much self-overlap (as we can clearly appreciate from the figures). For $d(\Gamma) = 2$, the probability of returning to a given area (no matter how small) is one. In three dimensions the situation is different; here $d(RW) = 2 < d(\Gamma)$ and the return probability is smaller than one. This is believed to be an explanation as to why many biological interactions (such as chemical reactions) occur on surfaces.

The implications of the presence of fractal behavior in ecology are easy to understand using the following example (Wilson, 1992). Consider again the beetle walking on the tree trunk. As it walks along the tree surface, it measures a circumference of, say, five meters. However, it cannot perceive the finer structure full of many dips and hollows in the bark. But in such irregularities other species of beetles exist, perceiving a different scale that the larger beetle ignores. The small details not preceived by larger organisms are actually the natural scale exploited by smaller creatures. To them, the tree trunk is about ten times larger and the surface is a hundred times larger too. Such a disparity has nontrivial consequences in relation to diversity: It translates into more niches. If we think about the different regimes of temperature and moisture present in different crevices, a huge number of microenvironments can host a plethora of fungi and algae species on which insects can feed. As discussed by Wilson, we can go further toward smaller scales (not perceived by the small beetles) and look at still smaller insects (such as oribatid mites) measuring less than one millimiter in length. Morse et al. have explored this idea by analyzing the question of why there tend to be so many more small-sized individuals (in a given habitat) than larger ones (Morse et al., 1985). Using the fractal dimension of various types of vegetation they studied the body size of arthropods. By measuring D (using box-counting techniques) of planar projections, they found a range of fractal dimensions $1.3 < D < 1.5$. It can be shown (using $L(\epsilon) \sim \epsilon^{1-d}$) that assuming $D \approx 1.5$, an order of magnitude decrease in ϵ leads to an increase in surface vegetation between 3.16 and 10 times. This study (consistent with allometric data) suggests that the steep increase in arthropod abundance with decreasing body size is a consequence of the fractal scaling of their habitat.

Important is the fact that the fractal behavior (and thus the estimated fractal dimension) of real objects can be to some extent dependent on the scale of observation. As previously discussed, a fractal object will have characteristic lower and upper cut-offs, since self-similarity is restricted to the scale of observation implicit in these limits. Another problem we shall explore is that related to the observation that many natural systems cannot be characterized by a single number such as D. Instead, they are so-called multifractal objects (see below), and so an infinite spectrum of dimensions must be introduced.

The definition of fractal dimension can be reviewed by considering it as defined in terms of r (Takayasu, 1990). So the previous definition would read:

$$D(r) = -\frac{\log N(r)}{\log r} \tag{4.15}$$

and is properly defined as long as $N(r)$ is a smooth function. If $N(r)$ follows a simple power law for all r, then $D(r) = D$. Solving the previous equation, we get

$$N(R) = N(r) \exp\Big(-\int_r^R \frac{D(s)}{s} ds\Big). \tag{4.16}$$

This can be applied to a random walk whose mean free path is finite. We have seen that the fractal dimension of the trajectory of an ideal RW is $D_{rw} = 2$, but for a finite free path we should expect a scale-dependent fractal dimension. Clearly, for a scale of observation much shorter than the mean free path, the trajectory will look like a straight line (thus $D \approx 1$) and will become more and more entangled as r grows.

FRACTAL TIME SERIES

The previous examples of fractal structures are mainly related to spatial patterns, but as mentioned in relation to RW, fractal structures can also be at work in time series. The presence of fractal structures in biological time series has been widely analyzed (West and Goldberger, 1987; Goldberger et al., 1990; Sugihara and May, 1990). The most popular tool is spectral analysis, based on Fourier transform techniques. Spectral analysis involves the decomposition of the variation in a given time series into components. The time series $\{x(t)\}$ can be considered a superposition of regular sinusoids (sine or cosine functions) of differing amplitudes, frequencies (f), and starting points (phases). In that case, we can represent $x(t)$ (defined in some interval $0 < t < T$) as:

$$x(t) = \sum_{n=0}^{\infty} a_n \cos(n\omega t) + b_n \sin(n\omega t) \quad (4.17)$$

where $\omega = 2\pi f$ and a_n, b_n are the amplitudes of each term. Standard methods allow us to calculate the appropriate coefficients associated with this series and thus to determine their contribution to the whole signal. The Fourier transform $F(f)$ of $x(t)$ is defined as:

$$F(f) = \int_0^T x(t)e^{-2\pi ift}dt. \quad (4.18)$$

Since the Fourier transform often turns out to be complex, a standard, real-valued quantity to be used is the power spectrum

$$S(f) = |F(f)|^2, \quad (4.19)$$

and is the quantity typically computed from real data. The power spectrum (PS) has typically one or more peaks, corresponding to the main frequencies present in the signal. Often, the PS is broad-band (fig. 4.6), and this is characteristic of chaotic attractors. However, the broad-band shape is also common to many types of stochastic processes. One particularly important family of stochastic processes is characterized by the presence of scaling in the PS

$$S(f) \sim f^{-\beta}, \quad (4.20)$$

with $0 < \beta < 2$. The two limiting cases correspond to white noise ($\beta = 0$) and brown noise ($\beta = 2$). White noise is associated with a complete lack of correlations: It is characteristic of the noise exhibited by a string of random numbers. Brown noise is highly correlated and corresponds to Brownian motion. The scaling behavior indicates that the time series

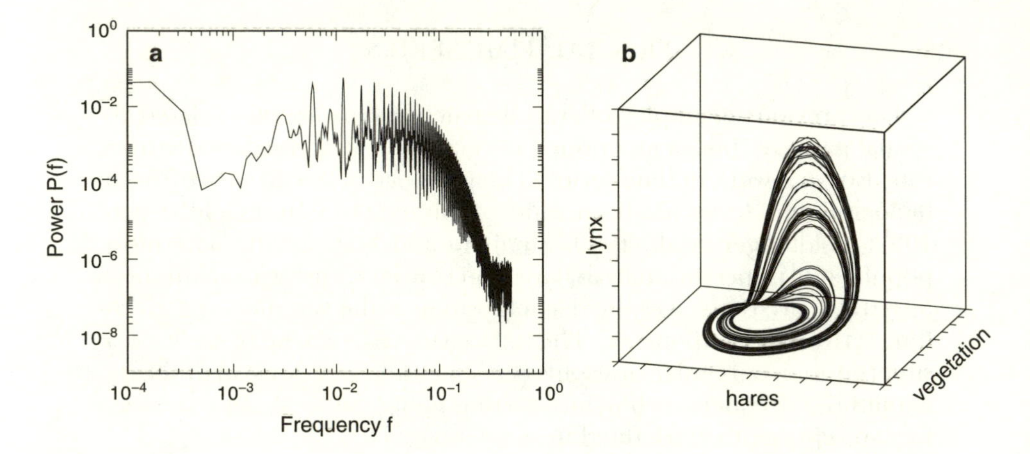

FIGURE 4.6. The power spectrum of a chaotic dynamical system. In (a) the spectrum is shown, corresponding to the strange attractor displayed in (b). Here one of the components (the simulated lynx population) is used for the computation of $P(f)$. We can see that multiple peaks emerge through a background of continuous spectrum, characteristic of strange attractors.

is fractal: When looking at different time scales, the same basic features will be observed. This type of PS is actually widespread in both biology and inanimate systems.

Fractal properties of time series can also be analyzed by means of Hurst's Rescaled Range Analysis (RRA) (Feder, 1988; Mandelbrot, 1983; Korvin, 1992). Let us consider a given time series $\{X_t\}$ with $t = 1, 2, \ldots, T$. This can be the records of discharges from a river, the number of sunspots, or the extinction size for a given group of organisms. The average of X_t over T time steps will be $< X >_T = (\sum_t X_t)/T$. The departure from the average over a t-year time horizon is given by

$$X(t,T) = \sum_{i=1}^{t}[X_i - < X >_T] = \left\{\sum_{i=1}^{t} X_i\right\} - t < X >_T . \qquad (4.21)$$

Obviously at the end of the period, we get $X(T,T) = 0$. $X(t,T)$ is usually calculated by dividing the time series into $M(T)$ adjacent segments of size T.

Two key quantities are computed from the previous time series. The first is the standard deviation, defined as $S(T) = [< (x_t - < x >_T)^2 >]^{1/2}$ and the so-called *range* of the time series, given by the difference between the maximum and the minimum over the period T:

$$R(T) = \max_{1 \leq t \leq T} X(t,T) - \min_{1 \leq t \leq T} X(t,T) \qquad (4.22)$$

Using these quantities, the rescaled range $F(T)$ is defined as (Feder, 1988): $F(T) = R(T)/S(T)$. In this way we have a measure that scales the range by taking the standard deviation as the unit of measurement.

Ordinary white noise corresponds to the case $F(T) \propto T^{1/2}$ where the values of the time series are uncorrelated with one another. In terms of forecasting, the best prediction is the last measured value. However, when Hurst analyzed the scaling in different natural systems, he found that, instead of the previous relationship, a more general scaling was present—$F(T) \propto T^H$ where H was shown to be differing greatly from 1/2. Here H is often larger than 1/2 for systems as diverse as market-price fluctuations, light intensity curves from quasars, or precipitation data (Mandelbrot, 1983; Korvin, 1992). It can be easily shown that the fractal dimension of the signal D is related to the Hurst exponent by a linear relation $D = 2 - H$.

When the Hurst exponent is greater than 1/2, the system shows *persistence* on all time scales: On average, an increasing trend in the past implies an increasing trend in the future. If $H < 0.5$ then we will see an opposite effect: An increase in the past implies a decrease in the future; the local trend will be reversed and the predicted value tends to the mean value over the interval (antipersistence).

The previous definitions are useful in problems related to the presistence of rare species (Sugihara and May, 1990b). In particular, there seems to be a link between the vulnerability to extinction and the value of the Hurst exponent. High values of H (or lower fractal dimension) imply higher vulnerability whereas higher stability would be associated to lower H. As an example, numerical estimation of H from the population fluctuations of the American redstart (*Setophaga ruticilla*) and the least flycatcher (*Empidonax minimus*) gave $H = 0.28$ and $H = 0.56$, respectively. These results suggest that the latter is more prone to local extinction and would be less density-dependent controlled. The antipersistent pattern associated with the American redstart suggests the opposite and indicates density dependence.

PERCOLATION

Let us consider the simplest and clearest example of statistical fractal behavior on a two-dimensional spatial landscape. This is an important example not only because it illustrates several relevant aspects of fractal behavior, but because it is an essential component of most of the problems arising in natural systems dealing with scaling phenomena.

Let us consider a two-dimensional, $L \times L$ square grid Γ where each site can be occupied by a tree with some probability p. The location of a given site will be indicated as $(i,j) \in \Gamma$, and its state by $S(i,j) \in \{0, 1\}$.

Here 0 and 1 stand for empty and occupied sites, respectively. Each lattice point has four nearest sites: The *neighborhood* of (i,j) will be indicated by $n(i,j)$. Since the occupied sites are randomly distributed, they do not display any kind of patchiness or spatial correlations. Now let us ask the following question: What is the probability of finding a path of neighboring trees connecting the bottom to the top? In other words, what is the probability $P(p,L)$ of finding a connected cluster of trees with a characteristic length of order $O(L)$? Our intuition suggests that such probability will increase in some monotonous fashion as p increases. We will see that such intuition is totally wrong.

When a cluster of connected sites spanning the entire lattice is present, we say that *percolation* takes place, and such cluster is called the *percolation cluster* (Stauffer and Aharony, 1985; Schroeder, 1991). Percolation is important because it underlies the presence of a spatial pattern that emerges at large scales from local interactions. In order to detect the presence of percolation, we can use a simple burning algorithm (Peitgen et al., 1992) where a limited fraction of trees is burned and fire spreads to the nearest trees. This example is not arbitrary: Forest fires are known to play a relevant role in maintaining diversity. They are both a threat and an essential ingredient to many ecosystems. In this context, along with climate and soil, fire is one of the most important forces affecting forest growth.

Let us assume that the trees at the bottom row are burned (their "burning" state will be indicated as $S(i,j) = 2$). If a nearest tree has a burning neighbor, it also burns, and in this way the fire propagates. The resulting set of burned trees,

$$\Omega(p) = \{(i,j) \in \Gamma \mid S(i,j) = 2\} \tag{4.23}$$

is our object of interest. Now the question is to compute $P(p,L)$. One initial observation is that percolation seems to occur only after a given threshold p_c is reached (fig. 4.7). As L is increased, this transition becomes sharper. For large lattices, no percolation occurs for $p < p_c$, and it always takes place for $p > p_c$. Another striking observation concerns the set of trees that is burned, as shown in insets of figure 4.7. For small p values (a), the fire is rapidly extinguished and thus only a few trees are marked. For large p values (c), the fire spreads burning almost all trees. The unexpected result is found at $p = p_c$, where we can see that the burning cluster $\Omega(p_c)$ percolates the lattice (fig. 4.7b) and displays a complex structure. This structure is a fractal: We can see islands of nonburned trees spanning many different sizes and the box-counting algorithm gives a non-integer dimension between one and two. Besides, if we compute the frequency of connected, nonburned clusters of trees of size S, it reveals itself to be a power law: $N_b(S) \propto S^{-\eta}$. In

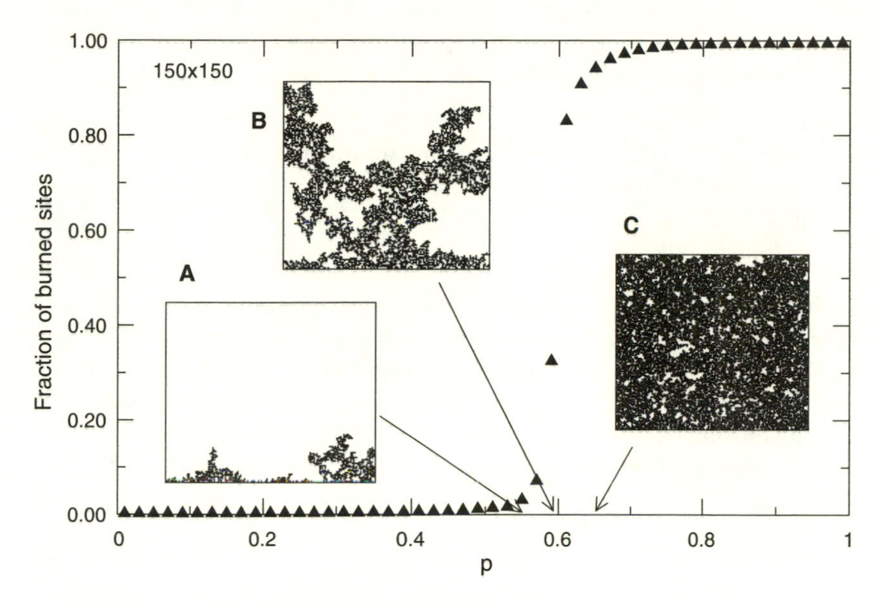

FIGURE 4.7. Phase transition in percolation processes. The main graph shows the order parameter for the transition against p, displaying a sudden increase at the critical point $p_c \approx 0.59$. The three insets show the resulting clusters of burning trees for different values of p close to the transition.

summary, this simple numerical experiment reveals three basic aspects of percolation that will play an important role in our thesis:

1. Percolation involves a threshold phenomenon, separating two well-defined domains or *phases*. The first, subcritical phase (for $p < p_c$) is characterized by short-distance interactions: the fire (or any other type of information) cannot spread to far-away places. The second phase is characterized by the free propagation of interactions thoughout the entire set of units (trees in our previous example). The two phases are separated through a well-defined boundary or critical point p_c. For very large L, $P(p, \infty)$ behaves as

$$P(p, \infty) \sim (p - p_c)^{\beta} \tag{4.24}$$

   for $p > p_c$ (and zero for $p < p_c$). Here $\beta \approx 1.16$ for a two-dimensional lattice.
2. At criticality, the cluster of connected units propagating the fire is fractal: It reveals the presence of long-range correlations with a nontrivial structure arising from short-term interactions.
3. Scaling laws arise at criticality: Looking at the distribution of connected clusters, a power law distribution is observed. Most clusters

have small size but a few have a size of order $O(L)$. The same basic process generates small and large structures.

A different characterization of the percolation transition is the divergence of the correlation length ξ. This quantity is defined as the mean distance between two sites on the same finite cluster. Close to criticality, it is found that

$$\xi \sim |p - p_c|^{-\nu} \tag{4.25}$$

with the same exponent ν above and below the threshold. A fundamental result of percolation theory is the *universal* character of the scaling exponents β and ν. In spite of the fact that p_c is different if different neighborhoods are used, the exponents defining the transition are universal, only dependent on the lattice dimension. Additionally, the fractal dimension of the percolation cluster is given by (Stauffer and Aharony, 1985):

$$d_{pc} = d - \frac{\beta}{\nu}, \tag{4.26}$$

which is $d_{pc} = 91/48$ in two dimensions (i.e., for $d = 2$, where $\beta = 5/36$ and $\gamma = 4/3$.

The fact that percolation takes place suddenly has many implications. For a system with short-range interactions, large-scale structures (and thus emergent phenomena) will arise only if some well-defined thresholds are reached. Percolation thresholds are recognized in landscape ecology as an important ingredient in understanding forest growth dynamics. In this sense, studies of forest spread in a tallgrass prairie–dominating landscape have shown the influence of percolation thresholds in accelerating the invasion process (Loehle et al., 1996). But percolation also has implications in the other direction: For a given ecosystem, where high diversity needs a minimal spatial habitat, random habitat loss can lead to percolation, which means breaking available space into many patches and triggering ecosystem collapse (Bascompte and Solé, 1996; Solé and Alonso, 2000; see chapter 5). Some theoretical computations on percolation thresholds can be performed by means of the so-called renormalization group analysis (appendix 2).

Percolation on Bethe Lattices

There is one particular example of percolation process that allows one to derive a number of relevant statistical features. It occurs in a rather particular tree structure, known as the Cayley tree or the Bethe lattice (Stauffer and Aharony, 1985), but it serves as a simple (mean field) model of percolation. This type of lattice has a tree structure

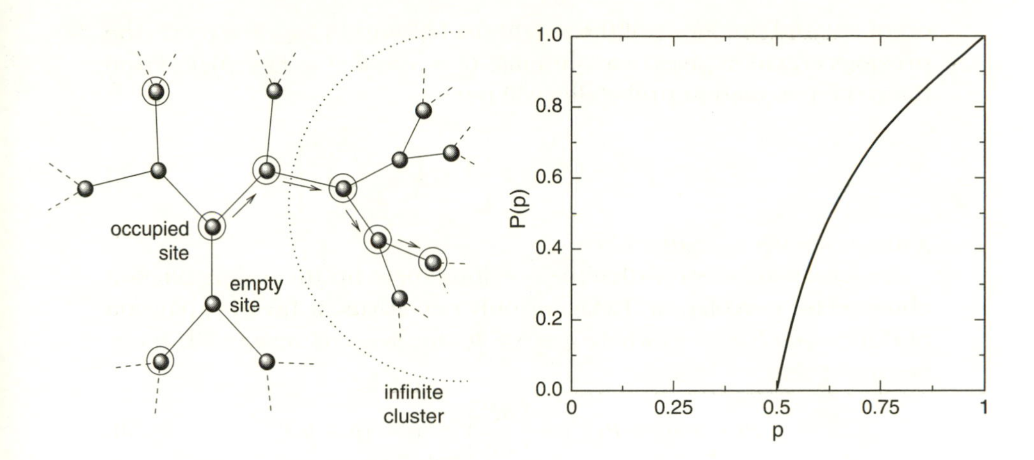

FIGURE 4.8. (Left) Bethe lattice: Here each node has exactly three bonds. In (right) the mean-field calculation of the phase transition is shown for $z = 3$.

(fig. 4.8a) with no loops and has been used in epidemics as a simple illustration of disease propagation.

Let us consider the nodes of this lattice occupied or free (thus similar to the previous example) and assume that disease or fire can propagate through from site to site if and only if the sites are occupied. The branching is assumed to be directional, that is, although each node has z links (the so-called coordination number) only $z - 1$ of them can actually propagate the signal. If p is the probability of occupation of a site, the average number of sites that can propagate epidemics will be $p(z - 1)$. The critical condition for propagation to define the percolation threshold would be $p_c(z - 1) = 1$ and thus the critical occupation is:

$$p_c = \frac{1}{z - 1}. \tag{4.27}$$

Assuming that the lattice is infinite, we can also compute the percolation probability $P(p)$ that, given one infected site, the disease will propagate (percolate) to infinity. For $p < p_c$ we have $P(p) = 0$, but for $p > p_c$ the specific form of $P(p)$ must be derived. If we indicate as C_∞ an infinite (percolating cluster) and defining Q as the probability that a randomly chosen site site s_i does not belong to C_∞, it is not difficult to see that:

$$Q = P[s_i \notin C_\infty] = (1 - p) + pQ^{z-1}. \tag{4.28}$$

The first term on the right-hand side corresponds to the probability that the node is empty. The second term is the probability that s_i is occupied and that none of the neighbors belongs to C_∞. For $z = 3$, the previous equation gives two solutions: $Q = 1$ and $Q = (1-p)/p$. From them the percolation probability will be:

$$P(p) = p(1 - Q^z) = p\left[1 - \left(\frac{1-p}{p}\right)^3\right] \tag{4.29}$$

and it is shown in figure 4.8b.

It is easy to demostrate that the previous curve fits the scaling relation obtained for percolation. In fact we only need to use a Taylor expansion of $P(p)$, using $P(p_c + \eta)$ where $\eta \equiv p - p_c$, for $p - p_c$ is very small. It can be shown that

$$P(p_c + \eta) \approx P(p_c) + \left[\frac{\partial P}{\partial \eta}\right]_{\eta=0} \eta \approx (p - p_c)^\beta \tag{4.30}$$

with $\beta = 1$. This result is certainly far away from the exact value $\beta = 0.15$ and illustrates how mean-field approximations fail to give satisfactory quantitative agreement. This is compensated by the considerable insight they can provide and by the fact that mean field results become exact as the dimension $d(\Gamma)$ grows. It also means that when interactions become increasingly nonlocal, mean field approximations (where no correlations among nearest sites are taken into account) are good descriptions of macroscopic phenomena.

NONEQUILIBRIUM PHASE TRANSITIONS

The percolation transition is an example of a so-called critical phenomenon and the shift from no propagation to propagation occurs at a so-called critical point. Critical points are associated with the presence of a narrow transition domain separating two well-defined phases, and we generally speak of critical phenomena to refer to such transitions. These two phases are characterized by the presence of macroscopic features linked to the capacity of the system to propagate some type of signal (such as fire, epidemic, or information). The term "phase transition" is well known from the physics of critical phenomena and the theory of phase transitions has been successfully applied to such diverse phenomena as polymerization, gelation, or turbulence, and not surprisingly (largely due to the presence of universal phenomena) to many other domains outside physics. This also includes ecological applications, particularly extended within landscape ecology (Milne, 1998). A critical phase-transition is characterized by some order parameter $\Phi(\mu)$ that depends on some external control parameter that can be continuously varied. For the percolation problem, $P(p)$ acts as an order parameter and p as a control parameter. In critical transitions Φ varies continuously at $\mu = \mu_c$ (where it takes a zero value) but the derivatives of Φ are discontinuous at criticality. For the so called first-order transitions (such as the water-ice phase change), there is a discontinuous jump in Φ at the critical point. First-order transitions do not exhibit the scaling features displayed by critical transitions and their mean field theories correspond to threshold phenomena in catastrophe theory (see chapter 2).

Some standard models of phase transitions in physics involve equilibrium. This is the case for example of the so-called Ising model (Binney et al., 1992), which has been used in a number of contexts, including model neural networks, flocking birds, forest dynamics, and beating heart cells (see for example Katori et al., 1998). Although the Ising model provides a powerful illustration of critical-phase transitions, it is not representative of the related phenomena observed in nonequilibrium systems. Most complex systems are in a nonequilibrium state: They are subject to a constant flow of energy, matter, and/or information (de Groot and Mazur, 1984).

One of the characteristic defining features of equilibrium systems is the fulfillment of the so-called detailed balance condition. Consider a system describable in terms of a set of states $\mathcal{A} = \{\alpha\}$ such that each state $\alpha \in \mathcal{A}$ occurs with a probability P_α (so that $\sum_\alpha P_\alpha = 0$). For each possible pair of states $\alpha, \alpha' \in \mathcal{A}$, a transition probability $\omega(\alpha \to \alpha')$ can be defined, such that

$$\sum_{\alpha'} \omega(\alpha \to \alpha') = 1. \tag{4.31}$$

The detailed balance condition is given by:

$$P_\alpha \omega(\alpha \to \alpha') = P_{\alpha'} \omega(\alpha' \to \alpha), \tag{4.32}$$

which can be easily interpreted: The intuitive meaning of detailed balance is that the balance between the probability of "leaving" a state and arriving in it from another state holds not just overall, but in fact individually for any pair of states. This condition allows one to properly define a Markov process.

Nonequilibrium systems are not determined (solely) by external constraints and violate the detailed balance condition. As a given control parameter is changed, a given stationary state can become unstable and be replaced by another. This new state might involve an oscillatory or a chaotic attractor, and transitions to these new states are usually path dependent. Nonequilibrium phase transitions (NPT) have been identified in numerous systems, mainly in physics and chemistry (Nicolis and Prigogine, 1977; Haken, 1977, 1983). How are both types of transitions related? Since detailed balance is not operating in NPT, their macroscopic properties are no longer obtainable from a stationary probability distribution. Besides, in systems out from equilibrium no free energy can be defined. In equilibrium systems, the phase diagram can be determined on the basis of free energy but this approach becomes useless far from equilibrium. However, phases are recognizable in terms of differential qualitative features exhibited by the system's macroscopic behavior, which arise when key parameters are changed.

An important subclass of NPTs is provided by lattice models known as interacting particle systems (Marro and Dickman, 1999), which can also be labeled stochastic cellular automata. This class of models will be widely explored in the following pages in different contexts. Although they might seem too simple and the discrete character of the unit interactions too simplistic, they succesfully reproduce macroscopic aspects displayed by real systems.

In the following section, the most simple example of these transitions, the so-called branching process (BP), is considered. It allows us to derive some analytic results and is the backbone of different important applications.

THE BRANCHING PROCESS

Many dynamical systems can be described in terms of the extinction and birth of given objects. These can be individuals, names, species, or particles (Harris, 1963). In its simplest form, branching processes (BP) are models in which the members of the population are assumed to reproduce independently of one another (so that distinct lines of descent are independent). Branching processes have been widely used in biology, particularly in genetics.

Consider a simple stochastic process with no spatial structure in which a population of individuals reproduces and dies at given, constant rates (i.e., the BP is named homogeneous). Specifically, if $n(t)$ is the number of individuals, the following transition rates are defined:

$$\omega(n \to n+1) = \lambda n \tag{4.33}$$

$$\omega(n \to n-1) = \mu n \tag{4.34}$$

where λ and μ are birth and death rates (per individual), respectively. Using these rates, a dynamical equation for the probability $P_n(t)$ of having n individuals at time t can be derived. The first step is to write the discrete-time dynamics of P_n:

$$P_n(t+1) = P_n(t) + \omega(n+1 \to n)P_{n+1}(t) + \omega(n-1 \to n)P_{n-1}(t)$$
$$-P_n(t)\left[\omega(n \to n+1) + \omega(n \to n-1)\right] \tag{4.35}$$

if we make the approximation: $dP_n/dt \approx P_n(t+1) - P_n(t)$; then we can write the so-called master equation for the branching process:

$$\frac{dP_n(t)}{dt} = \omega(n+1 \to n)P_{n+1}(t) + \omega(n-1 \to n)P_{n-1}(t) \\ -P_n(t)\left[\omega(n \to n+1) + \omega(n \to n-1)\right]; \qquad (4.36)$$

in our case, this equation reads:

$$\frac{dP_n(t)}{dt} = \mu(n+1)P_{n+1}(t) + \lambda(n-1)P_{n-1}(t) \\ -[\lambda+\mu]nP_n(t) \qquad (4.37)$$

by using $\mu = 1$ we have:[1]

$$\frac{dP_n(t)}{dt} = (n+1)P_{n+1}(t) + \lambda(n-1)P_{n-1}(t) - (1+\lambda)nP_n(t). \qquad (4.38)$$

In order to solve this equation, the so-called generation function $G(x,t)$ can be used:

$$G(x,t) = \sum_{n=0}^{\infty} x^n P_n(t). \qquad (4.39)$$

Using this function and its derivatives,

$$\frac{\partial G(x,t)}{\partial x} = \sum_n nx^n P_n(t) \qquad (4.40)$$

$$\frac{\partial G(x,t)}{\partial t} = \sum_n nx^n \frac{dP_n(t)}{dt}, \qquad (4.41)$$

the master equation for the BP can be written as

$$\frac{\partial G(x,t)}{\partial t} = (1-x)(1-\lambda x)\frac{\partial G(x,t)}{\partial x} \qquad (4.42)$$

with $G(1,t) = 1$. If a single individual is present at $t = 0$, then $G(x,0) = x$, and we obtain, by solving the last equation,

$$G(x,t) = \frac{(1-\lambda x)e^{(1-\lambda)t} - (1-x)}{(1-\lambda x)e^{(1-\lambda)t} - \lambda(1-x)}. \qquad (4.43)$$

Now, for $x = 0$, we have $P_0(t) = G(0,t)$ (the extinction probability), and since $P(t) = 1 - P_0(t)$, we obtain the survival probability associated to the BP (see fig. 4.9b):

$$P(t) = \frac{\lambda - 1}{\lambda - e^{(1-\lambda)t}}. \qquad (4.44)$$

[1] Without loss of generality: This is actually equivalent to rescaling the coefficients by dividing the master equation by μ.

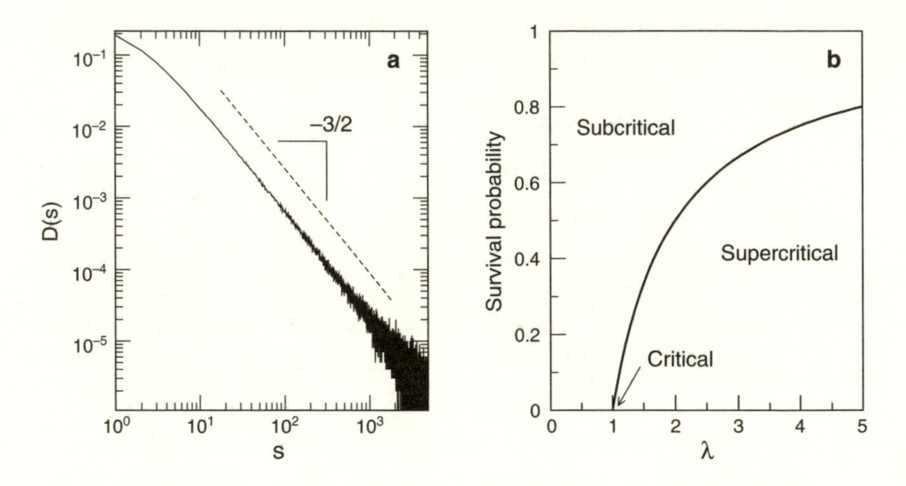

FIGURE 4.9. Scaling in the homogeneous branching process. In (a) an example of the power law distributions arising at criticality. Here a simple BP has been simulated starting from a single particle and averaged over 10^6 runs with $\lambda = \mu = 0.25$. In (b) the phase transition curve for the survival probability is shown (here for $\mu = 1$).

This time-dependent solution has different behaviors for $\lambda < \lambda_c = 1$ and for $\lambda > \lambda_c$. By taking the limit for $t \to \infty$ on the last equation, we obtain the asymptotic survival probability P_∞:

$$P_\infty = \lim_{t\to\infty} P(t) = \begin{cases} 1 - \frac{1}{\lambda} & : \lambda > 1 \\ 0 & : \lambda < 1. \end{cases} \tag{4.45}$$

The critical point $\lambda_c = 1$ defines the boundary between extinction and survival. Here P_∞ is our order parameter and displays a sudden transition at λ_c. At criticality, we have

$$G(x, t; \lambda_c) = \frac{1 + (1 - x)(t - 1)}{1 + (1 - x)t}, \tag{4.46}$$

which gives a survival probability $P(t) = 1/(1 + t)$. Several scaling laws are obtained close to criticality for the BP. Examples are the lifetime distribution (i.e., the time required until the population goes extinct) or the size of the trees generated before extinction takes place. Another is the probability $D(s)$ that the BP creates a tree with exactly s individuals. For large s, it follows a scaling law

$$D(s) \sim a^{-s} s^{-3/2} \tag{4.47}$$

where a is a given constant. An example of this scaling law is shown in figure 4.9a. Here a BP has been simulated and averaged over 10^6 replicas (using $\lambda = \mu = 0.25$).

THE CONTACT PROCESS: COMPLEXITY MADE SIMPLE

The branching process has a number of limitations when applied to real systems—(1), it allows an infinite number of elements to be present (at the supercritical phase), and (2) interactions have a nonlocal character. In this section we will consider an important class of spatially explicit, stochastic models widely used in spatial ecology and epidemiology. It is known as the contact process (Marro and Dickman, 1999) and was first proposed by Harris (1963) as a toy model of epidemics, though it has since been often used to model epidemiological dynamics (Durrett and Levin, 1974).

As with the Ising model, the contact process is defined on a discrete lattice of sites that can be in one of a finite number of states. In the easiest case, each site can be in either state 0 (vacant) or state 1 (occupied by a "particle"). This is a general model, and so one can think of different meanings for a "particle." In the last chapter we dealt with local populations, and so states 0 and 1 were associated to the presence and absence of a local population, respectively. Within the context of epidemics, states 0 and 1 would represent susceptible and infected individuals, respectively. The dynamic of the contact process is specified by a probabilistic function of the current state of the considered site, and the value of a set of neighboring sites (see fig. 4.10).

Let us first consider the one-dimensional CP (Marro and Dickman, 1999). At a given step t, the system is defined as a string $\{S_1(t), \ldots, S_N(t)\}$ of N sites where $S_i(t) \in \{0, 1\}$. Here 0 and 1 stand for "healthy" and "infected" units, respectively. The rules for this model are:

1. Infection: A healthy site becomes infected proportionate to the number of infected neighbors. So we have:

$$0 \xrightarrow{\lambda n_i} 1 \tag{4.48}$$

   where $n_i = S_{i+1}(t) + S_{i-1}(t)$ and λ is the probability of infection in a single contact.
2. Recovery: An infected individual can recover with probability μ,

$$1 \xrightarrow{\mu} 0. \tag{4.49}$$

The first rule specifies the propagation of the infection and the second involves the decay (assumed to be independent of the neigh-

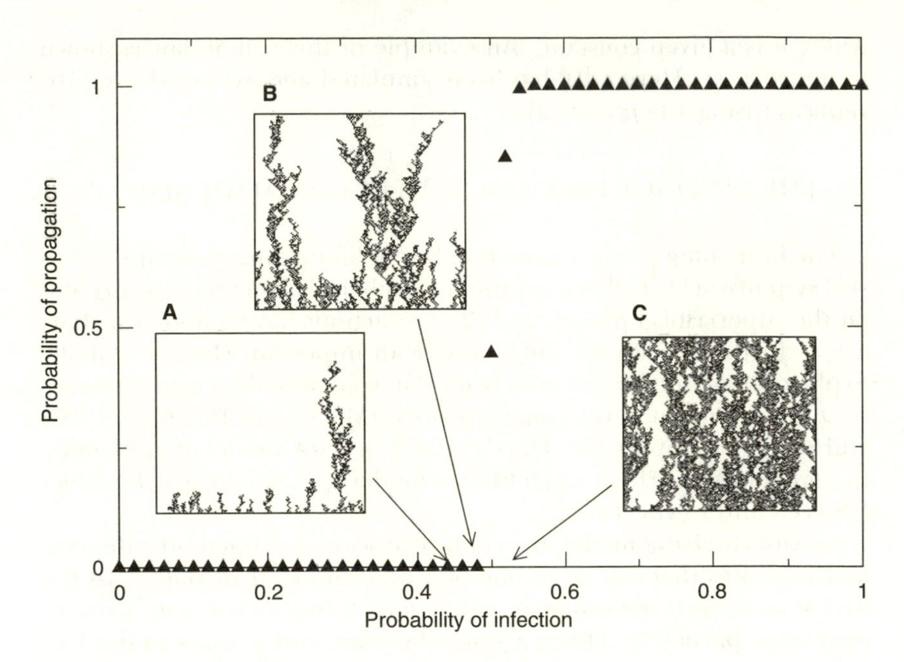

FIGURE 4.10. Phase transition in the contact process: Here a string of 150 sites has been used and a decay probability $\mu = 0.5$. The black triangles indicate the probability that active particles are still observed after $\tau = 10^3$ steps. A sharp transition is observed at the probability of infection $\lambda_c = 0.5$. Three space-time diagrams of the process are also shown at three different points in parameter space.

borhood). The system is updated synchronously: All sites are simultaneously changed at $t + 1$, following the previous rules.

One of the main questions to ask is wether the epidemic will persist indefinitely or become extinct. Computer simulation helps to answer this question. If $\rho(x, t)$ indicates the probability of infection at a given point $x \in \mathcal{L}$, that is

$$\rho(x, t) = P[S_x(t) = 1], \tag{r}$$

then $\rho(x, t)$ will evolve in time following:

$$\frac{d\rho(x,t)}{dt} = -\mu\rho(x,t) + \frac{\lambda}{q}\sum_{<u>}^{q} P\left[S_x(t) = 0, S_u(t) = 1\right] \tag{4.50}$$

where $< u >$ indicates the sum over the set of q nearest neighbors. For one dimension, this set is simply $\{i - 1, i + 1\}$. The previous equation is exact, but its computation would require knowledge of the probabilities associated with the interactions between nearest sites. The simplest approximation ignores such correlations and leads to the logistic equation:

$$\frac{d\rho}{dt} = \phi(\rho) = -\mu\rho + \lambda(1-\rho)\rho \tag{4.51}$$

where global spatial mixing is assumed (i.e., $\rho = \rho(x), \forall x$). The first term $-\mu\rho$ gives the recovery rate of infected units and the second term $\lambda(1-\rho)\rho$ gives the rate at which new infections occur. This equation has two equilibrium points: $\rho_0^* = 0$ and $\rho_1^* = 1 - \mu/\lambda$. It is not difficult to show that the epidemics will spread (and thus ρ_0^* will be unstable) if $\partial\phi(\rho_0^*)/\partial\rho < 0$, that is, if $\lambda < \mu$. For $\lambda > \mu$ propagation occurs and the second fixed point is reached. The critical point $\lambda_c = \mu$ separates the two phases of this model: (1) the subcritical phase, where the epidemics dies out and (2) the supercritical phase, where a steady average number of infections is always present (so that the epidemic self-maintains). The contact process has been used as a model for a large number of phenomena, particularly in epidemics and ecology. The CP and its variants have been used in: (a) understanding the coupling between spatial structure and fluctuations in metapopulations (Snyder and Nisbet, 2000), (b) modeling competition between annuals and perennials (Crawley and May, 1987), (c) defining resource-biodiversity relations in complex ecosystems (Weitz and Rothman, 2004), and (d) modeling competition in spatially explicit landscapes (Bengtsson, 1991).

RANDOM WALKS AND LEVY FLIGHTS IN POPULATION DYNAMICS

As mentioned previously, one of the simplest but most relevant examples of statistical fractal behavior is displayed by random walks. Together with population changes, the spatial spread of individuals has important consequences for ecosystem structure (see chapter 3). Deviations from this simple type of RW behavior have been reported as common in ecology. What is the importance of such deviations? And in a related context: What are the consequences of fractal landscape structures on the movement of organisms?

When building models of population dynamics, an important theoretical consideration is the underlying phenomenon taking place at the microscopic level. The study of pattern movements is a relevant aspect of landscape ecology (Crist et al., 1992). These patterns are influenced by different scales and are thus likely to be analyzed through the tools of fractal geometry. A null hypothesis assumes that the d-dimensional landscape is homogeneous and that individuals perfom independent random walks.

These assumptions, together with constant population size, give a macroscopic, dynamical description of population dispersal in terms of a simple diffusion equation (Murray, 1989):

$$\frac{\partial N(r,t)}{\partial t} = D\nabla N(r,t) \tag{4.52}$$

where $N(r,t)$ is the local density of individuals, D is their diffusion rate, and ∇^2 is the diffusion operator that in two dimensions reads: $\nabla^2 \equiv (\partial^2/\partial^2 x, \partial^2/\partial^2 y)$.

Two sources of deviation emerge from the constraints imposed by real landscapes and by behavioral rules (Johnson et al., 1992). The first has to do with the fact that movement is effectively limited or facilitated by landscape features. In this context, landscapes very often display fractal properties (Rodriguez-Iturbe and Rinaldo, 1997) and thus the pattern of movement displayed by individuals is nonrandom. The second is associated to intrinsic departures from randomness in animal movement (Bascompte and Vilà, 1997): Optimal foraging might be associated with special behavioral patterns of movement resulting in optimization through evolution.

In relation to the first constraint imposed by landscape heterogeneity, it can be shown (see Johnson et al., 1992) that non-Fickian diffusion is common in animal movements in real landscapes. The mean squared displacement scales as:

$$< R(t) > \sim t^{\nu} \tag{4.53}$$

where ν is related to the fractal dimension of the landscape and is $\nu = 0.71$ for a random walk restricted to jumping within a percolation cluster. Studies on *Eleodes* beetles in short-grass prairies have shown that the movement paths of individuals are strongly influenced by vegetation structure (Crist et al., 1992). They also indicate that individual movements would likely be controlled by different processes at different spatial and temporal scales. Such differences may be reflected in community patterns of species distribution and abundance linked to the fractal structure of vegetation (Morse et al., 1985; Williamson and Lawton, 1991).

Lévy Flights and Search Patterns

Observations of search patterns in different species has shown that previous assumptions from foraging theory about the existence of characteristic scales of search are wrong. Actually, the statistical analysis of the distribution of flight lengths L_j is long-tailed, that is, a power law

$$P(L_j) \sim L_j^{-\mu} \tag{4.54}$$

that corresponds (for $1 < \mu \leq 3$) to a so-called Lévy flight (Lévy, 1965).

LF have been found to be of great relevance in many areas (Klafter et al., 1996). Detailed studies on the behavioral patterns of search in protozoa, insects, mammals (both herbivores and carnivores), and birds have shown that their statistical pattern of search is of LF type with a well-defined scaling exponent consistent with an optimal strategy (see below).

It is useful to compare LF distributions with those generated from Brownian motion. Let us consider a random walk with the length of step following a probability distribution $p(L)$. The mathematician Paul Lévy asked what type of distribution $P_N(L)$ for the sum lengths over N steps, that is,

$$L = \sum_{j=1}^{N} L_j \tag{4.55}$$

would have the same form as $p(L)$. We recognize in this question the problem of how fractals are generated, this time for a random trajectory. Although the standard answer is that a Gaussian distribution will follow this rule (due to the central limit theorem), Lévy found other solutions. For a Gaussian distribution, the sum of $p(L)$ is a Gaussian but with N times the original variance. But the solutions found by Lévy have no finite variance. This lack of characteristic scale makes LF scale-invariant fractals.

A simple model has been developed to analyze the possible role of scale-free distributions of search paths in foraging (Viswanathan et al., 1999). The model assumes that target sites are randomly scattered through a given spatial domain, and that foragers behave as follows (see fig. 4.11a–b):

- If the target site is at a distance below the so-called direct vision distance r_v, then the forager travels the distance to the target in a straight line. The presence of r_v introduces thus a constraint on the maximum perception of target sites.
- If no target site is available for $r < r_v$, then the forager travels a distance L_j in a random direction. The distance is obtained from the previous probability distribution. It progressively approaches the random location and keeps going unless a new target is detected. In the latter case, it proceeds in a straight line.

The model can be solved analytically (Viswanathan et al., 1999). If the density of targets is ρ, the mean free path λ between two target sites explored consecutively by a given forager will be, in two dimensions (we assume that the distances between any two successive sites is λ)

$$\lambda = \frac{1}{2r_v\rho}. \tag{4.56}$$

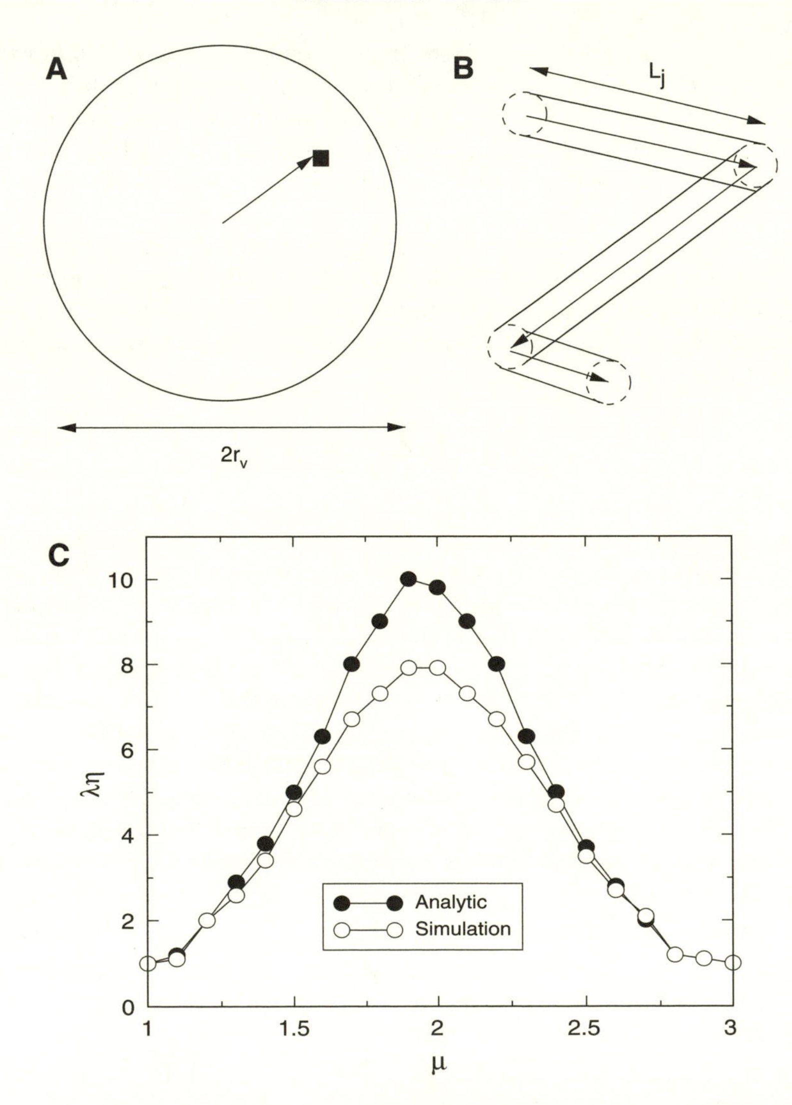

FIGURE 4.11. Foraging strategies: (A) It is assumed that a given individual has a characteristic, direct vision range r_v. Here a given target is indicated as a black square. In that case, the forager moves in a straight line to the target. If no target is available within r_v, the forager performs a random search by traveling in straight lines through varying distances L_j, where the distribution $P(L_j)$ is of Lévy type (redrawn from Viswanathan et al., 1999). In (C) the fitness of foraging strategies with different Lévy parameters is shown, with a maximum at $\mu_o = 2.0$.

Assuming that the distance between any two successive sites is λ, the average flight distance is thus

$$< L > \approx \frac{1}{\int_{r_v}^{\infty} L^{-\mu} dL} \left(\int_{r_v}^{\lambda} L^{1-\mu} dL + \lambda \int_{\lambda}^{\infty} L^{-\mu} dL \right) \tag{4.57}$$

and after performing the integrations, it reads

$$< L > \approx \left(\frac{\mu - 1}{2 - \mu} \right) \left(\frac{\lambda^{2-\mu} - r_v^{2-\mu}}{r_v^{1-\mu}} \right) + \frac{\lambda^{2-\mu}}{r_v^{1-\mu}}. \tag{4.58}$$

This mean field model involves a truncated Lévy distribution. Truncation occurs at the cutoff distance λ. By defining a search efficiency function as the ratio,

$$F(\mu) = \frac{1}{< L > N}, \tag{4.59}$$

that is to say, the ratio between the number of target sites visited to the total distance traversed by the forager. Here N is the mean number of flights taken by an individual while traveling between two successive target sites.

Using the definition for $< L >$, it can be shown that the search efficiency reaches an optimal value at $\mu_o \approx 2$. This is shown in figure 4.11c (white circles). Here the product $\eta\lambda$ is shown against the Lévy parameter μ. The robustness of this result (based on simple mean field equations) is shown by using a simulation model in which the previous rules are explicitly introduced for foragers moving on a two-dimensional system. This model considered the case of nondestructive foraging, which occurs when target sites are only temporarily depleted or fall below some fixed concentration threshold but it also applies if foragers become satiated and leave the area. Although the specific values obtained for the spatially explicit simulation are lower than the mean field prediction (as one might expect), they show a maximum in efficienty at the same point. These results indicate that optimal flights appear to occur in a way that avoids the overlapping common to standard random walkers but also the too-straight paths present at higher μ values.

Comparison with real data of foraging strategies in bees, albatross, deer, jackals, and spider monkeys seems to confirm the theoretical prediction. When nectar availability is low, the flight-length distribution $P(L)$ in bees decays as a power law with $\mu \approx 2$, consistent with theory. The same is found in albatross exploring large oceanic areas, and deer (in both wild and fenced areas), and a value of $\mu \in [2.0, 2.5]$ found in amoebas is also in agreement with the prediction.

In all these examples individuals are assumed to have no information about their resources, which are assumed to be scarce. Further theoretical studies (Bartumeus et al., 2002, 2003) suggest that changes in LF distributions are key to optimizing the long-term statistics of encounter rates under different scenarios. Scale-free patterns of foraging would appear under the situations explained above, which are common in nature (exploration of new habitats, unpredictable resource distributions, etc.). Both theoretical and field analyses support the idea that foraging patterns of very different organisms share universal laws related to optimal search strategies.

PERCOLATION AND SCALING IN RANDOM GRAPHS

The concept and consequences of percolation can be extended in many ways to ecological systems. One particularly interesting application concerns the connectivity of landscapes and ecosystems. A connected landscape would be roughly characterized by the existence of paths connecting most patches among them. Since the connectivity of patches or sites in a landscape has a strong (if not leading) effect on species distributions, whether or not a landscape is connected is a key issue. Hereafter, connectivity will refer to the accessibility to a given species to the whole set of patches or a fraction of them. As we have seen previously, percolation involves a threshold phenomenon that implies a transition from an essentially disconnected to a connected lattice. This picture can be generalized to other types of systems (see, for example, percolation in food webs, chapter 6) including some particular cases, such as the Bethe lattice.

The Bethe lattice is an abstract picture of connected systems composed of many parts. We used this simplistic picture in order to obtain some insight and analytic results from mean field models of percolation. But actually many interesting systems already exhibit some underlying graph structure describing interactions among their components. Examples in biology abound: Genetic, neural, metabolic, and ecological networks are obvious examples (Bornholdt and Schuster, 2002). In all these cases, two given components (genes, neurons, metabolites, or species) interact in some way and with some strength. The whole network can be extremely complicated and evolutionary processes are reflected in different ways in the topological properties of these networks. The natural formalism in which these properties can be analyzed is graph theory.

Graph theory has a long history in mathematics and applied sciences (Bollobás, 1985). A graph $\Omega(V,E)$ is defined as a finite set V of N vertices and a finite set E of edges such that each edge $e_{ij} \in E$ is associated with a pair of vertices $\omega_i, \omega_j \in V$. The simplest assumption in modeling a real graph is random wiring. The random graph is a null model that can be used as a first approximation to an arbitrary complex network. Early models of complex ecosystems used such approximation (May, 1972, 1973). The standard model considers a fixed number N of nodes (these can be species or habitat patches) and also a probabilistic rule defining the number and position of connections (edges). Specifically, one way of defining a random graph is to sequentially add edges by randomly choosing pairs of nodes and linking them. If we use a finite set of $V < E^2$ links, then at the end of this process we have a random graph with an average number of connections $z = V/E$. If we look at the so-called degree distribution, the frequency distribution of nodes with k links, $P(k)$, this particular example of random graph has a Poisson distribution: $P(k) = e^{-z}\lambda^k/k!$ Here

$$\bar{k} = \binom{S-1}{k} P_r^k (1 - P_r)^{S-k-1} \tag{4.60}$$

where P_r is the probability that two nodes are connected.

The theory of random graphs was widely developed in different directions and several relevant applications have been developed, mainly in communication networks and optimization. Perhaps one of the most spectacular findings of graph theory is the phase transition displayed by these graphs, as formulated by the famous Erdös-Renyi theorem. Roughly, this theorem establishes that, for an undirected graph and assuming a very large number of nodes, as z reaches a critical (percolation) threshold $z_c = 1$, a giant component emerges in the graph including a large fraction of nodes. The number of unconnected (isolated) nodes n_u rapidly decreases as: $n_u \propto O(1/N)$. Below the percolation threshold the graph is formed by a number of separated components. The overall features of this transition are summarized in figure 4.12. Here the fraction of nodes S belonging to the giant component is plotted against the average degree z.

Detecting Critical Scales in Fragmented Landscapes

A perfect example of the use of random graph theory in ecology is provided by the work of Bruce Milne and coworkers within the context of conservation ecology (Milne et al., 1996; Keitt et al., 1997; Bunn et al., 2000). The problem involves a given landscape where a set of habitat patches is accessible to a given species or set of species. In fragmented habitats (see chapter 5) where islands of available habitat are often punctuated by stretches of destroyed or poor habitats, species will perceive the habitat in different ways depending upon their range

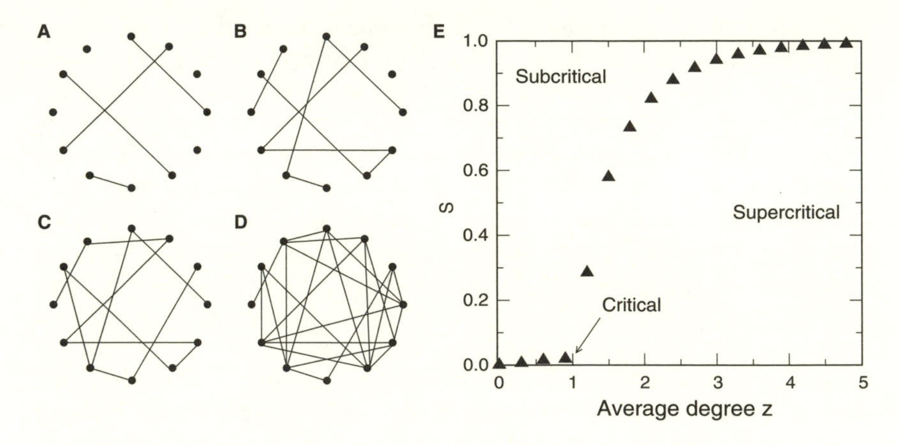

FIGURE 4.12. Phase transitions in random graphs: In (A–D) four different states of the wiring process are shown for a small graph with $n = 12$ nodes. Here (A) $z = 0.22$, (B) $z = 0.6$, (C) $z = z_c = 1$ and (D) $z = 2$. A phase transition occurs at $z_c = 1$, where the fraction γ of nodes included in the giant component rapidly increases. This picture corresponds to a very small system. Appropriate characterizations of the phase transition require large network sizes. An example is shown in (E), where the fraction S of nodes belonging to the giant component is plotted against z. Here the transition has been computed for graphs of size $N = 10^3$ and averaging over 10^4 replicas.

of dispersal. Long-range dispersors will "see" the landscape as more connected than a species with a shorter range.

Consider a given habitat composed of a set of patches $\{\omega_1, \ldots, \omega_N\}$. The area of these patches will be indicated as $\mu(\Omega_i)$. The total set $\Omega = \bigcup_{i=1}^{N} \omega_i \subset \Gamma$ defines the potential, suitable habitat available to a given species on a surface Γ. It is assumed that we have a resolution defining a lower bound: the pixels of our lattice. This habitat can correspond to a homogeneous or heterogeneous set with different vegetational composition, so that each ω_i can be different in size and internal organization. The problem considered by several authors (Milne et al., 1996; Keitt et al., 1997; Bunn et al., 2000) is the following: Given a species that can occupy Ω with a dispersal range δ, which fraction of the current habitat is effectively available to colonization?

This question can be answered using graph theoretic concepts. The first step is to construct a landscape graph $\Omega(V, E(\delta))$ defined as follows. The set of vertex is simply the set of patches $\{\omega_i\}$. This set is the same for any δ. However, the edges $e_{ij} \in E(\delta)$ connecting different patches define a set that is dependent upon dispersal range. Here link $e_{ij} \in E(\delta)$ exists if the distance between the centers of ω_i and ω_j are such that $d(\omega_i, \omega_j) < \delta$ (fig. 4.13a). The distance is calculated using the

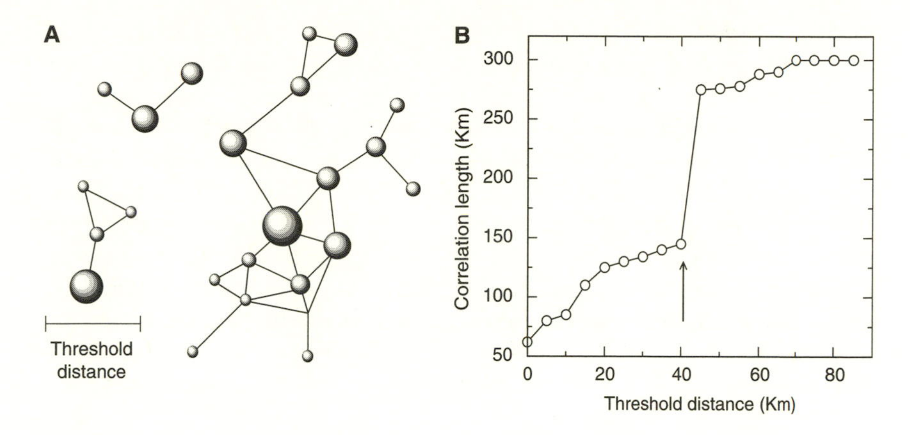

FIGURE 4.13. Critical scales in fragmented landscapes. (A) An example of a graph connecting all patches (irrespective of their size, here indicated by different radius) for a given threshold distance. (B) Correlation length of the habitat distribution versus threshold distance.

center of mass of each patch. It is clear that species with long-range dispersal will perceive a given habitat Ω as more connected than will a species with short-range dispersal. In this sense, the landscape pattern acts as a scale-dependent "filter" acting differently on the movement of species with different degrees of dispersal (Keitt et al., 1997). In more or less continuous landscape mosaics, the effect of filtering will also be smooth. But landscapes showing a high degree of fragmentation will introduce strong filtering. For these landscapes the filter response will be thresholded: Species with small δ will have access only to a small fraction of Ω and thus become isolated, while high-δ species can percolate through Ω. This is actually a key issue (see chapter 5): Fragmentation is a common situation enhanced by increased human activity during the last century. An appropriate characterization of the landscape graph can greatly help to identify the presence of critical scales in landscape connectivity and provide useful guides to conservation (Milne et al., 1996).

Let us consider a case study (Keitt et al., 1997) on a digital map involving a landscape Ω with two types of forest covers (mixed conifer and ponderosa pine). The study area Γ included a physiographic region comprising desert, rocky canyons, semiarid grassland, and piñon juniper woodland, interspersed with numerous mountain ranges supporting both conifer and deciduous hardwood forests (McLaughlin, 1986). The specific choice of Ω was not arbitrary: It approximates the distribution of suitable nesting habitat for Mexican spotted owls

(*Strix occidentalis lucida*) within the southswestern United States. The data used were in fact gathered in a recovery plan for this species (see chapter 5).

A first statistical measure to be defined on the thresholded graph $G(V, E(\delta))$ involves connectivity. Specifically, we can ask which critical dispersal range δ_c will be such that the whole habitat will be available to the species under consideration. To this end, using the previous data set, it was observed that, for $\delta = 20$ km, the landscape was largely formed by disconnected patches and thus a set of m small habitat clusters. These clusters can be characterized by their size, measured in terms of the so-called radius of gyration. For the k–th cluster it is defined as

$$R_k = \frac{1}{M_k} \sum_{i=1}^{M_k} \left[(x_i - < x >_k)^2 + (y_i - < y >_k)^2 \right]^{1/2} \tag{4.61}$$

where $< x >_k$ and $< y >_k$ are the mean x and y coordinates of the sites in the cluster and $\{x_i, y_i\}$ the coordinates of each grid cell in the cluster. This quantity measures cluster size taking into account the shape. The correlation length ξ of the landscape is then computed using

$$\xi(\delta) = \frac{\sum_{s=1}^{m} n_s R_s}{\sum_{s=1}^{m} n_s} \tag{4.62}$$

where n_s is the number of grid cells in cluster s (Creswick et al., 1992). The interpretation of $\xi(\delta)$ is clear: It is the average distance an individual is capable of dispersing, and it gives a measure of landscape connectivity. In figure 4.13b we can see the behavior of this quantity for increasing threshold distances. The landscape is basically disconnected for $\delta < \delta_c \approx 40$ km. Species with a dispersal range below δ_c will be subjected more to stochastic effects and extinction due to isolation. But the graph becomes almost totally connected at greater distances, and a species with $\delta > \delta_c$ can spread through the entire system by way of many alternative paths.

This analysis has been extended (considering, for example, the quality of each patch) and used in comparative studies (Bunn et al., 2000). The graph theoretic approach allows one to compare two species sharing the same habitat but displaying different dispersal ranges. Take for example the American mink (*Mustela vison*) and the prothonotary warblers (*Protonotaria citrea*). In a case study in a large area involving many patches of hardwood forest (the focal species habitat) it was shown that the same landscape is connected for mink and unconnected for prothonotary warblers (Bunn et al., 2000). The previous method (and some refinements of it) was applied to this system, showing that a

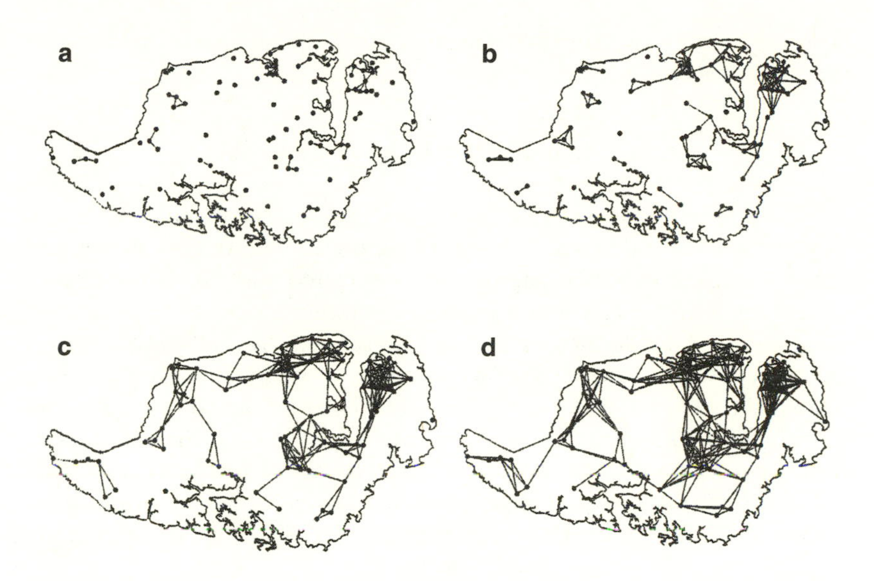

FIGURE 4.14. Edge thresholding in a study area in North Carolina, involving diverse vegetation types, including freshwater swamps, pine woods, and coastal vegetation (from Bunn et al., 2000). A well-defined transition occurs for $\delta \approx 20$ Km, where the landscape would be perceived as being connected.

critical distance $\delta_c \approx 19$ km exists separating a fragmented graph from a single large component. In figure 4.14, the available area and the patches (black circles) are indicated. The edges of the different thresholded graphs are also shown for increasing distances of $\delta = 5, 10, 15$, and 20 km. We clearly appreciate that no isolated patches exist at $\delta = 20$ km.

These analyses illustrate the power of graph and percolation theory for conservation biology. We can go further by performing patch removal experiments in which the effect on ξ of removing a patch from the landscape is analyzed. By eliminating each patch and recording the change in correlation length, it was shown that (as one might expect) the landscape is almost insensitive to patch removal for $\delta > \delta_c$. But close to percolation, it was also found that a number of small patches can have an important effect: These correspond to stepping-stone patches connecting larger areas of habitat, thus playing an especially relevant role. The graph theoretic approximation to conservation biology is thus a valuable method for understanding population survival strategies of patch preservation. These issues will be further discussed (using metapopulation dynamics together with implicit and explicit spatial habitats) in the next chapter.

ECOLOGICAL MULTIFRACTALS

We have previously defined the concept of fractal dimension and how to measure it through box-counting methods. We also mentioned that the presence of intrinsic cut-offs associated with the finite nature of real ecosystems can initiate some scale dependence in the measured dimension. Here we further expand these ideas by analyzing a generalization of the concept of fractal measure. Many fractals have a scaling behavior that is far more complex than the ones described at the beginning of this chapter. They all include intricate structures with more than one scaling exponent. These objects can be properly described by means of multifractal formalism, which provides global information on self-similarity (Schroeder, 1991; Halsey et al., 1986; Peitgen et al., 1992; Drake and Weishampel, 2000).

Considering a given set $\Omega \subset \Gamma$, the multifractal nature of this set is practically determined by covering the system with a set of boxes $\{B_i(r)\}$ $(i = 1, \ldots, N(r))$ of side length (size) r. These boxes are non-overlapping and such that

$$\Omega \subset \bigcup_{i=1}^{N(r)} B_i(r). \tag{4.63}$$

This procedure fully corresponds to the box-counting method, but now a measure $\mu(B_n)$ for each box is computed. This measure can correspond, for example, to the total population or biomass contained in B_n. In general, this measure will scale as

$$\mu(B_i(r)) \sim r^{\alpha} \tag{4.64}$$

where for a large class of systems α is restricted to a finite interval $[\alpha_{min}, \alpha_{max}]$, where $0 < \alpha_{min} < \alpha_{max} < \infty$. Now the idea is that the sets of boxes defined by the previous equation will themselves be fractals, but their fractal dimension is not necessarily the same. More precisely, we would like to determine the number of boxes $N_r(\alpha)$ of size r having an exponent (the so-called Hölder exponent)

$$\alpha = \frac{log\, \mu(B_i(r))}{log\, r}. \tag{4.65}$$

The fractal dimension of such a set, to be indicated as $f_r(\alpha)$ (or simply $f(\alpha)$) will be

$$f_r(\alpha) = -\frac{log\, N_r(\alpha)}{log r}. \tag{4.66}$$

The definition of $f(\alpha)$ means that, for each α, the number of boxes increases, for small r, as

$$N_r(\alpha) \sim e^{-f_r(\alpha)} \tag{4.67}$$

with f a continuous function of α.

In order to obtain the basic expressions for the multifractal dimensions, let us first expand the definition of fractal dimension D. The fractal dimension already defined is actually one of an infinite spectrum of so-called correlation dimensions of order q. These are defined as follows (Shuster, 1984):

$$D_q = -\lim_{r \to 0} \frac{1}{q-1} \frac{log\left[\sum_{i=1}^{N(r)} p_i^q\right]}{log\, r} \tag{4.68}$$

where $p_i \equiv \mu(B_i)$ and measure normalization is assumed:

$$\sum_{i=1}^{N(r)} p_i = 1. \tag{4.69}$$

For $q = 0$ we have the familiar definition of fractal dimension. To see that this is true, we need only perform the limit $q \to 0$ in D_q, obtaining:

$$D_0 = -\lim_{r \to 0} \frac{N(r)}{log\, r}. \tag{4.70}$$

It can be shown that the inequality $D_{q'} \leq D_q$ holds for $q' \geq q$. The sum

$$M_q(r) \equiv \sum_{i=1}^{N(r)} [\mu(B_i(r))]^q = \sum_{i=1}^{N(r)} p_i^q \tag{4.71}$$

is the so-called moment (or partition function) of order q. Varying q allows one to measure the non-homogeneity of the pattern: The moments with large q will be dominated by the densest boxes. Using different values of q the relevance of the contributions of the p_i's is modified: For $q > 0$, the main contribution to the moments is determined by the largest p_i's. Conversely, for $q < 0$, the main contribution will come from small probabilities.

Two particularly important cases are $q = 1$ and $q = 2$. The result for $q = 1$ is the Shannon entropy of the set measure:

$$D_1 = -\lim_{r \to 0} \sum_{i=1}^{N(r)} p_i \, log\, p_i \tag{4.72}$$

and the second is the so-called correlation dimension, defined as:

$$D_2 = -\lim_{r \to 0} \frac{log\left[\sum_{i=1}^{N(r)} p_i^2\right]}{log\, r}. \tag{4.73}$$

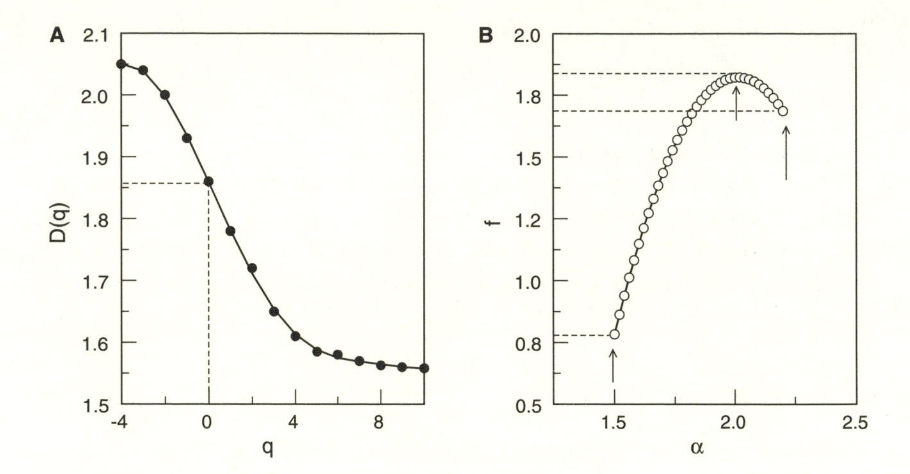

FIGURE 4.15. Multifractal spectrum for a rain-forest map from BCI. Here the $D(q)$ spectrum is shown in (A) for $H = 10$ m threshold canopy height. The broken line indicates the position of the fractal dimension D_0 within the spectrum. In (B) the corresponding $f - \alpha$ spectrum is shown. The broken lines indicate the three characteristic values of α and $f(\alpha)$ for this plot.

An example of the $D(q)$ spectrum is shown in figure 4.15a. It has been computed for the canopy gap distribution of the Barro Colorado Island rain forest plot (see fig. 4.4 right). Using a given predefined height H, the set $\Omega(h)$ of low-canopy points (i.e., places where $h < H$) will change with H. The different sets that can be defined are in fact different views of how the canopy is organized through space and the nature of correlations between trees. Here the measure is obtained by computing the number of low canopy points (black squares) inside each box. For $q = 0$, the fractal dimension of the plot is shown to be $D_0 = 1.86 \pm 0.03$ (Solé and Manrubia, 1995a, 1995b; Solé et al., 2005). If the BCI plot were a pure fractal, a straight line would be observed (i.e., $D(q) = D_0$ for all q). The wide variation indicates that several nested fractal objects are present. This observation can be used to constrain the type of models able to reproduce field observations.

The set of generalized dimensions D_q allows one to calculate the complete spectrum of fractal dimensions for the set Ω. Specifically, the fractal dimension $f(\alpha)$ for the set with scaling exponent α is obtained from (Halsey et al., 1986; Solé and Manrubia, 1995a, 1995b):

$$\alpha(q) = \frac{\partial}{\partial q}[(q-1)D_q] \tag{4.74}$$

$$f(\alpha) = q\alpha - [q-1]D_q. \tag{4.75}$$

In plotting $f(\alpha)$ against α we represent the complete multifractal spectrum, which typically appears as a $\cap$-shaped curve, with a maximum at $\alpha(0)$ so that the maximum in f corresponds to the box-counting dimension. If the set Ω were a single-scaled fractal, then the $f - \alpha$ spectrum would collapse into a single point. Again, a well-defined curve indicates that different regions have different fractal dimensions.

The numerical computation of the $f - \alpha$ curve can be done by first computing the set of mass values for each box, that is, $\{\mu(B_i)\}, (i = 1, \ldots, N)$. Afterward, the Höllder exponent is measured (a finite range of values will be available for the BCI case):

$$\alpha_i = \frac{log\, \mu(B_i)}{log\, r} \tag{4.76}$$

and since we expect $N(\alpha) \sim e^{-f(\alpha)}$, plotting $-log N(r)/log\, r$ versus α gives the $f - \alpha$ spectrum. Further extensions of these ideas have been explored concerning their relationships with species-area laws and diversity indices (Borda-de-Agua et al., 2002).

SELF-ORGANIZED CRITICAL PHENOMENA

Since fractals are so widespread in natural systems, one obvious question to formulate concerns their dynamical origin. We have seen that percolation processes or the Ising model can display fractal patterns by tuning the approriate control parameters to the critical point. However, this is far from satisfactory when exploring real systems, where no one (in principle) is tuning parameters externally. Having in mind that several mechanisms can lead to scaling and fractal behavior (some to be discussed in this book), let us consider one particularly appealing theoretical approximation first suggested by Per Bak and coworkers (Bak et al., 1987; Jensen, 1998). This idea has inspired many developments in a plethora of scientific disciplines, including biology (Bak and Sneppen, 1993; Kauffman, 1993; Pietronero, 1995; Gisiger 2001; Solé et al., 1999).

The basic idea in Bak's theory is that some far-from-equilibrium systems formed by many interacting units can spontaneously drive *themselves* into the critical point, thus spontaneously leading to spatial and temporal fluctuations of all sizes. This might seem a difficult task, since the critical point is a well-defined place, but as we will see, it naturally arises as a consequence of an appropriate feedback between control and order parameters. It is important to mention, as shown by Mercedes Pascual and coworkers, that scaling in spatially extended systems can be obtained in ecological models with no critical dynamics involved (Pascual et al., 2002).

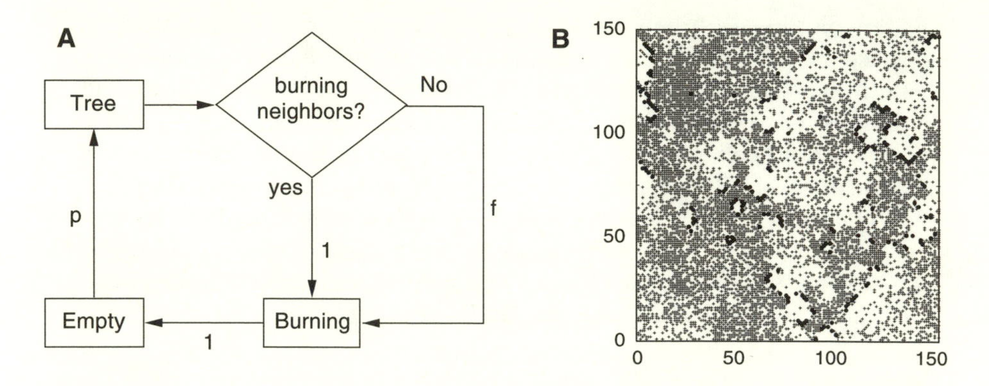

FIGURE 4.16. (A) Basic rules of the forest-fire cellular automaton model (as defined by Drossel and Shwabl, 1992). In (B) a snapshot of the forest-fire simulation is shown (see text).

Let us consider one of the first proposed examples of a self-organized critical system: the forest fire model (Bak et al., 1990; Drossel and Schwabl, 1992; Clar et al., 1999), which has been successfully applied to the propagation of epidemics on small islands (Rhodes and Anderson, 1996) and the dynamics of measles spreading (Grenfell et al., 2001). Once again, a square lattice (with periodic boundary conditions) is used, and the elements (say trees) of the lattice $S_i(t)$ can have three allowed states: $S_i \in \Sigma = 0, 1, 2$ corresponding to empty (burned), green, and burning trees, respectively. The basic rules for this model (Drossel and Schwabl, 1992, figure 4.16a) are:

1. Burning into ashes: A burning tree at t becomes an empty site (ashes) at $t + 1$,

$$1 \longrightarrow 0. \tag{4.77}$$

2. New tree is born, with probability p:

$$0 \xrightarrow{p} 1. \tag{4.78}$$

3. Spontaneous lightning: A green tree burns with probability f,

$$1 \xrightarrow{f} 2 \tag{4.79}$$

4. Fire propagation: If at least one nearest neighbor of a green tree is burning, the green tree also burns with probability one.

In the double limit $p \to 0$ and $f/p \to 0$, the system self-organizes into a state in which very complex fluctuations and scale-free properties spontaneously emerge. An example of the patterns obtained is shown in fig. 4.16b. Here empty, green, and burning are indicated by means of white spaces, crosses, and black circles, respectively. We can see that the fronts propagate through the system (thus introducing activity and information transfer) and create a complex landscape with green and empty sites forming clusters of many sizes. The activity is measured in terms of burning trees and it is clearly inhibited by the rapid propagation of the fire. Typically, after a large fire, growing clusters of trees will burn if one of their components starts to burn, thus generating a small event. But from time to time a very large, percolating fire will be observed. The spontaneous lightning rule allows burning to restart in case fire goes totally extinct. This is the so-called driving term, which provides a source of activity to the system. Then the rules of propagation provide the mechanism to quickly transfer the initial perturbation into the whole system. In order to have a self-organized critical state, a well-defined separation of time scales is necessary (Jensen, 1998; Sornette 2000; see also Adami, 2000). Specifically, the driving must be slow compared to the dynamics of propagation. These are rather constraining assumptions that will be discussed throughout this book.

Computer simulations of the FFM allow one to calculate the scaling relations involved, and it was found that the number of trees burned, s, follows a power law $P(s) \sim s^{1-\tau}$ with $\tau \approx 3$. Some analytic results are derived following mean field arguments. One first quantity to be estimated is the average mean time between two lightning strikes hitting a tree. For a square lattice with L^2 sites with an average density $\bar{\rho}$ of trees, the rate R_f at which trees are hit will be

$$R_f = f\bar{\rho}L^2 \tag{4.80}$$

and the time between hits is simply $T_f = 1/R_f$. The average number of trees $\bar{s}$ born between two fires is computed from: (i) the average time T_f between two strikes and (ii) the average number of empty sites transformed into green trees, $p(1-\bar{\rho})L^2$. Thus we have:

$$\bar{s} = \frac{p(1-\bar{\rho})L^2}{f\bar{\rho}L^2} = \frac{p}{f}\frac{1-\bar{\rho}}{\bar{\rho}}. \tag{4.81}$$

At the steady state this also corresponds to the distribution of trees burning. We have seen that the spatial pattern generated by the FFM involves clusters, and that their average size will also be $\bar{s}$.

From the previous results the frequency distribution of burning clusters $N(s)$ can be derived (Adami, 2000). The total number of burning trees is given by

$$N_b = \sum_{j=1}^{S} sN(s) \tag{4.82}$$

where S is the maximum cluster size (to be limited only by system size L) and the probability of burning for a given site (in a cluster of size s) is simply $P(s) = sN(s)/N_b$. Thus the mean number of burning trees will be

$$\bar{s} = \sum_{j=1}^{S} sP(s) = \frac{\sum_{j=1}^{S} s^2 N(s)}{\sum_{j=1}^{S} sN(s)}. \tag{4.83}$$

Assuming a scaling relation $N(s) \sim s^{-\tau}$, we obtain: $\bar{s} \propto S^{3-\tau}$ if $2 < \tau < 3$ and $\bar{s} \propto \frac{S}{log\, S}$ if $\tau = 2$ and consequently $S \rightarrow \infty$ as the system size grows: Clusters of all sizes (up to the system size) are expected to be observed.

The presence of scaling behavior in real forest fires has been shown through the analysis of the frequency-area distribution of fires from several geographic regions with different vegetation types and climates (Malamud et al., 1998; Ricotta et al., 1999; Caldarelli et al., 2001).

COMPLEXITY FROM SIMPLICITY

Understanding the origins of scaling can provide valuable information about the origins of ecological complexity. Sometimes it is asked why scaling laws are relevant. They are in fact the key to understanding a large number of phenomena using simple and clever theories. There is no reason, in principle, why scale-free, power-law relations should apply to biological systems. But actually they do and are widespread. The presence of fractal patterns reveals the nonequilibrium conditions under which structures at different levels are created and how large-scale patterns are generated from local interactions. They have provided the clues to understanding the origins of allometric scaling in biology (Brown and West, 2000) and are necessary ingredients of any theory of ecological organization. As Brown and coworkers have indicated, the self-similar structure of many ecosystem properties reflects the constraints on their organization, which stem from basic physical, biological, and even mathematical principles (Brown et al., 2002). The self-similar character of these systems, sometimes spanning many orders of magnitude, can be exploited in order to extrapolate between different scales.

The lessons from the physics of nonequilibrium phenomena (particularly of critical phenomena) are important here. When looking at real ecosystems, we see an enormous diversity of structures and interactions unfolding over disparate spatial and temporal time scales. It is difficult to not get trapped in a dilemma when dealing with simple models: Which features can be safely discarded and why? In statistical physics, many examples provide support to the idea that when looking at macroscopic patterns (in our context, species abundances, diversity patterns, or spatial distributions of individuals), the small details do not matter.

The success of CA models in describing reality stems from the existence of several levels of description when studying a complex system (Chopard and Droz, 1999). The two main levels are roughly describable as the microscopic and macroscopic, and the same system will typically look different at these two scales. At the smallest scale, interactions among the basic units (and the units themselves) will be described by means of complicated sets of rules. This is specially true for behavioral components in ecology, but not less true when dealing with the difficulties arising from, say, the quantum description of physical systems. What we learn from complex systems theory and statistical physics is of great value when looking at the macroscale: At this level, the features displayed by the system result from the aggregate effect of (all) microscopic events. In this sense, the macroscale phenomenology is only dependent upon the generic features of the microscopic interactions. Put more strongly: The complexity of the macroscale appears *disconnected* from that of the microscale (Chopard and Droz, 1999). This might seem surprising, since the former is driven by the latter. But theory and experience reveal that the fine details implicit in the microscopic rules vanish when looking at the large picture.

This separation of scales is strongly tied with the concept of emergent phenomena (Solé and Goodwin, 2001). New properties emerge at the macroscale and only make sense at that level: They simply cannot be inferred from the properties of the microscopic details. The Ising model is a good illustration from physics: It offers an extremely poor description of the quantum, atom-level reality, but totally retains the relevant aspects that show up when the system is analyzed at the macroscale. A similar situation is provided by hydrodynamics: The so-called Navier-Stokes equations give a complete description of fluid behavior in spite of a lack of description of the type of molecules that form the fluid.

Biological systems are different from physical ones in a number of ways (Hopfield, 1994). The existence of functionality (or purpose) is an obvious difference but biosystems are also far-from-equilibrium entities and the scale separation that applies for physics should also be at work in ecosystems. It might be difficult to know a priori what ingredients from the microscopic must be included to get a good macroscopic understanding (see, for example, Wootton, 2001) but the lessons from the past indicate that the approach is feasible. As Einstein noted: "Everything should be made as simple as possible, but no simpler." No successful approach to the real world can ignore the multiple faces of biocomplexity, but no theory can hope to succeed by only looking at the small details.

CHAPTER FIVE

Habitat Loss and Extinction Thresholds

HABITAT LOSS AND FRAGMENTATION

Current rates of habitat destruction are extremely high. According to the Food and Agriculture Organization of the United Nations (FAO), deforestation produced a net loss of some 180 million hectares between 1980 and 1995, that is, an annual average loss of 12 million hectares (Food and Agriculture Organization of the United Nations, State of the World's Forests 1997, FAO, Rome, 1997, p. 16). Extensive losses have been observed in all continents (see, for example, fig. 5.1a). As early reviewed by Sanders et al., the physical changes associated with habitat loss and fragmentation include:

1. A reduction of total area and productivity of native areas.
2. Increased isolation of forest remnants (and their local populations).
3. Changes in the physical conditions of the remnant fragments. These include soil composition, water flux, or solar radiation (see Laurance and Bierregaard, 1997).

These anthropogenic changes trigger further community responses that sometimes end in a biotic collapse. The sequence of biotic decay (Wilcove, 1987) includes initial exclusion of some species, deleterious effects of isolation, and ecological imbalances. The latter involve nonlinearities and cascade effects implicit in ecosystem structure. The loss of some given species can promote subsequent loss of their predators, parasites, or mutualists. One dramatic example is the so-called ecological meltdown observed in predator-free forest fragments. The loss of predators generates strong imbalances shown by the disproportionate increase in the densities of prey and severe reductions of seedlings and saplings of canopy trees (Terborgh et al., 2001).

It is well known that such a human-induced habitat alteration is the major variable leading to the loss of biodiversity (Wilcox and Murphy, 1985; Barbault and Sastrapradja, 1995). Some authors have estimated an increase of the extinction rates by 1,000 times during the last 300 years. These current rates of extinction are comparable in magnitude

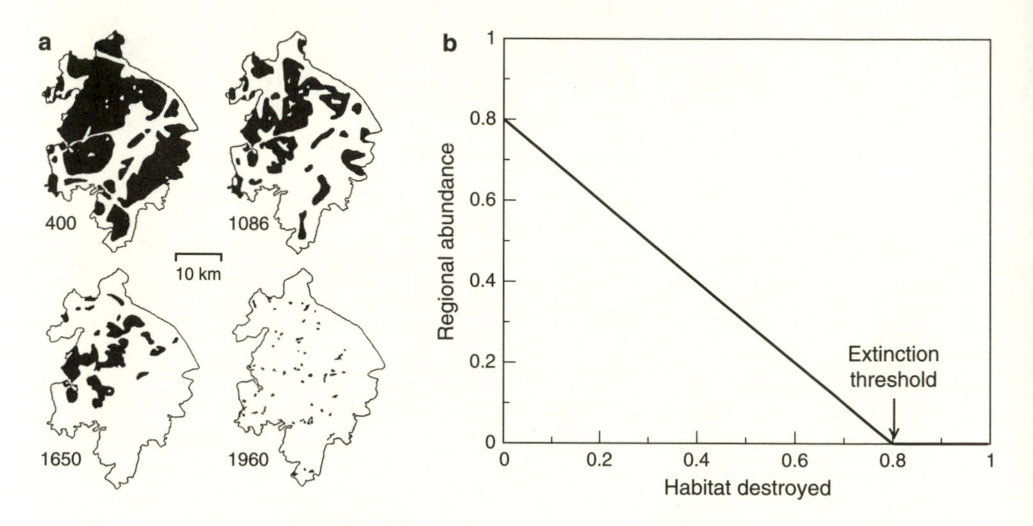

FIGURE 5.1. Real and simulated habitat destruction. In (a) an example of habitat loss and fragmentation in a real ecosistem is shown through time. This example corresponds to the deforestation process experienced by native forests in Warwickshire, England, between 400 and 1060 C.E. (adapted from Wilcove, 1987). In (b) the metapopulation regional abundance is plotted as a function of the amount of habitat loss for the Levins model. An extinction threshold takes place for a critical amount of habitat loss (D_c) even when some habitat is still suitable.

to one of the five big mass extinction events (Lawton and May, 1995; see chapter 7). If these estimations are correct, a species disappears from Earth on average every fifteen minutes. This means that at this rate, 1 million species may become extinct during the next twenty years.

Given the magnitude and consequences of habitat destruction, it is imperative to get enough insight to understand the effects of habitat loss on species survival, and predict its further consequences. Since economic trade-offs are at play, scientists are oftentimes faced with the question of how much habitat can be destroyed before a certain species goes extinct. We start to get some data on the consequences of habitat loss, using both field observations (Andrén, 1994, 1996) and experiments (Holt et al., 1995; Robinson, et al., 1992; Debinski and Holt, 2000). While this approach is fundamental, it is hard to get enough information about the long-term consequences of habitat loss due to the large spatial and temporal scales at which this process takes place. For example, we can record the loss of species a few years after a human alteration, but there may be time lags. Thus, other species still present may go extinct during the next years and so we may underestimate the effect of habitat destruction. In order to fully understand the consequences of habitat loss we have to develop a useful body of theory.

In this chapter we review how the theory on spatial ecology addressed in chapter 3 has been adapted to study the consequences of habitat loss. In particular, we want to characterize extinction thresholds, that is, the critical amount of habitat destroyed at which a metapopulation goes extinct. We will start by introducing spatially implicit models of habitat loss. We will follow by incorporating explicit space. The distinction between habitat loss and habitat fragmentation (that is, the breakdown of a continuous habitat into smaller, isolated patches surrounded by destroyed habitat) can only be accounted for by means of spatially explicit models. As we shall see, the concept of percolation is particularly relevant in relation to the biological problem of habitat fragmentation. As a consequence, the critical fraction of habitat destroyed at which some characteristic landscape indices show sharp transitions can be anticipated by a general theory on complex systems.

The last goal of this chapter is to use theoretical tools to better understand how ecosystems will respond to the destruction of pristine habitat. Some relevant concepts are extinction thresholds, nonlinearities, and time delays. We will review these concepts here and discuss their relevance for theoretical ecology and conservation biology. Our strategy in this chapter is the same as throughout the book, that is, to use simple models in the hope that they capture the essential, while omitting the irrelevant.

EXTINCTION THRESHOLDS IN METAPOPULATION MODELS

As seen in chapter 3, a metapopulation can be defined as a set of geographically distinct local populations maintained by a dynamical balance between colonization and extinction events. Let us start this section by revisiting the Levins (1969a) model, which captures the dynamics of a metapopulation

$$\frac{dx}{dt} = cx(1-x) - ex, \tag{5.1}$$

where x is the fraction of patches occupied, and c and e are the colonization and extinction rates, respectively. This model has a nontrivial solution given by $x^* = 1 - e/c$. The colonization rate has to be larger than the extinction rate for the metapopulation to persist.

One can easily introduce habitat loss into the framework of model (5.1). If a fraction D of sites is permanently destroyed, this reduces the fraction of vacant sites that can potentially be occupied. Model (5.1) becomes

$$\frac{dx}{dt} = cx(1 - D - x) - ex. \tag{5.2}$$

Similarly, one can use a discrete time version of the Levins model. In this case we can write:

$$x_{t+1} = c(1 - D - x_t) + (1 - e)x_t. \tag{5.3}$$

The long-term fraction of occupied sites, that is, the metapopulation regional abundance, is now given by $x^* = 1 - D - e/c$ in both the continuous time and discrete time model. In figure 5.1b we plot x^* as a function of D. As can be seen, there is a linear decrease in metapopulation abundance as more and more sites are destroyed, until a certain point, at which the metapopulation becomes extinct. Let us call the extinction threshold (D_c) this critical fraction of sites destroyed at which the metapopulation goes extinct. Interestingly, the metapopulation goes extinct even when some habitat is still available. This is an interesting and counterintuitive result. We can understand the existence of such extinction thresholds because the metapopulation is maintained by a balance between colonization and extinction events. Extinction events imply that some patches are always non-occupied. Even when all habitat is available, the fraction of sites occupied is less than one. The extinction threshold is given by $D_c = 1 - e/c$. This corresponds to the fraction of habitat occupied when all habitat is suitable ($x^*(D = 0) = 1 - e/c$). This is an interesting result. It means that we can estimate the extinction threshold without any knowledge of demographic parameters such as c and e. All we need is to measure the fraction of sites occupied when all the habitat is available. This will be our estimate of D_c. Hanski et al. (1996) have coined the term Levins rule for this prediction. In a more general context, Nee (1994) was the first to emphasize that the eradication threshold corresponds to the used amount of the limiting resource (this being habitat or nonvaccinated human hosts). Note that we can consider either the fraction of habitat destroyed (D) as we do here, or the fraction of habitat available ($h = 1 - D$). In the latter case, the eradication threshold corresponds to the unused amount of the limiting resource as originally stated by Nee (1994).

We have started with the Levins model for a pedagogical purpose, but the first metapopulation model predicting an extinction threshold was that of Lande (1987), who generalized the concept of a metapopulation by incorporating life history, territoriality, and dispersal behavior. He applied his model to the northern spotted owl (*Strix occidentalis caurina*). This is an illustrative example of the predictive power of simple models and their potential use in conservation biology. Let us briefly review Lande's model.

The Case of the Northern Spotted Owl

The northern spotted owl is a highly endangered subspecies living in old coniferous forests in northern California, Oregon, and western Washington. Its abundance has been drastically reduced due to intense logging, and it has become an icon for conservation biologists. Each pair of spotted owls needs a territory of about one to three square miles of old forest (more than about 250 years old). Lande (1987) first estimated the geometric growth rate of the subspecies using analytical tools. Second, he predicted the effects of future habitat loss on metapopulation abundance. In this case, a patch corresponds to a territory, and these can be either suitable (a female can reproduce) or unsuitable (destroyed). The basic model deals now with the demography of females, as in similar demographic models.

Let us denote by h the fraction of suitable territories ($h = 1 - D$); l_x is the probability of surviving to age x, and the rate of production of female offsprings per unit time by female aged x is b_x. Juveniles have to find a territory before reaching the reproductive age (a). They can either inherit their natal territory with a constant probability ϵ, or search randomly for a new, vacant, suitable territory. Let us imagine that the juveniles can check up to a number m of territories before dying. The probability that a randomly found territory is not good for settlement of the juvenile (it is either nonsuitable or occupied by a female) is $D + ph$, where p denotes the fraction of suitable territories already occupied by a female. Then, following Lande (1987), the probability of a surviving juvenile finding a suitable, non-occupied territory is $1 - (1 - \epsilon)(D + ph)^m$.

If l'_x is the probability of survival to age x conditional on finding an available territory, then we can write the following expression for the probability of survival to age x (older than the age of first reproduction):

$$l_x = l'_x[1 - (1 - \epsilon)(D + ph)^m]. \tag{5.4}$$

At demographic equilibrium, the net lifetime production of female offspring by female (R_0) is by definition one. This can be expressed as the product of the probability of finding a suitable territory times the total production of female offspring when a vacant territory has been found. If p^* denotes the equilibrium fraction of suitable territories occupied by a female, we can write:

$$[1 - (1 - \epsilon)(D + p^*h)^m]R'_0 = 1, \tag{5.5}$$

where

$$R'_0 = \int_a^{\infty} l'_x b_x dx. \qquad (5.6)$$

The equilibrium occupancy of suitable territories is then

$$p^* = 1 - (1-k)/h \quad \textit{if} \quad h > 1-k, \qquad (5.7)$$

$$p^* = 0 \quad \textit{if} \quad h \leq 1-k, \qquad (5.8)$$

where

$$k = \left(\frac{1 - 1/R'_0}{1-\epsilon} \right)^{1/m} \qquad (5.9)$$

is a demographic parameter determined by the life history and dispersal properties of the species; k is named the demographic potential of the population, and it gives the maximum occupancy of territories at equilibrium when all sites are available.

For the cases in which $0 < k < 1$, the survival of the metapopulation is conditional on a certain fraction of habitat being preserved. For this case, one can see from equation (5.7–5.8) that an extinction threshold occurs when suitable habitat is reduced to a critical fraction given by

$$h_c = 1 - k. \qquad (5.10)$$

Figure 5.2 plots p^* as a function of habitat loss ($D = 1 - h$). We use D instead of its inverse h (as in the original paper) to better emphasize the consequences of habitat destruction, and to allow comparison with other examples. The threshold occurs when the fraction of habitat destroyed reaches the critical value $D_c = k$. That is to say, despite the additional complexities of the life cycle and demographic behavior considered by Lande, the Levins rule still applies. None of the species-specific demographic details is important to estimate the extinction threshold. All that one has to know is the fraction of territories occupied at equilibrium when all territories are available.

The last result has important implications in relation to the role of simple models in ecology (Bascompte and Solé, 1998a; Bascompte, 2003). A common objection to models such as Levins's (equation 5.2) and Lande's (equation 5.5) is that they are too simplistic to have any predictive value in the real world. However, as pointed out by Nee (1994), demographic details, or other species-specific details are here irrelevant, since they do not appear in the estimation of the extinction threshold. This justifies simple approaches like the ones described throughout this book. This also makes possible the use of general

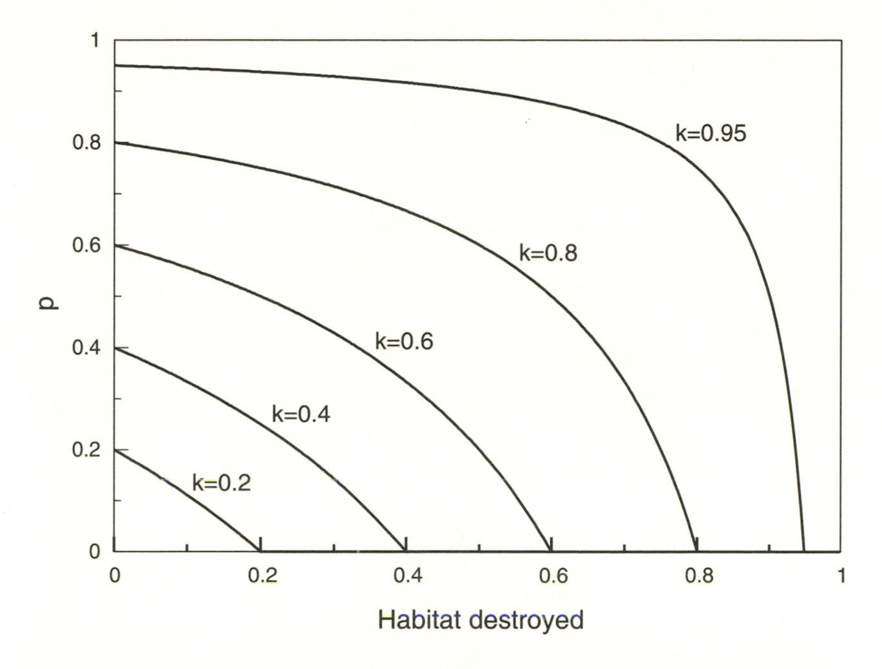

FIGURE 5.2. The equilibrium occupancy of suitable territories (p) as a function of the fraction of habitat destroyed. Here k is the so-called demographic potential of the population, that is, the fraction of suitable territories when all territories are available. Modified from Lande (1987).

models in different fields, such as epidemiology and conservation biology to find generalizations.

Lande (1987) goes a step further by applying this result to the conservation of the northern spotted owl. He estimated the parameters in equation (5.7–5.8). National forests in the Douglas Fir region of western Washington and Oregon currently contain about 38% old forest greater that 200 years old. Thus $h \approx 0.38$. One can also estimate the fraction of suitable territories occupied by a reproductive female. Different surveys discussed by Lande (1987) give an average annual occupancy of $p \approx 0.44 \pm 0.04$. One can now substitute these two values into model (5.7–5.8), and estimate k. For this particular case, $k \approx 0.79 \pm 0.02$. That is, the metapopulation will become extinct when the fraction of suitable territories decreases to 79%. Lande (1987) thus proved that the effect of implementing the forest service guidelines suggesting reducing h to values between 0.07 and 0.16 would lead the northern spotted owl toward extinction. This example shows how simple models can have important practical applications in conservation biology.

EXTINCTION THRESHOLDS IN METACOMMUNITY MODELS

Until now we have considered single metapopulation models. However, we are often more concerned about preserving diversity than about preserving particular species. This is more relevant as one moves from temperate countries to tropical ones. We have to include species interactions in our models. In this section we will consider extensions of the Levins model in which species interactions are taken into account.

Competition and the Extinction Debt

Nee and May (1992) were among the first to extend the Levins model to study the effect of patch removal on the coexistence of two competing species. Here we will summarize their model and their most important results.

Consider a scenario with an infinite number of patches of the same quality. Each one of such patches can be destroyed, available but empty, occupied by species A, or occupied by species B. Let us imagine that species A is the superior competitor, and thus can invade either an empty but suitable patch, or a patch occupied by species B. If A occupies a patch already occupied by B, then B is excluded from it. Similarly, species B can only colonize empty but suitable patches. If we denote by x, y, and z the proportion of empty patches, A-only patches, and B-only patches, and if a fraction h of the habitat is available, then the dynamics of such a metacommunity is described by the following model:

$$\frac{dx}{dt} = -c_a xy + e_a y - c_b xz + e_b z, \tag{5.11}$$

$$\frac{dy}{dt} = c_a xy - e_a y + c_a zy, \tag{5.12}$$

$$\frac{dz}{dt} = c_b zx - e_b z - c_a zy, \tag{5.13}$$

where c_a and c_b are the colonization rates of species A and B respectively, and e_a and e_b are the corresponding extinction rates.

This system has the following nontrivial solution:

$$x^* = \frac{1}{c_b}(hc_a - e_a + e_b), \tag{5.14}$$

$$y^* = h - \frac{e_a}{c_a}, \tag{5.15}$$

$$z^* = \frac{e_a(c_a + c_b)}{c_a c_b} - \frac{e_b}{c_b} - \frac{hc_a}{c_b}. \tag{5.16}$$

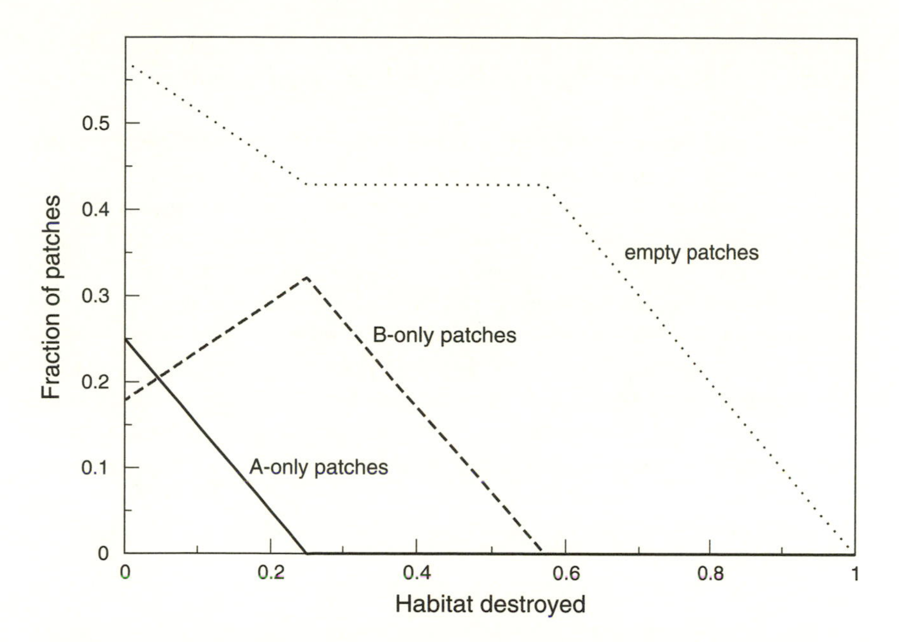

FIGURE 5.3. Regional abundance of two competing species plotted as a function of habitat loss. A-only patches are those occupied exclusively by species A, that is, the superior competitor, inferior disperser. B-only patches are those occupied exclusively by species B. Available, nonoccupied patches are empty patches. As noted, the regional abundance of the inferior competitor increases with habitat destruction due to the decrease of the superior competitor. Thus, habitat loss can change community composition in remaining patches. Here $e = 0.3, c_a = 0.4, c_b = 0.7$. Modified from Nee and May (1992).

As noted by Nee and May (1992), the inferior competitor (species B) can persist only if the following inequality holds:

$$\frac{c_b}{e_b} > \frac{c_a}{e_a}, \tag{5.17}$$

which happens if the inferior competitor has a fugitive lifestyle with a high colonization rate. In other words, the inferior competitive species can only persist by acting as a "weed."

Here we follow Nee and May's example by assuming that both species have the same extinction rate ($e_a = e_b = e$). Figure 5.3 shows the equilibrium abundance of all kind of patches as habitat is destroyed (as in the previous example by Lande, the original paper was using the inverse variable, that is, the fraction of patches that are suitable). As can be seen by looking at figure 5.3, coexistence occurs for $0 < D < e/c_a$. In this domain, the proportions of the three types of patches are given by (5.14–5.16). The extinction threshold for the superior competitor takes place at $D_c = e/c_a$. In the domain $e/c_a > h > e/c_b$, only the inferior competitor, the better disperser can persist. Finally, both species are extinct when $D > e/c_b$.

One would assume that the inferior competitor would be the first species to go extinct since it persists by colonizing empty patches. However, habitat loss increases the inferior competitor's regional abundance, reducing the number of patches occupied by the superior competitor. Thus, habitat destruction can change community composition in the remaining, nondestroyed patches.

Tilman et al. (1994) built on the previous model to consider a set of n competing species in an ideal forest. They again assumed a trade-off between competitive and dispersal abilities. The best competitor species is also the poorest disperser. As before, we assume that a superior competitor can instantly colonize a patch occupied by an inferior competitor, excluding it from such a site. If such a set of species is ranked from the best to the poorest competitor, we have the following equation for species i:

$$\frac{dp_i}{dt} = c_i p_i \left(1 - D - \sum_{j=1}^{i} p_j\right) - m_i p_i - \sum_{j=1}^{i-1} c_j p_i p_j. \tag{5.18}$$

The first term in equation (5.18) describes how species i can colonize any nondestroyed site that is not occupied by superior competitors or individuals of its own (species 1 to i). The last term indicates that species i can be driven extinct from any site by any superior competitor colonizing such a site. As in the case of the Levins model (5.2) and its extension by Nee and May (5.11–5.13), p_i describes the regional abundance of species i (i.e., the fraction of patches occupied by species i), and c_i, e_i are the colonization and extinction rates, respectively, of species i.

The solution of model (5.18) giving the long-term abundance of species i is:

$$p_i^* = 1 - D - \frac{m_i}{c_i} - \sum_{j=1}^{i-1} p_j^* \left(1 + \frac{c_j}{c_i}\right). \tag{5.19}$$

The previous equation (5.19) has to be solved from species 1 to i with $c_i > 0$ and $p_i^* \geq 0$ for all i. Then, we see that the solution for the superior competitor is the same as the solution for the Levins model (5.2), that is, $p_1^* = 1 - D - m_1/c_1$. In other words, as noted by Tilman et al. (1994), the Levins rule is again at work. The superior competitor goes extinct when the fraction of habitat destroyed equals the fraction of habitat occupied when all habitat was pristine.

Tilman et al. (1994) assume that species abundance in a pristine habitat follows a geometric series, that is $p_i = q(1 - q)^{i-1}$ where q is the abundance of the best competitor. Assuming that extinction rates are the same for all species ($m_i = m_j = m$), we found that the required colonization rates compatible with the geometric series are given by $c_i = m/(1-q)^{2i-1}$. Tilman et al. (1994) substitute this result into equation (5.19) to obtain the extinction thresholds, that is, the critical values of habitat destruction at which any given species becomes extinct. For the *ith* species, we have

$$D_{c_i} = 1 - (1-q)^{2i-1}. \tag{5.20}$$

As can be observed from this last result, species go extinct from the best competitor to the poorest as more habitat is destroyed, which confirms the result by Nee and May (1992).

One important question we can address with this model is how diversity is reduced as more habitat is destroyed. The number of species driven extinct by habitat loss (the extinction debt) can be calculated by solving (5.20) for i, which yields:

$$E = \frac{ln[(1-D)(1-q)]}{2ln[1-q]}. \tag{5.21}$$

Figure 5.4a plots the number of extinct species as a function of the fraction of habitat destroyed based on equation (5.21). As can be seen, the response is highly non-linear. The number of extinct species increases sharply as more habitat is destroyed. That is to say, habitat destruction threatens more species in previously destroyed habitats. The shape of this function depends on the assumed ranking of species abundances. The lower the best competitor's abundance, the higher the number of extinct species for a given value of habitat loss (fig. 5.4a). Since abundance of the best competitor is lower in tropical forests than in temperate ones, one can predict that the effects of habitat loss will be more detrimental in the former.

Another important result from Tilman et al. (1994) is that there are long time lags between habitat destruction and the consequent reduction in diversity. These time lags can range from 50 to 400 years. We saw in chapter 3 the relevance of long transients in spatially extended ecological models. This is an important result. It suggests that the effects of habitat loss can be worse than previously thought. Even when some species are still present, they may be "ecological ghosts" doomed to an unavoidable extinction. Thus, habitat loss creates a "debt" that comes due in the future (Tilman et al., 1994). The robustness of this simple model has been explored by Tilman et al. (1997) using more complicated assumptions as well as spatially explicit simulations. The main results hold. Similarly, Loehle and Li (1996) relate Tilman et al.'s model to real situations of habitat loss and find that the extinction debt is a robust property.

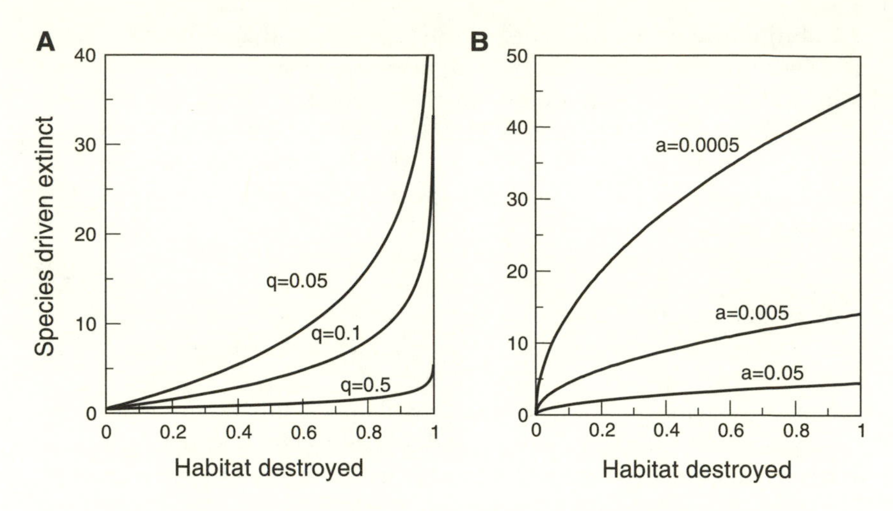

FIGURE 5.4. Habitat destruction and biodiversity loss. The number of species driven extinct as habitat is destroyed is plotted for a forest model (A), and for a model of a coral reef (B). While in the forest model a larger number of species is driven extinct as more habitat is destroyed, the coral reef is more fragile, with a large number of extinctions for moderate values of habitat loss. In (A), q is the abundance of the best competitor (a geometric series of abundance is assumed). In (B), a represents the ratio between the abundance of two successive species. As noted, more species are driven extinct when the dominant competitor is rare (e.g., tropical forests), and when abundances between species are more equifrequent. Modified from Tilman et al. (1994), and Stone (1995).

As we have pointed out, the results by Tilman et al. (1994) are based on a specific ranking of species and competition-colonization trade-offs that are representative of forests. Building on that, Stone (1995) applies the model to coral reefs, and compares his results with Tilman et al.'s (1994) for the forest. Stone summarizes some observations from field studies on coral reefs in Eilat, Israel, and incorporates this information into a model like (5.18). Some differences between forests and coral reefs are as follows. In coral reefs, the best colonizer has now the largest abundance. Also, interspecific competition is no longer an important force, so population abundance is largely correlated to colonization rates. Finally, mortality rates are no longer assumed constant but are ranked in a way in which the superior colonizers have the highest mortality rates. After incorporating these life-history properties into the model, Stone (1995) gets the following expression for the number of extinct coral species as a function of habitat loss:

$$E = \sqrt{D/\alpha}, \tag{5.22}$$

where α is the ratio between the abundance of two successive species used to set the initial distribution of abundances.

E is plotted as a function of the amount of habitat destroyed in figure 5.4b. The figure can be compared with the one corresponding to the forest (fig. 5.4a). As noticed, habitat destruction is even more detrimental for coral reefs. Very small amounts of habitat destruction can now lead to the extinction of a large number of species. A destruction of only 10% of habitat can induce the extinction of up to 50% of the initial species. For comparison purposes, the same amount of destruction in the forest leads to extinction of only about 5% of the initial species. As in the case of the forest model, the effect of habitat destruction is more detrimental in those locations in which there is less dominance (i.e., a higher similarity between species' abundances; low α-values, fig. 5.4b).

Collapse under Recruitment Limitation: Neutral Model

The previous models are characterized by a well-defined hierarchical ordering. Competitors are defined in terms of their extinction and colonization rates. The competition-colonization trade-off seems to be fundamental among competing species: the ability to persist on a site (to be a good competitor) versus the ability to get to a new site (to be a good colonizer). But to what extent are actual hierarchies structuring natural competitive communities?

Some recent studies strongly indicate that competitive hierarchies are not always at work in natural conditions. Rather than competitive local exclusion under a strict hierarchy, studies on tree diversity in neotropical forests seem to indicate that recruitment limitation appears to be a major source of control of tree diversity (Hubbell, 2001). In other words, the winners of local competition are not necessarily the best (local) competitors but those that happened to colonize that particular site first. The power of so-called neutral models has been shown to be high. To a large extent, the approach to community structure involving neutrality seems to be a rather good one. Following this idea, in this section we consider a hierarchy of competitors, where species are ecologically equivalent in their colonization ability (Solé et al., 2004). In a metapopulation context, such an equivalence regarding the fraction of colonized habitat and extinction thresholds is achieved by an equal c/e ratio (Hanski, 1999). In this way, we consider a community where species may just colonize empty sites and share the same c/e ratio (Solé et al., 2004). This constancy can be seen as another possible expression of the fundamental competition-colonization trade-off. The better colonizer a species is, the worse ability to persist on a site it must have. The model is as general as that of Tilman, but instead of assuming a strict competitive hierarchy, it assumes strict neutral competitive interactions. In looking at such extreme cases, we can gain some insight into how other scenarios of community organization can influence the response against fragmentation.

A spatially implicit, mean field metapopulation model of such recruitment limitation process can be easily formulated by means of a generalization of the Levins model (Levins, 1969a) as:

$$\frac{dp_i}{dt} = f_i(p) = c_i p_i \left(1 - D - \sum_{j=1}^{S} p_j\right) - \phi(c_i) p_i, \quad i = 1, \ldots, S, \qquad (5.23)$$

including habitat fragmentation. Here the extinction rate m_i has been replaced (last term) by $m_i = \phi(c_i)$, which gives the functional form of the trade-off. This trade-off is chosen to render all species ecologically equivalent in their colonization ability, where as in Hanski (1999) this equivalent ability means the same ability to colonize and persist if they were in isolation. So, we assume the same c/e ratio for all species. Equivalently, we can write the functional form of the trade-off then as:

$$\phi(c) = \alpha c. \qquad (5.24)$$

We assume $\alpha < 1$, which means that no species would go to extinction in isolation. Therefore, such trade-off implies that high (low) colonization rates are linked to high (low) local extinction rates and can be seen as a particular expression of the fundamental competition-colonization (persistence ability versus colonization ability) trade-off (Lehman and Tilman, 1997).

As a result of this choice, it is easy to see that the total fraction of occupied patches $P = \sum_i p_i$ will evolve in time as:

$$\frac{dP}{dt} = < c > (1 - D - P - \alpha) P \qquad (5.25)$$

where $< c >= \sum_i p_i(t)\, c_i / P(t)$ is the average colonization rate at time t, and the average extinction rate is $< \phi(c) >= \alpha < c >$ under the assumption of a linear trade-off given by (5.24). Hence, (5.25) has a unique positive stationary value, namely, $P^* = 1 - D - \alpha$, which is globally asymptotically stable for any initial condition $P(0) > 0$. This implies that the main result of this mean-field model (where spatial correlations are not taken into account) is that the *global* population will experience a collapse at a critical threshold of habitat destruction $D_c = 1 - \alpha$ since, for this value of D, $P^* = 0$. And thus a diversity collapse will occur at D_c, instead of at increasing values for D, as was shown by Tilman et al. (1994).

Equation (5.23), endowed with a positive initial condition $p^0 = (p_1^0, \ldots, p_S^0)$, has as a unique stable equilibrium p^*, which is given by

$$p_i^* = p_i^0 \exp\left(c_i \int_0^\infty (1 - D - \alpha - P(t))\, dt\right), \qquad i = 1, \ldots, S, \qquad (5.26)$$

with $P(t)$ a solution of (5.25) with initial condition $P^0 = \sum_i p_i^0$. However, such an equilibrium is not asymptotically stable. Notice that for any $D < D_c$, any initial condition p^0 with $P^0 = 1 - D - \alpha$ is also an equilibrium of the model since, in this case, $P(t) = P^0$ for all $t > 0$ and, from (5.26), $p_i^* = p_i^0$. In other words, model (5.23) has an *infinite* set of steady states p^* given by the solutions compatible with the set of conditions

$$p_i^* = 1 - D - \alpha - \sum_{j \neq i} p_j^*, \quad i = 1, \ldots, S. \tag{5.27}$$

When $P^0 \neq 1 - D - \alpha$, it follows from (5.23) that, for $D < 1 - \alpha$ and $P^0 < P^* = 1 - D - \alpha$ ($P^0 > P^*$), all fractions of occupied habitat $p_i(t)$ increase (decrease) monotonously to p_i^* given by (5.26) as $t \to \infty$ since $P(t) < P^*$ ($P(t) > P^*$) for all $t > 0$. Instead, for $D \geq 1 - \alpha$, all fractions of occupied habitat tend uniformly to 0 as $t \to \infty$ and, hence, diversity collapse occurs.

It is important to mention that the existence of a nontrivial equilibrium of coexistence depends crucially on the assumption of this ecological equivalence in the way we have defined it, that is, as a linear trade-off function $\phi(c)$. We could have defined an even more strict version of ecological equivalence, as in (Hubbell, 2001), as an equivalence in per capita rates among all individuals of every species. Even such a strict ecological equivalence, as Hubbell states, "permits complex ecological interactions among individuals so long as all individuals obey the same interaction rules." In particular, notice that under such a strict equivalence, our model predicts as well a biodiversity collapse at the same threshold $D_c = 1 - e/c$.

It is also worthy of mention that ecological equivalence must be assumed, at least in the form of a linear trade-off (5.24), in order to have a nontrivial equilibrium coexistence of all species. When $\phi(c)$ is nonlinear, a model where species just colonize empty sites like ours (5.23) does not have a positive equilibrium solution different from the one with only one competitor, namely, the best competitor, that one with the lowest ratio $\phi(c_i)/c_i$. In this case, equations (5.23) do not support diversity. The underlying hierarchy that can be seen now as a ranking of species after colonization ability (the ratio e_i/c_i) leads to the extinction of all species but one, the best competitor. Once this point is reached the known critical threshold of habitat destruction for monospecific metapopulations remains true but now reads $D_c = 1 - \min_i\{\phi(c_i)/c_i\}$. For instance, if $\phi(c) = \alpha c^2$, the best competitor, the one that survives, will be the one with the lowest c_i. However, if $\phi(c) = \alpha\sqrt{c}$, then the survival will be the competitor with the highest c_i.

FOOD WEB STRUCTURE AND HABITAT LOSS

Until now we have considered metacommunities formed by competing species. Similar work has also considered trophic and mutualistic interactions (May, 1994; Kareiva and Wennergren, 1995; Holt, 1997; Nee et al., 1997; Bascompte and Solé, 1998a, c; Melián and Bascompte, 2002a; Bascompte, 2003). Now we will briefly consider the predator-prey metapopulation model by Bascompte and Solé (1998c). Results are similar to those obtained by other authors (May, 1994; Holt, 1997; Nee et al., 1997).

Consider a metacommunity defined by a predator and a prey. We assume that the predator is a specialist, that is, it can only live in patches occupied by prey. If x and y denote the fraction of patches occupied by prey and predator, respectively, the following model describes the trophic interaction:

$$\frac{dx}{dt} = c_x x(1 - x - D) - e_x x - \mu y, \tag{5.28}$$

$$\frac{dy}{dt} = c_y y(x - y) - e_y y. \tag{5.29}$$

As usual, c and e denote extinction and colonization rates, respectively. Prey mortality is decomposed in two terms. Prey have a mortality rate e_x at patches where only prey is present (a fraction $x - y$ of sites). Prey mortality rate is increased to $e_x + \mu$ in those patches where prey and predator coexist (a fraction y). Then, total mortality for prey is $e_x(x - y) + (e_x + \mu)y = e_x x + \mu y$. Available sites for predators are now the nondestroyed, empty sites occupied by prey (i.e, $x - y$) because predators are considered specialists (see below), and so they cannot live without their prey.

One interesting advantage of model (5.28–5.29) is that it can describe different trophic interactions: donor control ($\mu = 0$) and top-down control ($\mu > 0$). This is important for two reasons. First, because theoretical models of trophic interactions have generally assumed scenarios in which natural enemies have a significant impact on prey populations (top-down control). Second, because donor control may operate in half of the parasitoid-host interactions, which in turn contain roughly half of the world's multicellular species (Hawkins, 1992).

Model (5.28–5.29) has three different solutions depending on the values of habitat destruction. Let us refer to D_{c_1} and D_{c_2} as the critical values of habitat loss at which predator and prey become, respectively, extinct. We have the following domains:

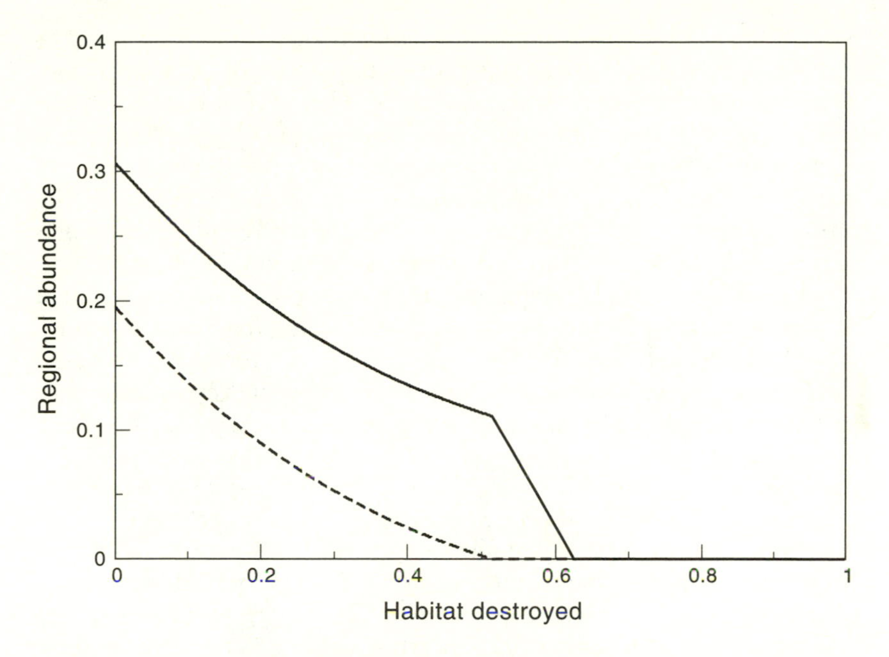

FIGURE 5.5. Prey's (solid line) and predator's (broken line) regional abundance as a function of habitat loss. Here $\mu = 0.2, c_x = 0.4, c_y = 0.9, e_x = 0.15, e_y = 0.1$. Modified from Bascompte and Solé (1998c).

1. $D < D_{c_1} = 1 - e_x/c_x - e_y/c_y < D_{c_2} = 1 - e_x/c_x$. The nontrivial solution of model (5.28–5.29) is given by:

$$x^* = \frac{1}{2c_x}\left[\Gamma + \left(\Gamma^2 + 4c_x\mu\frac{e_y}{c_y}\right)^{1/2}\right], \tag{5.30}$$

$$y^* = x^* - \frac{e_y}{c_y}, \tag{5.31}$$

   where $\Gamma \equiv c_x(1 - D) - e_x - \mu$.
2. $D_{c_1} \leq D < D_{c_2}$. Now $y^* = 0$ and $x^* = 1 - D - e_x/c_x$.
3. $D \geq D_{c_2}$. Both species are extinct in this domain, that is $x^* = y^* = 0$.

We plot prey and predator densities as a function of habitat destruction in figure 5.5. Two interesting conclusions can be drawn. First, extinction takes place earlier for the predator than for the prey. If predators are specialists we can expect a well-defined order in the pattern of extinctions, starting with the highest trophic level species going down through the trophic chain. This may change if the predator is a generalist (results by Swihart et al., 2001, building on Bascompte and

Solé, 1998c). This means that habitat destruction is not only going to reduce biodiversity, but will also reduce the length of the food chain. The reduction of species is highly biased. This may have great implications for ecosystem functioning and biological control. Krues and Tscharntke (1994) have found that habitat loss can lead parasitoids toward extinction, freeing phytophagous insects from their control. Big outbreaks can now be observed, and since these phytophagous insects may be agriculture pests, this may have large economic implications. Although biological control depends largely on top-down attributes such as parasitism rate, Kruess and Tscharnke (1994) show that bottom-up considerations may also be relevant.

The second result stressed by figure 5.5 is that the response of prey to habitat loss depends on the fraction of the habitat that has been already destroyed. There are two different domains separated by D_{c_1}, the extinction threshold for the predator. When predators are present, prey abundance decays slower than when predators are extinct. This is because the response to habitat loss is a trade-off between different trends depending on the trophic position. Thus, while habitat loss has only a negative effect for predators, prey face two opposite trends. On one hand their own habitat is reduced, which tends to decrease their regional abundance. On the other hand, habitat loss affects largely the predators, and so, reduces predation pressure, which tends to increase prey regional abundance. For some scenarios, it may even happen that habitat loss increases the abundance of the prey, as happened with the abundance of the inferior competitor in the model by Nee and May (1992, equations 5.11–5.13). After predators go extinct, the trade-off between the two opposite forces disappears, and prey abundance decreases much faster with additional habitat loss. Thus, the rate of decrease depends on the amount of habitat already destroyed.

The colonization rate has an important role in the response to habitat loss for all species, but this effect is more accentuated for species located at higher trophic levels. Predator regional abundance is plotted in figure 5.6 as a function of its colonization rate. As can be seen, there is a critical colonization rate below which the metapopulation can no longer persist. This threshold is given by

$$c_{y_c} = \frac{e_y}{x^*}. \tag{5.32}$$

After the above threshold is reached, a slight increase in colonization rate is translated into a high increase in abundance. A plateau is then reached, and regional abundance is no longer increased by additional increases in colonization rate. Abundance may even slightly decrease for additional increases in colonization rate (for high $\mu - values$). This

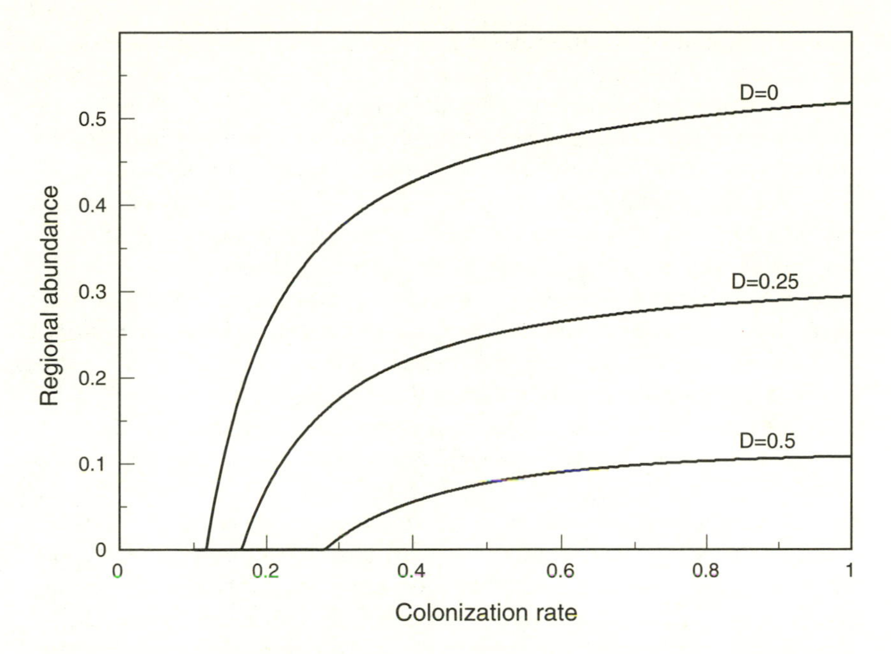

FIGURE 5.6. Predator's regional abundance as a function of its colonization rate, for three different values of habitat destruction D. Here $\mu = 0.2, c_x = 0.7, e_x = e_y = 0.1$. Modified from Bascompte and Solé (1998c).

is due to the negative feedback between prey and predator. If predators can easily colonize patches, prey abundance is decreased, which in turn decreases predator abundance. Similar thresholds in colonization rates were studied by Hastings (1977).

Melián and Bascompte (2002a) have studied how food web structure modifies the extinction threshold of the top predator. They considered four trophic web modules: simple food chain, omnivory, apparent competition, and intraguild predation. Each module contained three trophic levels with three or four species. If we represent by r, c, and p the regional abundance of resource, consumer, and predator, respectively, we can write an extension of the Levins model for each trophic structure. For the simplest case, the simple food chain, we can write:

$$\frac{dr}{dt} = c_1 r(1 - r - D) - e_1 r - \mu_1 rc, \tag{5.33}$$

$$\frac{dc}{dt} = c_2 c(1 - c - D) - e_2 c - \psi_1 c(1 - r) - \mu_2 cp, \tag{5.34}$$

$$\frac{dp}{dt} = c_3 p(1 - p - D) - e_3 p - \psi_2 p(1 - c), \tag{5.35}$$

where the bulk of parameters are as above. The new parameters ψ_i and μ_i can be interpreted in the following way. Until now we have been considering a specialist predator, that is, a predator that could not exist without its main prey. Now, colonization of a patch by the predator occurs independently of patch occupancy by its main prey. Therefore, in patches without prey, intermediate and top species pay an added cost ψ_i in terms of an increase in the rate of local extinction for mistakenly colonizing an inferior resource patch. In the case of a perfect generalist species, ψ_1 and ψ_2 would be equal to 0 (Swihart et al., 2001); μ_i represents the increase in mortality due to predation. This last parameter allows us to consider two scenarios: donor control ($\mu_i = 0$), and top-down control ($\mu_i > 0$).

From this simple food chain model we can build more complexity and introduce the other food web structures (see Melián and Bascompte, 2002a, for details). For example, in omnivory, the predator consumes both consumer and resource; in apparent competition, there are two consumer species sharing the same predator; in intraguild predation, a trophic link is introduced between the two intermediate species (see inserts in fig. 5.7). Figure 5.7 shows the reduction in the number of species as habitat is destroyed. As noted, food web structure alters the response of top species to habitat loss. This means that extinction thresholds are not only determined by life history traits, competitive-colonization abilities, and landscape properties, but also by the complexity of the food web. Omnivory confers the higher persistence to the top species, while the type of interactions between the two intermediate species determine their extinction thresholds. Only the extinction threshold of the resource is constant and identical to that predicted by the Levins rule (since when all other species have gone extinct, we are dealing with a single species metapopulation model).

In summary, in this section we have reviewed simple metacommunity models of habitat loss. These models provide us with some interesting results about the ways communities may respond to this destructive process. Among these results we find nonlinearities in the decrease of diversity with habitat loss (that is, the rate of decrease in biodiversity depends on the amount of habitat already destroyed); the existence of time lags between habitat destruction and the subsequent reduction in biodiversity; and the counterintuitive effects of habitat loss in competitive and trophic interactions (habitat loss can in fact increase the abundance of the poorer competitor species, or of a prey species).

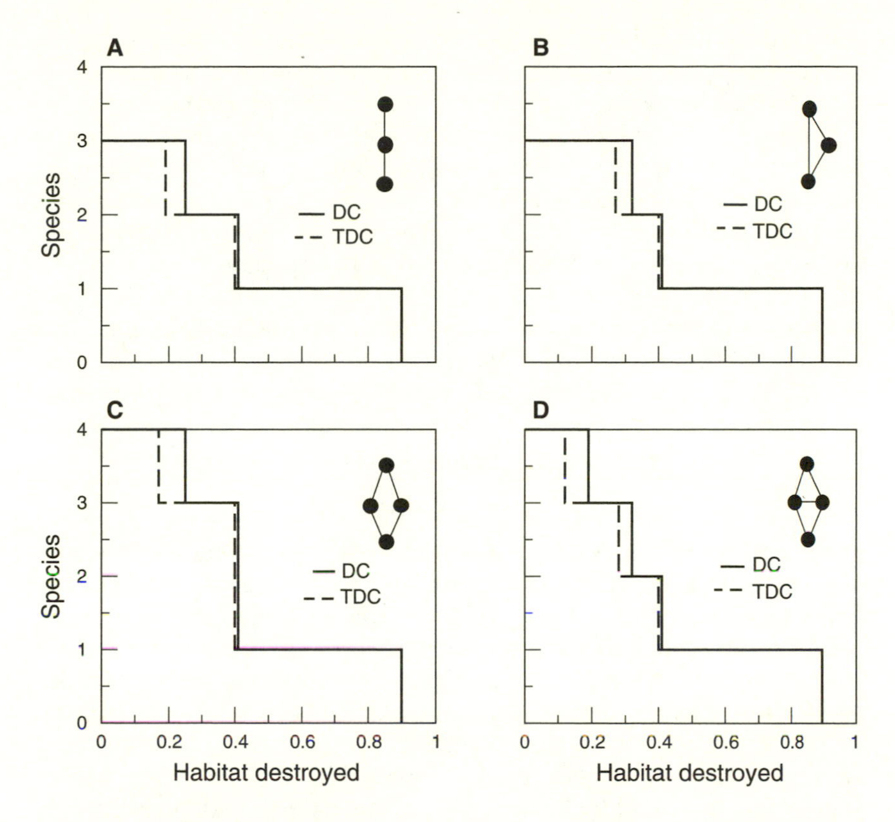

FIGURE 5.7. The effect of habitat loss on species richness depends on the structure of the food web. Each panel represents a specific food web type, as represented by the insert: (A) simple food chain; (B) omnivory; (C) apparent competition; (D) intraguild predation. In all cases, number of species is plotted versus habitat destroyed. The continuous and discontinuous line represents a donor and top-down control, respectively. Modified from Melián and Bascompte (2002a).

PERCOLATION IN SPATIALLY EXPLICIT LANDSCAPES

Up to this point we have reviewed spatially implicit models of habitat loss. These are mean-field models assuming global mixing. The probability of being colonized is the same for all patches regardless of distance. This simplification does not allow us to study habitat fragmentation and disentangle the effects of habitat loss and habitat fragmentation on metapopulation persistence. As noted by some ecologists, "models that gloss over spatial details of landscape structure can be useful for theoretical developments but will almost always be misleading when applied to real-world conservation problems" (Fahrig and Merriam, 1994).

Since we are now interested in fragmentation, let us start by identifying this process, using a simple null landscape model. We will consider only the properties of a landscape, and after understanding how the properties of such a landscape change with random habitat loss, we will proceed by introducing a metapopulation into this spatially explicit world.

There is a long tradition in ecology for using null landscape models (Turner, 1989; Turner et al., 1989; Lavorel et al., 1993; Gustafson and Parker, 1992). The idea is to differentiate patterns that are the result of simple random processes from patterns that arise from more complex processes. For a general overview of null models in ecology, see Gotelli (2001).

Let us consider a spatial grid of cells. Each site can be either available or destroyed. We shall proceed by randomly destroying a fraction D of sites. In figure 5.8a we plot examples of such ideal landscapes for different values of habitat loss. When D is very small, all available sites are connected in a single cluster, that is to say, all available sites share at least one edge. One can imagine that an organism with poor dispersal abilities could move through the entire network of available sites without going through a destroyed site. But as more and more sites are destroyed, some clusters become isolated from the rest, and so our ideal organism would be constrained to live within the domains of the isolated cluster. The number of distinct clusters of continuous habitat first increases with D, it reaches a peak, and finally decreases (fig. 5.8b). When almost all habitat is destroyed, there are only small clusters (of one or two cells) scattered around an ocean of destroyed sites.

As can be seen in figure 5.9a, there are two very different phases of reduction in the size of the largest cluster of habitat as more sites are destroyed. At the beginning, the decrease is slow. In this stage, the only process taking place is habitat loss. As a new patch is destroyed, the single big cluster decreases its size by a unit. This is a quantitative process. However, there is a big transition after a critical amount of habitat loss has been reached. There is a discontinuity in the distribution. At this point, an additional destruction has more implications than before. It does not only reduce the size of the largest cluster by a unit, but it breaks the cluster into two or more smaller clusters. The largest cluster of available habitat suddenly breaks down (fig. 5.9a). In other words, we face habitat fragmentation.

We may want to separate the two effects of habitat destruction, that is, habitat loss and habitat fragmentation. We can do this by using the

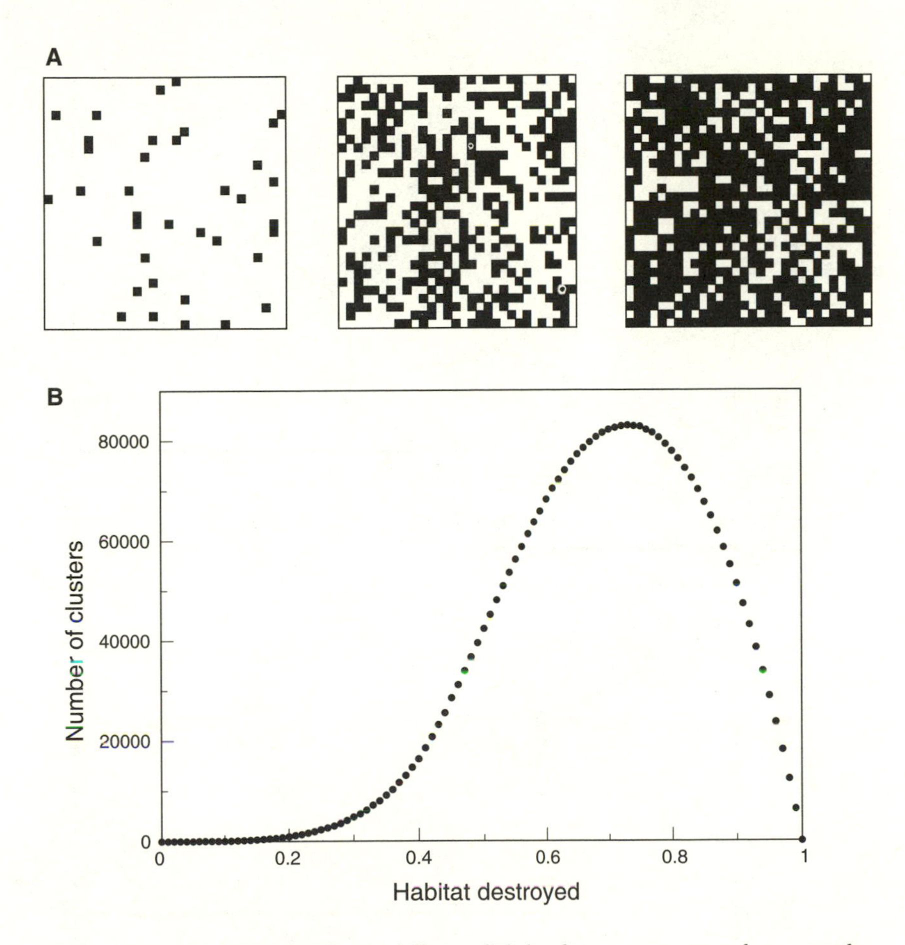

FIGURE 5.8. (A). Example of a spatially explicit landscape represented as a regular lattice. Available sites are represented in white; permanently destroyed sites are represented in black. From left to right, the fraction of sites destroyed is 0.05, 0.40, and 0.7, respectively. (B) The number of distinct clusters of sites as a function of the fraction of habitat destroyed. Lattice size is 800 × 800 and each point is the average of ten replicates. Destroyed sites are randomly scattered. Modified from Bascompte and Solé (1996).

following order parameter (Bascompte and Solé, 1996):

$$\Omega = \frac{S_{max}}{\sum_{k=1}^{N} \Theta(k)}, \tag{5.36}$$

where S_{max} is the size of the largest cluster, and $\Theta(k)$ is one if site k is available, and zero if it is destroyed. As defined, the denominator in

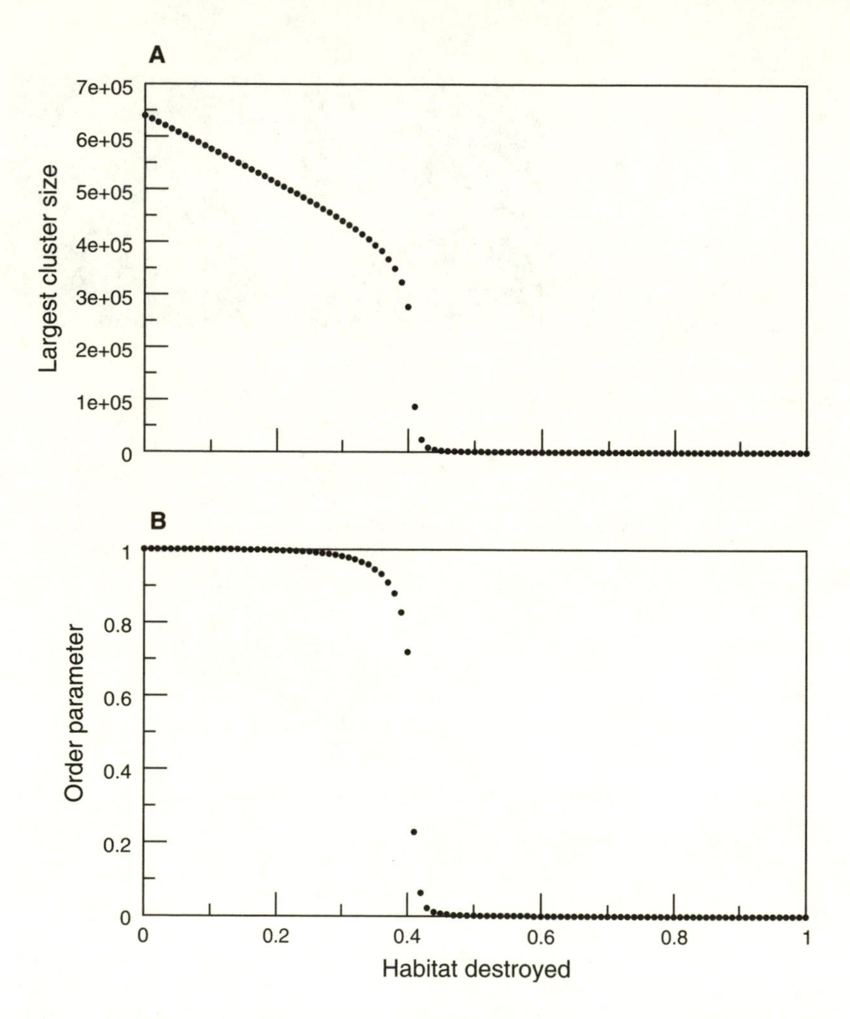

FIGURE 5.9. Largest cluster size (A) and order parameter (B) as a function of random habitat destruction. For low values of destruction, all sites belong to the same cluster, and so the decline in its size is linear. But for a critical destruction value, fragmentation takes place and the largest cluster collapses in a myriad of small, disjunctive clusters. The order parameter in (B) disentangles the effect of habitat loss and habitat fragmentation. Its behavior is similar to a phase transition near a critical point. Modified from Bascompte and Solé (1996).

the equation (5.36) gives the total number of available sites. When all available sites belong to the same cluster, numerator and denominator coincide, and so Ω is equal to one. When habitat is fragmented, Ω is less than one. Figure 5.9b plots the behavior of the order parameter as a function of habitat destroyed. The phase transition indicates habitat fragmentation. Before then, habitat destruction only reduces the amount of continuous habitat. We are only facing habitat loss. Close to the transition, habitat destruction introduces fragmentation. This order parameter is equivalent to similar order parameters introduced in the physics of critical phenomena to identify a phase transition. The value of the order parameter is normally zero before the phase transition and one after it. In this case we are dealing with a particular example of phase transition called percolation (Schroeder, 1991).

Percolation (a concept already introduced in the previous chapter) is a concept widely used in the study of complex systems. It was first introduced in relation to the design of gas masks, where the problem was to find the optimum degree of porosity. Not too dense, since then it would be difficult for air to pass through, not too loose, since then the toxic gas is not filtrated out. Since then, the concept of percolation has been used in a multitude of examples (Schroeder, 1991). One of the best is called the so-called forest fire. We introduced the forest fire in the previous chapter. Let us summarize the main result because it will be very important for understanding the extinction thresholds in spatially explicit metapopulation models. As noted before, one would imagine that the probability of fire spreading through the entire systems increases linearly. Wrong. In fact, one observes that this probability of transmission is almost zero below a critical density of trees, and becomes almost one after such a critical density. This critical fraction of trees is called the percolation threshold, because at this density a lot of small clusters percolate into a big one, so one can find a path going from one extreme of the lattice to the opposite one. For a two-dimensional system with spread to the four nearest neighbors, the percolation thresholds is 0.5928. This threshold does not depend on further details. Similar systems sharing these basic principles percolate exactly at this density.

EXTINCTION THRESHOLDS IN SPATIALLY EXPLICIT MODELS

An order parameter in percolation can be the probability of spreading from one side of the lattice to the other. The transition in such a parameter from zero to one at the critical point is characteristic of a phase transition. How is this relevant to our example of habitat destruction?

As we have seen, in the Levins model there is linear decay in the regional abundance as habitat is destroyed (the slope of the curve is constant). This is because colonization is global. As stressed before, the Levins rule is a powerful result that may have practical implications as a rule of thumb for predicting the extinction threshold for a certain metapopulation even when demographic information is not available. The next question is, to what extend does the Levins rule hold when additional levels of realism are introduced? To begin with, let us consider how the Levins rule is modified when local dispersal (as opposed to global dispersal) is considered. This will be the only difference. Let us continue to assume that the pattern of habitat loss is random.

Bascompte and Solé (1996) extended Lande's (1987) concept of extinction threshold to a spatially explicit metapopulation model. The dynamic of the metapopulation is described by a stochastic cellular automata. At each generation, an occupied patch has a given probability of becoming extinct. Similarly, an empty patch has a given probability of being occupied by one of its four nearest neighbors. Colonizations are considered independent events, so the larger the number of nearest neighbors occupied, the larger the probability of a site becoming colonized. This is a simple model, but it is interesting to keep things simple in order to understand how the the extinction threshold is shifted when moving from global to local dispersal. The key point in Bascompte and Solé (1996) is to relate landscape ecology to metapopulation dynamics. Once we understand the changes undergone by the landscape as habitat is progressively destroyed, we want to see how these changes are translated into the decline of the metapopulation.

One technicality should be mentioned here. The Levins model is a continuous time model with colonization and extinction rates. Our simulation, on the other hand, is a discrete time process with extinction and colonization probabilities. In order to make the comparison possible, one can follow two different strategies. The first is to work with the discrete time version of the Levins model (equation 5.3). Alternatively, one can simulate a continuous time process. In this case, one commonly assumed strategy is as follows: One uses small differential time intervals (Δt). Each time step corresponds to one of such small time intervals. We assume that the probability of colonization (or extinction) follows a Poisson distribution. A site will be colonized if there is at least one colonization event in one of its nearest neighbors. Let us call p the probability of this happening. Then $q = 1 - p$ is the probability of no colonization events, that is, the null term in the Poisson distribution. So one can write

$$p = 1 - (e^{-c\Delta t})^n, \tag{5.37}$$

where n is the number of nearest neighbors occupied. The colonization rate here can be interpreted as the mean number of propagules produced by an occupied site. Similarly, we can translate from extinction rates to extinction probabilities. Other authors follow a different strategy (K. Sato, personal communication), that is, assuming that a time interval is such a small amount of time that only one event can take place (the probability of having two events is small, so it can be disregarded). One then picks up a site randomly and allows this site to change its status with the corresponding probability. Both strategies are similar and coincide in the limit.

This is not the only point to keep in mind when moving from a mean field model to a numerical simulation. In the mean field model, we assume that there is an infinite number of patches. The equation is totally deterministic. However, the simulation uses a finite lattice, and probabilistic rules. Stochastic events can be important here as we will see in the last section. If we want to isolate the effect of local dispersal, we have to make sure that these other factors are not relevant. One can check for the role of stochastic fluctuations by running the simulation with global dispersal. In this case, the mean field model perfectly describes the system in the limit of an infinite lattice size. For very small lattices, demographic stochasticity may lead the metapopulation into extinction before the extinction threshold predicted by the deterministic model. Up to a certain lattice size (around 60×60 is large enough), the simulation and the mean field model converge. Then one can use lattices larger than or equal to this size and introduce local dispersal. Now we are ready to look at the effect of local dispersal and to compare the extinction threshold with the one predicted by the Levins model.

Figure 5.10 shows the decline in the fraction of patches occupied as habitat is destroyed for the spatially explicit model. Each point is the average of ten replicates. The discontinuous line shows the behavior of the spatially implicit Levins model. As noted, both models behave similarly when almost all habitat is available. In this domain, in which habitat loss is the only process, the spatially homogeneous description of the mean field model is a good enough description. However, as more and more habitat is destroyed, both models start to diverge. The fraction of patches occupied is lower for the spatially explicit model than for the Levins model. And the decay is faster as more habitat is destroyed. The decay is no longer linear (the slope is no longer constant). This is because spatial details now become more and more relevant. Destroying a site can have larger effects than the habitat loss effect described by the Levins model. As more habitat is destroyed, an additional destruction can be amplified by isolating local clusters, reducing the probability of local colonizations. As can be seen, patch occupancy is lower than predicted by the spatially implicit model. Also, the extinction threshold takes place for lower values of habitat loss. The Levins rule is no longer valid. Things are worse than predicted by the Levins model.

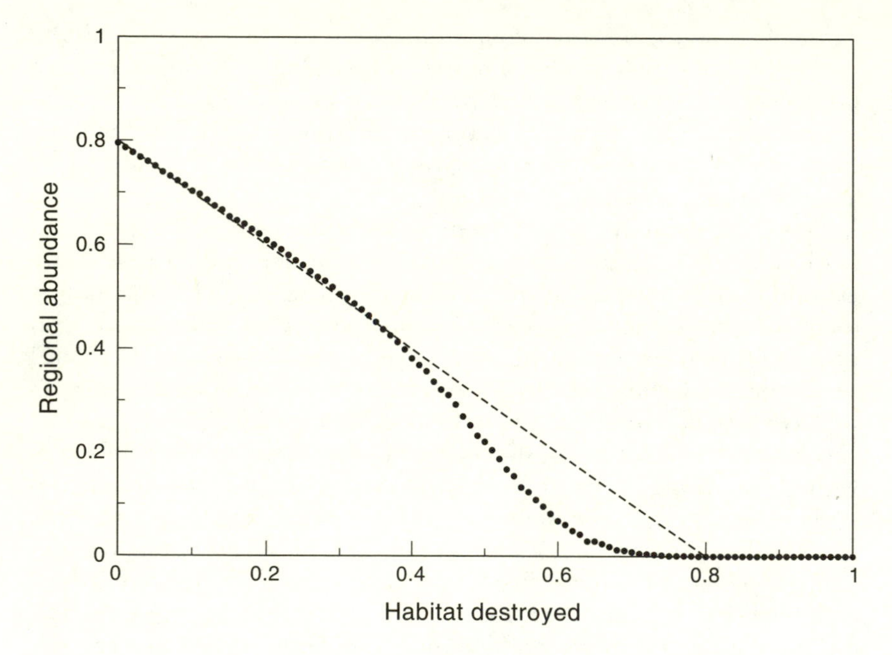

FIGURE 5.10. Regional abundance as a function of habitat destroyed for a spatially explicit simulation with local dispersal to one of the four nearest neighbors. Each point represents the average of ten replicates. Lattice size is 100×100, and extinction and colonization probabilities are 0.2 and 0.4, respectively. Broken line depicts the behavior of a spatially implicit model such as the Levins model. As noted, regional abundance is lower than predicted by a spatially implicit model. Modified from Bascompte and Solé (1996).

Similar comparisons between spatially implicit and explicit models looking at the effects of local dispersal have been done by Dytham (1995a,b) and Moilanen and Hanski (1995). These authors use the two-species competition model by Nee and May (1992) as the reference system. Dytham (1995a) uses a cellular automata model, while Moilanen and Hanski use the incidence function, a spatially explicit realistic model where patches have different sizes and distances from one another. Both sets of papers have similar results to the ones shown for the case of single species. The decrease in abundance is higher than predicted by the spatially implicit model (the number of empty patches increases with fragmentation, instead of remaining constant), and the extinction threshold takes place for lower values of habitat loss. Once more, the spatially implicit models underestimate the minimum patch density for metapopulation persistence. Due to the different modeling approaches, we can conclude that this is a general result of metapopulations with local dispersal.

So far, we have restricted our analysis to local dispersal and random habitat destruction. However, habitat loss is often far from the purely random scenario, so one important question is how the pattern in which the habitat is removed affects metapopulation regional abundance. Dytham (1995b) again took the two-competing-species model by Nee and May (1992) as his system of reference. Additionally to random habitat loss he considered the following patterns of destruction: destruction through a gradient, in blocks, and following a road. The order of extinction by the superior competitor, inferior disperser as more habitat is destroyed is as follows: First it becomes extinct when habitat loss is random; second when destruction is along a gradient; and finally when destruction takes place in blocks and lines. Similarly, both With and King (1999) and Hill and Caswell (1999) study spatially explicit models in which habitat loss takes place through a fractal pattern, and compare the extinction thresholds with the thresholds predicted by the Levins model. Both results coincide with Dytham's (1995b) finding, that is, the extinction threshold takes place for higher values of habitat destruction when the pattern of habitat loss is nonrandom.

ANALYTICAL MODELS OF CORRELATED LANDSCAPES

We started this section by asking whether or not spatial details were really important. We have seen that such spatial details shift the position of the extinction threshold toward lower values of habitat destroyed. As emphasized in chapter 3, a trade-off between simplicity and realism arises. Once again, the question is whether or not we can summarize spatial details by using aggregate measures. For the purposes of the present chapter, we want to find analytical results showing the effect of the pattern of habitat loss. Two recent developments give insight into this question. Both methods consider the spatial correlation in the pattern of habitat loss, and how equilibrium patch abundance can be related to the level of correlation for a specific fraction of habitat destroyed. Ovaskainen et al. (2002) compare both analytical approximations. One is based in the metapopulation capacity, a concept introduced by Hanski and Ovaskainen (2000), and the other in pair approximation (see chapter 3 and below for details).

The metapopulation capacity λ_M is a measure that describes the capacity of a fragmented landscape to maintain a viable metapopulation. Following Hanski and Ovaskainen (2000), consider the following spatially realistic version of the Levins model describing the rate of change in the probability of patch i being occupied:

$$\frac{dp_i(t)}{dt} = (c_i)[1 - p_i(t)] - (e_i)p_i(t). \qquad (5.38)$$

Although the theory does not depend on the particular form of the colonization and extinction functions, Hanski and Ovaskainen (2000) use the following functions for describing the extinction and colonization rates, respectively, of patch i:

$$e_i = \frac{b}{A_i} \qquad (5.39)$$

$$c_i = f \sum_{j=1, j \neq i}^{n} exp(-\alpha d_{ij}) A_j p_j(t), \qquad (5.40)$$

where A_i is the area of patch i, d_{ij} is the distance between patches i and j, $1/\alpha$ is the average migration distance, and b and f are constants.

We can now use matrix notation to describe the rates of change for the p_i values. Thus, we can define a matrix $\mathbf{M}$ containing the elements $m_{ij} = exp(-\alpha d_{ij})A_i A_j$ for $j \neq i$ and $m_{ii} = 0$. $\mathbf{M}$ is the so-called landscape matrix, which describes how the spatial configuration of the patches determines colonization and extinction rates. The metapopulation capacity λ_M is the leading eigenvalue of $\mathbf{M}$. Interestingly, this single parameter summarizes the spatial details of a complex landscape; λ_M captures the impact of landscape structure—the amount of habitat and its spatial configuration—on metapopulation persistence. It can be shown that λ_M is equivalent to the amount of suitable habitat $h = 1 - D$ in the Levins model (equation 5.2) and in the Lande's (1987) model (equation 5.7–5.8). So, the condition for metapopulation persistence is

$$\lambda_M > e/c. \qquad (5.41)$$

An appropriate weighted average of the p_i^* values can be approximated by the following equation (Hanski and Ovaskainen, 2000):

$$p_\lambda^* \approx 1 - \frac{\delta}{\lambda_M}, \qquad (5.42)$$

where $\delta = e/c$. One of the interesting aspects of the metapopulation capacity is that it can be used to rank different landscapes in terms of their capacity to support viable metapopulations, or it can be used to assess the effects of removing a specific patch or adding a new one. The contribution of fragment i to the metapopulation capacity, for example, can be given by $\lambda_i \equiv x_i^2 \lambda_M$, where x_i is the ith element in the leading eigenvector of matrix $\mathbf{M}$ (Hanski and Ovaskainen, 2000).

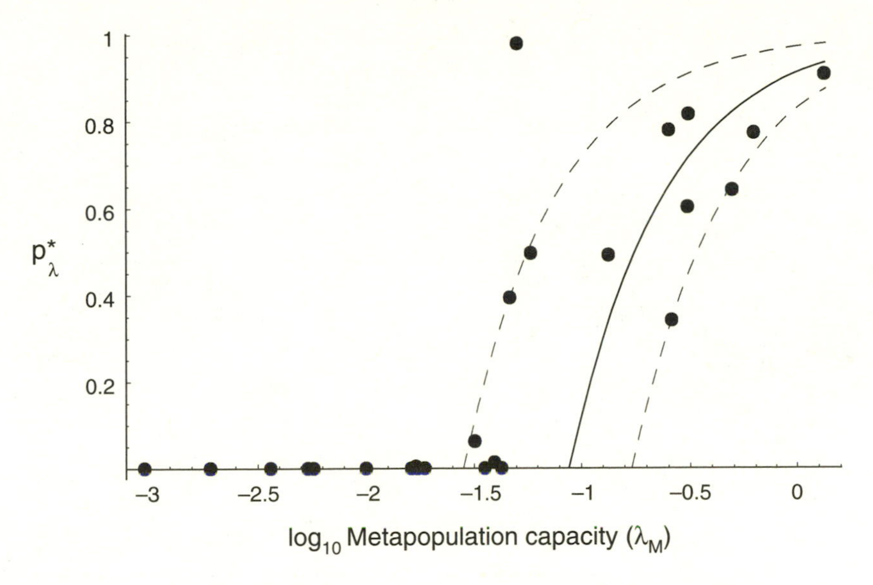

FIGURE 5.11. Metapopulation regional abundance (p^*_λ) plotted as a function of the metapopulation capacity of a fragmented landscape (λ_M). Dots correspond to empirical metapopulations of the Glanville fritillary butterfly. The continuous line represents the average estimate of the ratio extinction-colonization ($\delta = e/c$), and the broken lines represent the minimum and maximum values of the estimated δ values omitting the two networks yielding the most extreme values. This is empirical evidence of an extinction threshold. The figure also shows the remarkable prediction power of the metapopulation capacity. From Hanski and Ovaskainen (2000). Reproduced with permission of *Nature*. Figure courtesy O. Ovaskainen).

As noted in figure 5.11, the metapopulation capacity predicts an extinction threshold that matches well the observed threshold in a set of metapopulations of the butterfly *Melitaea cinxia*.

The metapopulation capacity approach can be used with landscapes composed of patches of different sizes and various distances from one another. Its potential importance as a realistic management tool does in fact rest in this property. However, since we want to compare the results obtained by this approach with the ones obtained by pair approximation, let us consider the common idealized landscape composed by a lattice with an infinite number of identical sites. Now we can explore how the metapopulation capacity changes with the pattern of correlation of habitat loss.

Ovaskainen et al. (2002) introduce spatial correlation of habitat loss and estimate the metapopulation capacity. They assume that spatial correlation in the landscape structure decays exponentially with a pa-

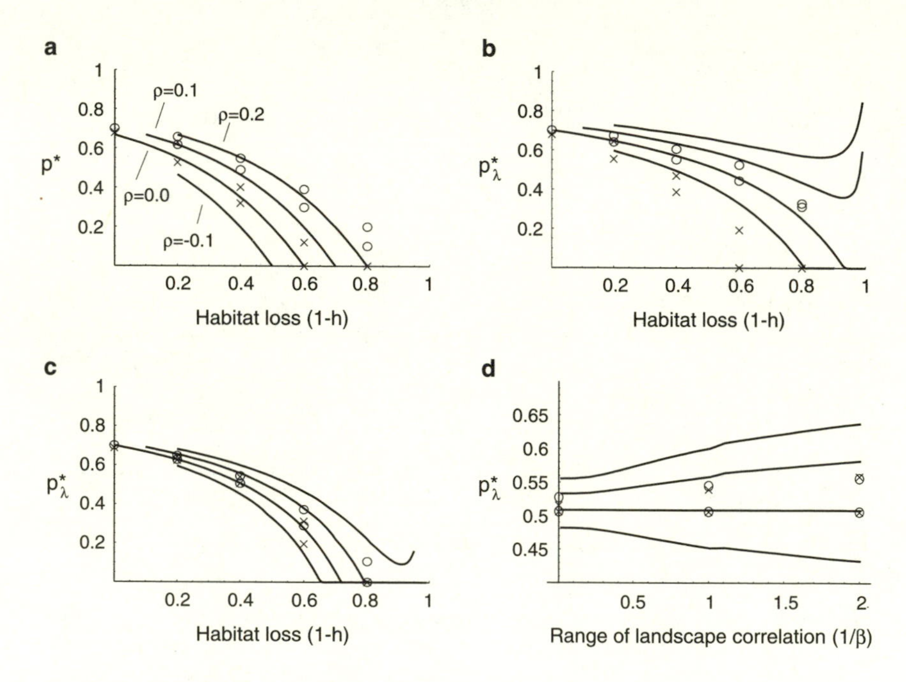

FIGURE 5.12. Effect of the pattern of habitat loss on metapopulation regional abundance (p^* and p^*_λ). On each panel, lines from top to bottom represent a spatial correlation in habitat loss from 0.2, 0.1, 0 and −0.1. Panel (a) corresponds to pair approximation. Dispersal is constrained to the nearest neighbors. Panels (b–d) are based on the metapopulation capacity of a fragmented landscape. Panel (b) corresponds to dispersal to the nearest neighbor (as in (a)), while panel (c) corresponds to a species with long-range dispersal. In panel (d) dispersal is again long-distance, but instead of varying the amount of habitat destroyed, one varies the range of correlation in habitat destruction. To illustrate the validity of the approach, circles and crosses represent the deterministic and stochastic simulation results for a set of 30 × 30 lattices with periodic boundary conditions (for the values $\rho = 0$ and $\rho = 0.1$). All things being equal, the least detrimental way of destroying a certain amount of habitat is in a positively correlated pattern. Based on Ovaskainen et al., 2002. Figure courtesy of O. Ovaskainen.

rameter β. Similarly, the dispersal kernel of the species also follows an exponential distribution with parameter α. Consequently, we can check the effects of the correlation in habitat loss, the spatial scale of such correlation (how fast correlation decays with distance), and the spatial scale of dispersal (for details, see Ovaskainen et al., 2002).

Figure 5.12b–c plots the metapopulation regional abundance as a function of habitat loss for different values of spatial correlation in habitat loss. In both panels we have the same range of spatial correlation constrained to the nearest neighbor. Increasing the spatial correlation of habitat loss increases metapopulation abundance for a specific value of habitat destruction (fig. 5.12b–c). Other things being equal, the less detrimental way of destroying habitat is by doing it in a correlated way, where sites near a destroyed site have a larger probability of being destroyed. This is because in correlated landscapes we leave large areas of continuous habitat. This is in agreement with previous results using the numerical simulations discussed above.

Both figures 5.12b and 5.12c differ in the scale of dispersal. While in figure 5.12b dispersal is constrained to one of the four nearest neighbors, figure 5.12c corresponds to a scenario with a long-range dispersal kernel. By comparing figures 5.12b and 5.12c, we can see that random habitat loss is even more detrimental for species with long-range dispersal, and that increasing correlation in habitat structure does not help as much the species with long-range dispersal as it helps the one with short-range dispersal (Ovaskainen et al., 2002). The reason can be understood as follows. Both species do equally well in an undisturbed environment ($D = 0$). Since the scale of habitat fragmentation is rather short, the remaining available habitat blocks in the fragmented landscape are higher as perceived by the species with short-distance dispersal. This is confirmed by figure 5.12d, where we plot the fraction of sites occupied by the long-range dispersal species for a given value of spatial correlation, as a function of the range of landscape correlation. As can be seen, abundance increases as the range of landscape correlation is increased.

The second analytical method used to explore the effects of spatial correlation is pair approximation (Hiebeler, 2000; Ovaskainen et al., 2002; see chapter 3). As explained in chapter 3, our system is defined by a lattice of infinite size with identical patches. Colonization is assumed to be local, to one of the four nearest neighbors. In addition to the two states each lattice site could be in the example of chapter 3—empty but suitable (1), and occupied (2)—a site can now also be destroyed (0). Then we indicate the fraction of sites destroyed, empty but suitable, and occupied by ρ_0, ρ_1, and ρ_2, respectively. Let us denote by ρ_{ij} the fraction of (ij) pairs of patches, where $i, j = 0$, 1, or 2. Finally, $q_{i/j}$ is the conditional probability of a site being in state i given that one of its nearest neighbors is in state j. By definition, $q_{i/j} = \rho_{ij}/\rho_j$. In a system with global mixing, $\rho_{ij} = \rho_i \rho_j$, and so $q_{i/j} = \rho_i$.

One problem with this notation is that ρ_0 and $q_{0/0}$ are not independent variables; $q_{0/0}$ increases with the density of destroyed sites. The effective correlation among near sites (a measure of spatial clustering seen in chapter 3 [see also Bascompte, 2001]) is defined as:

$$\rho = q_{0/0} - \rho_0. \tag{5.43}$$

The previous expression represents the increase with respect to what we would expect in a random situation (Ives et al., 1998; Bascompte, 2001). Spatial correlation is zero when $q_{0/0} = \rho_0$.

We normalize the colonization rate by setting the extinction rate as 1. Then the normalized colonization rate is $a = c/e$, where c and e are the colonization and extinction rates in the Levins model.

The time derivatives for singlet densities are now:

$$\frac{d\rho_2}{dt} = a\rho_2 q_{1/2} - \rho_2, \tag{5.44}$$

$$\frac{d\rho_1}{dt} = \rho_2 - a\rho_2 q_{1/2}, \tag{5.45}$$

$$\frac{d\rho_0}{dt} = 0. \tag{5.46}$$

Equation (5.46) stands for the fact that we assume that the fraction of destroyed sites is fixed, that is, that these patches are permanently destroyed. As in the example of chapter 3, doublet densities can now be defined as:

$$\frac{d\rho_{20}}{dt} = a\left(1 - \frac{1}{z}\right) q_{2/10}\rho_{10} - \rho_2 \tag{5.47}$$

$$\frac{d\rho_{10}}{dt} = \rho_{20} - a\left(1 - \frac{1}{z}\right) q_{2/10}\rho_{10} \tag{5.48}$$

$$\frac{d\rho_{00}}{dt} = 0 \tag{5.49}$$

$$\frac{d\rho_{11}}{dt} = 2\rho_{12} - 2a\left(1 - \frac{1}{z}\right) q_{2/11}\rho_{11} \tag{5.50}$$

$$\frac{d\rho_{22}}{dt} = 2a\left[\frac{1}{z} + \left(1 - \frac{1}{z}\right) q_{2/12}\right]\rho_{12} - 2\rho_{22} \tag{5.51}$$

$$\frac{d\rho_{12}}{dt} = \rho_{22} + a\left(1 - \frac{1}{z}\right) q_{2/11}\rho_{11} - a\left[\frac{1}{z} + \left(1 - \frac{1}{z}\right) q_{2/12}\right]\rho_{12} - \rho_{12}. \tag{5.52}$$

Once the previous system of equations is closed (see chapter 3 for details), we have only three independent variables. We can then solve the system and find the steady state. The analytical result for the equilibrium patch occupancy is

$$\rho_2^* = \frac{1}{2\{(z-1)a-1\}}\Bigg[a(z-1)\{(z+1)(1-\rho_0) - z\rho_0(1-q_{0/0})\}$$
$$-2z(1-\rho_0) - (z-1)\sqrt{a}$$
$$\sqrt{4z\rho_0(1-\rho_0)(1-q_{0/0}) + a\{(z-1)(1-\rho_0) - z\rho_0(1-q_{0/0})\}^2}\Bigg]. \tag{5.53}$$

Figure 5.12a plots metapopulation abundance as a function of habitat loss for different values of spatial correlation. As we saw for the case of the metapopulation capacity approach, spatial correlation increases metapopulation abundance for a specific value of habitat loss. The comparison between these two very different approaches suggests that this is indeed a very general result, independent of the details of the particular model used.

In summary, both numerical simulations using different patterns of habitat loss (Dytham 1995b), numerical simulations using fractal landscapes (Hill and Caswell, 1999; With and King, 1999); and analytical methods using pair approximation techniques (Hiebeler, 2000; Ovaskainen et al., 2002) and the metapopulation capacity of a fragmented landscape (Hanski and Ovaskainen, 2000; Ovaskainen et al., 2002) unambiguously suggest that not only the amount of habitat destroyed, but the pattern of habitat removal is important for metapopulation persistence. The Levins rule no longer applies when we consider local dispersal and spatial structure. When local dispersal is considered, the extinction threshold takes place sooner than predicted by the Levins rule. On the other hand, when we move from random habitat loss to correlated habitat loss, the extinction threshold takes place for higher values of habitat loss.

The analytical results provided by both the metapopulation capacity approach and pair approximation also confirm one of the points emphasized in chapter 3—that spatial detail is important, but that in some circumstances we can simplify such spatial detail by using aggregate statistical measures. Intermediate models between the simplicity of mean field models and the complexity of numerical simulations can be used, and by doing so, we can arrive at new insights into how to bridge these opposite approaches.

MORE REALISTIC MODELS OF EXTINCTION THRESHOLDS

Throughout this chapter, as in chapter 3, we have built complexity around the spatial dimension. We have also moved from metapopulations to metacommunities, toward a community approach that will be fully reviewed in the next chapter. However, there are many other elements of reality that have been added to the pioneering work by Levins. Among them are the incorporation of Allee effects, the rescue effect, transient dynamics, and stochasticity. We will now briefly review them.

The Allee and the Rescue Effects

The Allee effect was first described by Allee and colleagues (Allee et al., 1949), and describes those situations in which per capita growth rate is correlated with population density. In some extreme cases, growth rate may become negative below some density threshold, and so the population can become extinct. There are multiple mechanisms that can cause an Allee effect, such as difficulties in finding mates, lack of social cooperation, inbreeding depression, failure to satiate predators, and so forth. (Courchamp et al., 1999; Stephens and Sutherland, 1999). The Allee effect was only considered at the population level. However, it can also be considered at the metapopulation level, as done by Amarasekare (1998).

The Levins model can be modified to introduce an Allee effect that would describe a situation in which a metapopulation can go extinct when rare (low patch occupancy), even when it would persist as predicted by the amount of habitat destroyed. Thus, predictions based only on habitat availability may overestimate the ability for long-term persistence (Amarasekare, 1998). Field evidence for a role of the Allee effect in metapopulation dynamics includes several species in which the species is absent at places with few patches. Examples include *Daphnia* inhabiting rocky pools, and several butterfly species such a *Hesperia comma, Plebejus argus*, and *Melitaea cinxia* (see Amarasekare [1998] for details and data source).

Another factor that has been explored is the so-called rescue effect (Brown and Kodric-Brown, 1977; Hanski et al., 1996; Gyllenberg and Hanski, 1997). The rescue effect means that simultaneous extinctions and colonizations can take place within the annual cycle of events, and so the realized extinction probability is changed from e to $(1 - c)e$, where c is the colonization probability. Once the rescue effect is taken into account, the equilibrium fraction of empty but suitable sites is no longer constant, but increases with habitat loss. This is due to the fact that a large fraction of occupied patches can simultaneously colonize after an extinction event, reducing the effective extinction rate and so increasing metapopulation's regional abundance.

Hanski et al. (1996) provide clear experimental evidence for the existence of rescue effects in the Glanville fritillary butterfly. These authors compared observed and expected extinction probabilities and found that expected extinctions (predicted without allowing for rescue effects) were always much larger than the observed extinction probabilities (in which rescue effects can play a role). These authors also concluded that the only scenario in which the rescue effect can be ignored corresponds to metapopulations with low turnover rate and very low rates of habitat destruction. However, as noted by Hanski et al. (1996), the type of scenario faced by conservation biologists is the opposite one, with fast rates of habitat destruction. In this scenario, one has to consider the rescue effect, and so the amount of empty patches cannot be used to estimate the amount of habitat required for metapopulation persistence.

An additional problem is that the Levins model and its predictions deals with stationary regional abundances. But when one empirically measures a metapopulation's regional abundances, this may be out of equilibrium. This relates to the concept of supertransients described in chapter 3. Since for spatial systems transients may be large, it is not a surprise that actual metapopulation values may reflect more a transient than a stationary value. If this is the case, the situation is worse than expected, and a fraction of patches is doomed to extinction (Hanski, et al., 1996). Once more, several metapopulations may by only ghosts, already doomed to certain extinction because of previous habitat alterations.

The transient time before a metapopulation settles down to a new equilibrium regional abundance after a perturbation has been studied by Ovaskainen and Hanski (2002). These authors found that the time delay depends on several factors such as the strength of the perturbation, the characteristic turnover rate of the species, and how close the metapopulation is to the extinction threshold. Near this threshold, the transient time is longer, which is in agreement with the extinction debt reviewed in previous sections.

Similarly, in chapter 3 we discovered the existence of multiple attractors in simple spatially extended models. In the context of metapopulations, multiple attractors have been found in variants of the Levins model incorporating the rescue effect (Hanski and Gyllenberg, 1993). Although these papers have explored the effects of demographic variables such as colonization rates, one can also envision a scenario in which habitat destruction smoothly reduces the regional abundance until a critical point in which equilibrium regional abundance suddenly shifts to a much lower value. This scenario emphasizes the role of nonlinearities and multiple equilibria already seen in previous chapters. Once more, the consequences of habitat destruction would be more detrimental than expected by the predictions of the Levins model.

Extinction in Stochastic Environments

Throughout this chapter (as in the rest of the book), we have mainly dealt with deterministic models. This is because in this book we are emphasizing self-organization in simple, deterministic models. We are not implying that less work has been done with stochastic models. The interested author should look at the work by Gurney and Nisbet (1978), Lande (1993), Hanski et al. (1996), Lande et al. (1998), Saether et al. (1999), Frank and Wissel (2002), and Ovaskainen and Hanski (2002). In the text that follows, we will only provide a brief outline of how stochastic components may alter the extinction thresholds reviewed in this chapter.

The extinction threshold is a deterministic prediction. Even when we start with a large number of patches, as we are destroying more and more sites (that is, the number of available sites is reduced), stochastic events become more relevant. This means that because of demographic stochasticity, the metapopulation can become extinct before reaching the extinction threshold. How likely is this? The answer depends on the number of available patches, and thus, on the value of habitat loss. This source of stochasticity is called demographic stochasticity. Demographic stochasticity is caused by chance realizations of individual probabilities of colonization and extinction in a finite metapopulation (Lande, 1993). Because independent individual events tend to average out in large populations, demographic stochasticity is most important in small metapopulations. Another source of stochasticity is environmental stochasticity and random catastrophes. Environmental stochasticity arises from a nearly continuous series of small or moderate perturbations that similarly affect the colonization and extinction of all patches in a metapopulation (Lande, 1993). Catastrophes, on the other hand, are large environmental perturbations that produce sudden major reductions in metapopulation size. In contrast to demographic stochasticity, both environmental stochasticity and random catastrophes are important in large and small metapopulations (Lande 1993).

Gurney and Nisbet (1978) study a metapopulation model and find analytical estimates for the time to extinction and relative fluctuation size. Here we will assume a lattice with N patches and global colonization, that is, the situation described by the Levins model. If we denote by H_0 the number of occupied patches out of the total number of patches N, we can write the time to extinction (Gurney and Nisbet, 1978) as:

$$\tau = exp\left(\frac{H_0^2}{2(N - H_0)}\right). \tag{5.54}$$

The relative fluctuation size is:

$$\frac{\sigma_H}{H_0} = N^{-1/2}\left[\left(\frac{N}{H_0}\right)^2 - \frac{N}{H_0}\right]^{1/2}. \tag{5.55}$$

Let us assume now a stochastic version of the Levins model (equation 5.2). As we have seen, the equilibrium metapopulation abundance is $H_0^* \equiv Nx^* = N(1 - D - e/c)$ below the extinction threshold. If we substitute this into (5.54) we can write the following expression for the time to extinction from the steady state:

$$\tau^* = exp\left(\frac{[N(1 - D - e/c)]^2}{2N[1 - (1 - D - e/c)]}\right). \tag{5.56}$$

Similarly, the relative fluctuation size can be written as:

$$\frac{\sigma_{H^*}}{H^*} = N^{-1/2}\left[\left(\frac{N}{N(1 - D - e/c)}\right)^2 - \frac{N}{N(1 - D - e/c)}\right]^{1/2}. \tag{5.57}$$

Figure 5.13a illustrates how time to extinction from the steady state decreases with habitat loss. Time to extinction drops very fast as we are approaching the extinction threshold, and becomes zero at the threshold, where the metapopulation goes deterministically extinct. By using this alternative approach, we can assess the probabilities of a metapopulation becoming extinct due to demographic stochasticity when the amount of habitat available exceeds the minimum required for deterministic persistence. Figure 5.13a also illustrates how the time to extinction depends on the number of patches composing the metapopulation. Similarly, the relative fluctuation size is plotted in figure 5.13b.

As noted through this chapter, as well as in chapter 3, traditional metapopulation models assume an infinite number of patches. This approximation is unrealistic for a large number of species that live in a small network of patches. These species are oftentimes submitted to strong environmental fluctuations. In these cases, an interesting question is how the number of patches required for the persistence of a metapopulation depends on the level of environmental variability. Here we will summarize an approach taken by Bascompte et al. (2002).

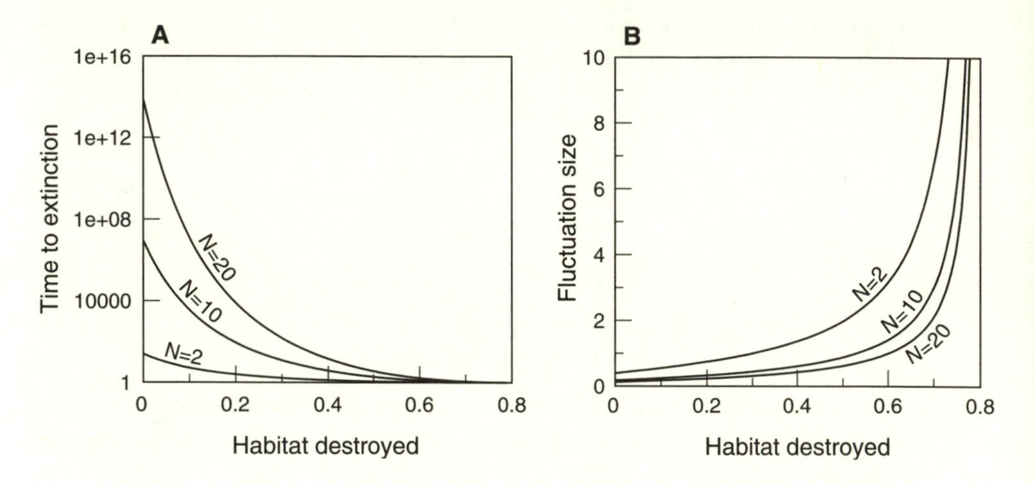

FIGURE 5.13. Time to extinction (A) and average fluctuation size (B) as a function of habitat destroyed in a stochastic metapopulation model. The three lines correspond to a different number of patches composing the metapopulation. The deterministic extinction threshold corresponds to a value of habitat destroyed $D_c = 0.8$. In a stochastic world, there is always a probability of extinction, and so a metapopulation may become extinct even when the deterministic model predicts persistence.

This approximation is interesting because persistence is now characterized from the point of view of the number of patches composing the metapopulation. We will now focus on the effect of environmental stochasticity on a small number of patches. The key of the approach is the concept of geometric mean fitness (GMF), a concept widely used in ecology and evolution to understand persistence in fluctuating environments (Lewontin and Cohen, 1969; Levins, 1969b; Gillespie, 1974; Kuno, 1981; Metz et al., 1983; Klinkhamer et al., 1983; Yoshimura and Jansen, 1996; Jansen and Yoshimura, 1998).

Imagine a metapopulation with density-independent dynamics, high dispersal, and strong environmental variability from year to year. Juveniles (we can think about larvae) disperse into a common pool, grow and move to a set of n patches for reproduction at the beginning of the next generation. The growth rate at each patch is a random variable with some stationary probability density function. This is the way environmental stochasticity is introduced into the model. We assume that environmental fluctuations can be spatially correlated but uncorrelated through time. Let R_{ij} be the growth rate at patch i at year j. If N_0 refers to metapopulation size at generation 0, we see that the metapopulation size after t generations will be (Bascompte et al., 2002):

$$N_t = N_0 \prod_{j=0}^{t-1} \bar{R}_j, \tag{5.58}$$

where $\bar{R}_j$ is the spatial arithmetic mean of the growth rates among the n patches at year j.

If we denote by $G(\bar{R}) = (\prod_{j=0}^{t-1} \bar{R}_j)^{1/n}$ the geometric mean (GM) of $\bar{R}$, equation (5.58) can be written as:

$$N_t = N_0 G(\bar{R})^t. \tag{5.59}$$

Then, the population will persist in the long term ($N_t \geq N_0$) (Kuno, 1981; Jansen and Yoshimura, 1998) if

$$G(\bar{R}) \geq 1. \tag{5.60}$$

In order to find analytical expressions for the minimum number of patches compatible with metapopulation persistence, we have first to approximate the GMF. For the case of a single patch, a well-known approximation is the so-called variance discount approximation (VDA) (Gillespie, 1974; Yoshimura and Jansen, 1996). Bascompte et al. (2002) extended the VDA to the metapopulation context and found a conservative estimate of $G(\bar{R})$ given by:

$$G(\bar{R}) \approx \bar{R} - \frac{\sigma^2 + (n-1)cov}{2n}, \tag{5.61}$$

where σ^2 is the variance of R_{ij}, that is, the range of environmental stochasticity, and cov is the covariance of the environmental variation among patches (correlation is assumed to be the same among all patches).

As can be observed by looking at the previous approximation, both the intensity of environmental fluctuations and covariance decrease the GMF, while the number of patches tends to increase it. This is somewhat intuitive, but expression (5.61) gives an exact account of how these variables are related. This dependence between the critical number of patches, $\bar{R}$, σ^2, and spatial covariance is illustrated in figure 5.14. Note that although dealing with environmental stochasticity, this approach is completely deterministic. This is a difference in relation to truly stochastic metapopulation models as reviewed above, which predict a time to extinction.

From the previous approximation we can write an estimate of the minimum number of patches required for persistence. Remember that the isocline for persistence is given when $G(\bar{R}) = 1$. By setting equation

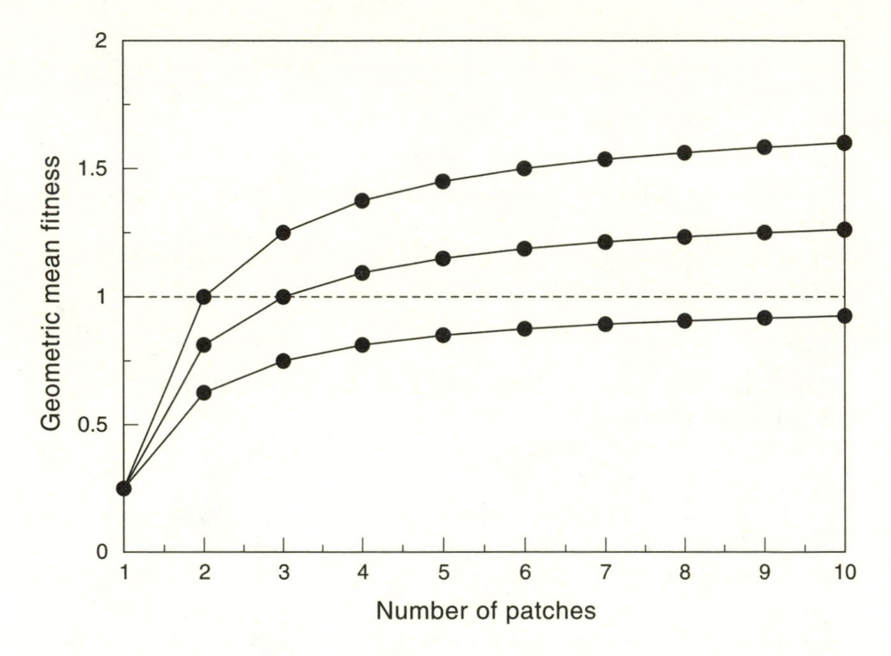

FIGURE 5.14. Dependence of the geometric mean fitness on the number of patches. As the number of patches increases, the sampling error in local growth rates decreases and so the GMF increases. Discontinuous line depicts the isocline separating extinction from persistence. $\bar{R} = 1.75$, $\sigma^2 = 3$, and spatial correlation is 0, 0.25, and 0.5 from top to bottom, respectively. Modified from Bascompte et al., 2002.

(5.61) equal to one, we can estimate the critical number of patches for persistence:

$$n_c = \frac{\sigma^2 - cov}{2\bar{R} - cov - 2}. \tag{5.62}$$

Imagine that the number of available patches is reduced due to habitat loss. The previous equation predicts how the extinction threshold changes with the level of environmental variability. As environmental fluctuations become stronger, the risk of metapopulation extinction becomes higher. Thus, we need a larger number of patches for metapopulation persistence.

To conclude, both demographic and environmental stochasticity can induce the extinction of the metapopulation before the deterministic extinction threshold is reached. This means that we should be more conservative than suggested by the deterministic approach of the previous sections. Habitat loss has different effects that are deeply interrelated. One one hand, the loss of habitat reduces the range of a metapopulation. On the other hand, habitat loss has an additional effect, the fragmentation of the remaining habitat. Additionally, reducing the amount of available habitat increases the potential role of demographic stochasticity, which superimposes on the deterministic threat by increasing the chances of random extinction. Continued work on integrating these different effects of habitat loss can provide new insight. Until now, however, we have learned that the effects of habitat loss can be more complex than expected. Nonlinearities seem to be common.

CHAPTER SIX

Complex Ecosystems

From Species to Networks

STABILITY AND COMPLEXITY

The analysis of model ecosystems including a small number of species and controlled experiments involving microecologies shows that low-diversity systems can be stable. One could imagine a simple biosphere inhabited by a few interacting species (such as an autotroph and an heterotroph) or even a planet fully covered by a layer of bacteria. Bacterial life was indeed the only dominant life form over 3,000 million years before multicellular life entered the scene, but even bacterial ecosystems are highly diverse (Wilson, 1992). Every place on Earth, at all scales, is shared by many coexisting organisms. This baroque of nature, as it was called by Ramon Margalef, is of course the result of the inevitable source of variation imposed by speciation events, but this is only one component of the whole picture.

In spite of the overwhelming complexity displayed by rich ecosystems, they also exhibit some well-defined regularities indicating universal laws of organization. When an empty field starts to be colonized by immigrant species (in a process called ecological succession), a new community gets formed and a pattern of species replacement develops. The transition from abandoned field to mature forest is one of the best-known examples of ecological sucession and is common in many places after the abandonment of agricultural land. In temperate climates, a mature forest is the end point of succession, in spite of the different potential beginnings (in a broad sense, the forest is here the final attractor of community development). Oportunistic species are the first to develop, sometimes modifying the landscape (such as some grasses invading sand dunes) and facilitating the establishment of further invasors. Eventually most if not all these species are replaced by longer-lived species.

One can get a glimpse of some universal principles of community organization by looking at the so-called rank-abundance diagram (May, 1975; Pielou, 1969). Here the relative abundance $N(r)$ of different species (in a set of S species) is plotted against their rank $r = 1, 2, \dots, S$ (i.e., ordered from the most frequent to the rarest). In figure 6.1 the time evolution of this diagram is shown for an old-field vegetation plot

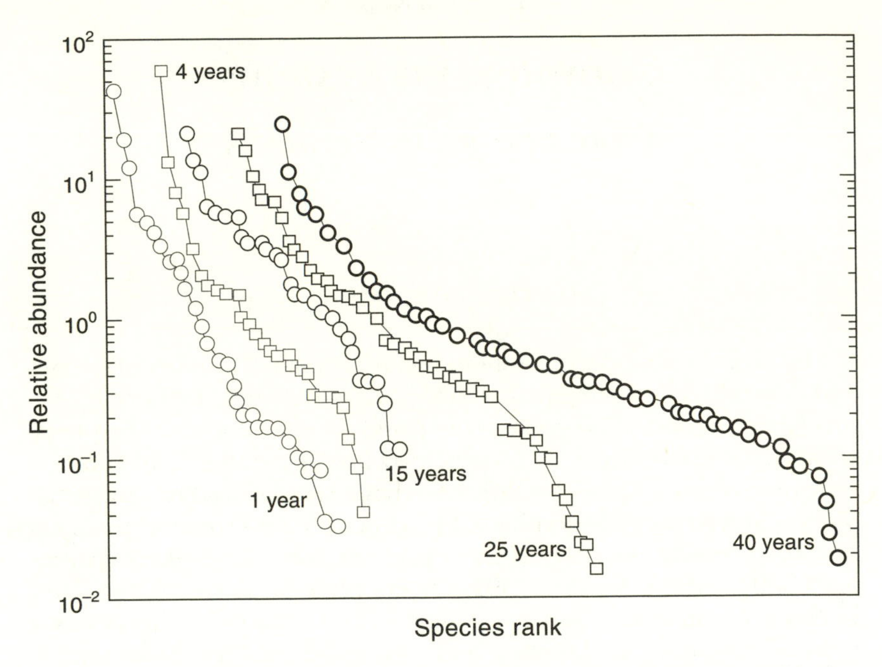

FIGURE 6.1. Species-rank distributions. Evolution of $N(r)$ for an old field plot, followed over a period of forty years (redrawn after Bazzaz, 1975).

(Bazzaz, 1975). Along forty years of fieldwork, different species of herbs, shrubs, and eventually trees entered into the study area. As the system becomes filled (how many species it can support is a key question analyzed in this chapter), a characteristic decay in $N(r)$ is observed (Pielou, 1969). These distributions tend to be long-tailed, with a few common species and a plethora of rare ones. Besides, the rank-abundance distribution observed in mature forests shows steeper decay in temperate forests and slower decay in rain forests (Hubbell, 1979, 1997, 2001).

If one of these systems is perturbed (by removing one species, for example), the distribution might be strongly modified, but in general, unless the available habitat has been destroyed too, a new community emerges and the same final distribution is recovered. An example of such a recovery pattern is the natural disappearance of a top predator in the Gulf of California (Dungan et al., 1982). This species (the large starfish *Heliaster kubiniji*), which preyed on sessile organisms, was decimated and almost disappeared throughout its geographic range in 1978. The rank-species plot changed through time after this event (Paine, 1966). Following the decline of *Heliaster*, many further events took place, such as a rapid increase in barnacles followed by their subsequent fall (as they became a common prey to other predators). In spite of the fact that the starfish remained absent, the distribution of abundances among species returned to the original values.

The previous example and others discussed below are related to another strong regularity: the presence of so-called keystone species. As this name implies, their removal can cause important effects on large parts of the community. When keystone species are returned to the system, it can sometimes return to its previous state, although the sequence of events is usually far from a simple reversal of the previous cascade of changes. Wilson cites the example of the human-driven depletion of the sea otters *Enhydra lustris* whose major prey were sea urchins (Wilson, 1992; Estes and Palmisano, 1974). The waning of this species was followed by the waxing of sea urchins, whose populations exploded and devastated large areas occupied by kelp and other seaweeds, which constituted a rich marine forest. The pressure exerted by sea urchins led to a very species-poor community. Some appropriate measures of conservation helped the reintroduction and waxing of otters and the waning of urchins. The original ecology eventually recovered, although such recovery has been only a transient phenomenon in some localities (Estes et al., 1998).

Keystone species (as we will discuss later) often have a high number of links with other species. At the other extreme of the spectrum, many species are specialized in such a way that they link to a few other species or even just one. This is the case, for example, of most parasitoids (Hawkins, 1994). They attack their single prey species (hosts in this case) and many species of them can have a common prey. In a similar way specialists often result from coevolutionary responses. A large body of literature has been mounting for decades concerning the mutual selection of adaptive responses between pairs of species. These include, for example, predator-prey or host-pathogen pairs (see chapters 2 and 3) or mutualism. The latter provides one of the best examples of coevolutionary dynamics, of which the mutually beneficial relation between flowering plants and their pollinators is one of the best studied. Here we would like to stress the fact that such coevolving pairs are commonplace in natural ecosystems providing a source of diversity associated to a small number of ecological links. As discussed below, the analysis of ecological food webs reveals that species with few links are common, but species with many links, although rare, are always present. This seems to occur under well-defined patterns of network organization.

A detailed description of a complex ecosystem is an impossible goal: Each species by itself involves too many variables to deal with. But as we have already discussed in other contexts, at some levels of description, statistical models of multispecies communities are able to explain the origins of ecological complexity and how it changes through time. In this chapter we explore the patterns emerging from network dynamics and topology. As discusssed below from different perspectives, ecological networks emerging in very different contexts share a number of (sometimes surprising) universal properties that can be explained by means of simple mathematical models.

N-SPECIES LOTKA-VOLTERRA MODELS

Mathematical models with three species are the simplest approximation to multispecies communities. They allow the introduction of several types of interaction patterns (such as omnivory) or explore the role of indirect effects. The simplest generalization is an ecosystem described through Lotka-Volterra (LV) equations, that is, a set of S differential equations:

$$\frac{dN_i}{dt} = \Phi_i(N_1, \ldots, N_S) \tag{6.1}$$

where $\{N_i\}$; $(i = 1, \ldots, S)$ is the population size of each species and the function $\Phi(N_1, \ldots, N_S)$ is defined as:

$$\Phi_i(N_1, \ldots, N_S) = N_i\left(\epsilon_i - \sum_{j=1}^{S} \alpha_{ij} N_j(t)\right) \tag{6.2}$$

where ϵ_i and α_{ij} are constants that introduce feedback loops and interactions among different species. The previous system will have several equilibrium points, $N_k^* = (N_{1k}^*, \ldots, N_{Sk}^*)$, obtained from:

$$\left(\epsilon_i - \sum_{j=1}^{S} \alpha_{ij} N_{jk}^*(t)\right) = 0. \tag{6.3}$$

The stability of these points (i.e., the stability close to equilibrium) is obtained from the Jacobi matrix

$$\mathbf{A} = (\alpha_{ij}) = \left(\frac{\partial \Phi_i}{\partial N_j}\right)_{N^*}, \tag{6.4}$$

which is assumed to have a connectivity C, defined as the (normalized) fraction of nonzero elements of $\mathbf{A}$.

By considering a small perturbation from the equilibrium point, $\mathbf{N}(t) = \mathbf{N}^* + \mathbf{n}(t)$, the time evolution of the small perturbation $\mathbf{n} = (n_1, \ldots, n_S)$ will be provided by the linearized system:

$$\frac{dn_i}{dt} = \sum_{j=1}^{S} \left(\frac{\partial \Phi_i}{\partial N_j}\right)_{N^*} n_j(t). \tag{6.5}$$

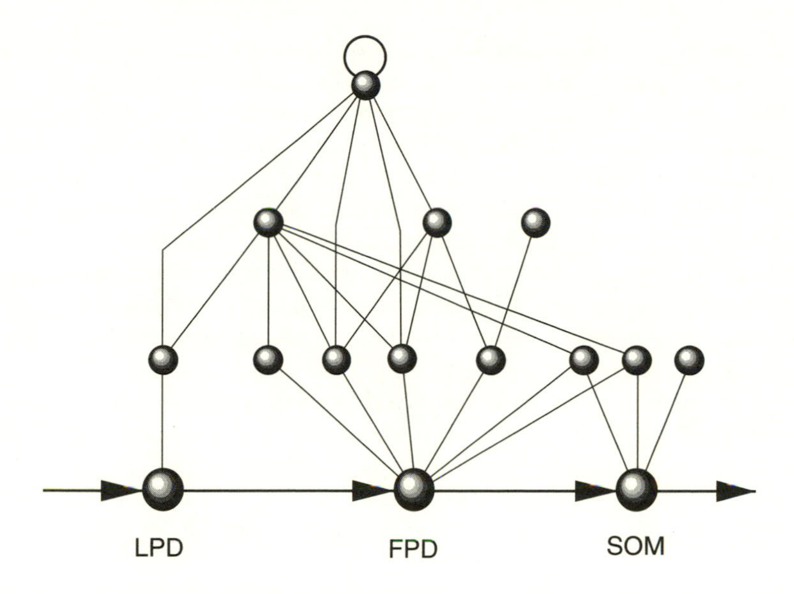

FIGURE 6.2. Complexity in food webs occurs at all scales. Here we show a schematic diagram of a water-filled tree-hole food web from southwest England (redrawn from Kitching, 2000). The bottom row corresponds to three connected components of the resources. Here LPD: large particle detritus; FPD: fine particle detritus; and SOM: suspended organic matter. Three levels of interacting species (including algae, insects, or mites) are observed.

The time evolution is thus given by a superposition (Strogatz, 1994)

$$n(t) = \sum_{j=1}^{S} \xi_j e^{\lambda_j t} \tag{6.6}$$

where ξ_j are constant (eigen)vectors and $\lambda_j (j = 1, \dots, S)$ are the eigenvalues of the Jacobi matrix (May, 1973). The longtime behavior of the perturbation will be dominated by the largest eigenvalue $\lambda_m = max\{\lambda_i; i = 1, \dots, S\}$. If $\lambda_m > 0$ then the system is unstable (in the linear stability sense; see chapter 2).

The underlying food web can be complex at all scales both in space and time. An example of these graphs is shown in figure 6.2. This diagram displays the web of interacting species in a water-filled tree-hole ecosystem (Kitching, 2000). These worlds within worlds are actually miniature ponds that can include many different groups of organisms, from bacteria to vertebrates. The tree-hole defines a microcosm where complex interactions are established on top of a given number of resources (mainly detritus coming fom leaf decay). When moving to larger scales, common patterns are observed thus indicating a nested hierarchy of structures (such as in fractal systems; see chapter 4).

Even for just three species, complex dynamics can be observed, as discussed in chapter 2. It can happen that the final species composition after a given transient is highly unpredictable. It can even be virtually impossible to predict the outcome of multispecies competition due to the intermingled fractal geometry of the basins of attraction. This seems to be the case for some models of phytoplankton ecology with multiple resources (Huisman and Weissing, 2001) and indicates that the nonlinearities present in these models makes forecasts of final species composition virtually impossible. But instead of studying the features of the attractors displayed by multidimensional dynamical systems (where the possible bifurcation patterns are likely to be extraordinarily rich and complex), in this section the global features of model ecosystems at equilibrium will be analyzed. In particular, let us consider the problem of how stability and complexity are related.

Stability and Complexity

A number of strong regularities have been reported from the analysis of biological networks (Strogatz, 2001), and several theoretical models have been proposed to explain them. One particularly influential work has been the theoretical analysis by Kauffman of several types of biological networks (Kauffman, 1993). Although the study of real biological nets has shown that they are not random, Kauffman's work revealed the importance of network-level properties beyond the specific features displayed by its component parts. This has obvious consequences for our understanding of evolutionary dynamics: Selective pressures might be operating not only at the level of species, but also at larger scales of ecosystem organization (Solé et al., 1996a; Solé et al., 1999). If true, then other principles beyond natural selection should be taken into account in long-term ecosystem evolution.

Stability seems to be associated with diversity, but the exact nature of such association has been a matter of debate for decades. Such debate involves many different aspects of ecological complexity, including structure, robustness, species richness, and fragility (MacArthur, 1955; Elton, 1958; Margalef, 1968, 1969; McCann, 2000). One important null model was presented by Robert May concerning the relation between stability and complexity in natural communities (May, 1972). In short, using Lotka-Volterra equations with random links among species, May analyzed the probability $P(S, \alpha, C)$ for an S-species model with connectivity C being stable. Here $C \in [0, 1]$ is defined as the fraction of nonzero elements in the community matrix A. Here stability is defined in terms of linear stability (as previously discussed): A given, randomly wired model ecosystem will be considered unstable if $\lambda_m > 0$. Using previous results from Wigner's semicircle law (Wigner, 1959), the May-Wigner theorem establishes that

$$\begin{aligned} &\text{if } \alpha^2 SC < 1 \text{ then } P(S,\alpha,C) \to 1 \text{ as } S \to \infty, \\ &\text{if } \alpha^2 SC > 1 \text{ then } P(S,\alpha,C) \to 0 \text{ as } S \to \infty. \end{aligned} \tag{6.7}$$

This result has a clear implication for the predicted (S, C) combinations to be observed in real ecologies: Those pairs such that they correspond to stable communities (i.e., those such that $\alpha^2 SC < 1$) should be linked to stable, real ecologies. The two "phases" displayed by the model are shown in figure 6.3a. The critical line $S \sim C^{-1}$ separates the two regimes. If the stability condition implicit in this approximation were at work in real ecologies, the upper part of the diagram would be empty and the lower part scattered by different combinations of (S, C) pairs. In figure 6.3b we also show a numerical calculation of the probability of having a stable system (our order parameter). Two different system sizes are considered, and α^2 has been fixed so that the transition occurs at $C = 0.5$. An interesting observation, to be discussed later in this chapter, is that real ecologies are actually scattered close to the critical transition. Although it is not easy to experimentally validate these predictions, some studies on microcosm communities seem to strongly support them (fig. 6.3c). Using ecosystems formed by bacteria, protists, and small metazoans, it has been shown that the probability of stable, feasible equilibrium declined sharply with complexity, as predicted (Fox and McGrady-Steed, 2002).

In spite of the oversimplified assumptions made by May's analysis (Cohen and Newman, 1984, 1985), further research revealed that the basic critical relation, described in terms of a hyperbolic law $S \sim C^{-1}$, appears to hold in more realistic models (Pimm, 1982). It should be noted that the stability criterion is based on linear stability grounds, and thus it labels as unstable those dynamical patterns involving oscillations or chaos. In this sense it is limited if we take into account the observation of high fluctuations exhibited by most field populations. However, this relation seems to be at work in real ecologies, although it deviates from the strict hyperbolic law and takes the form $S \sim C^{-1+\epsilon}$, with $0 < \epsilon < 1/2$. The underlying reason for the general validity of this result is that increasing connectivity provides increasing chances of positive feedback loops that introduce instabilities, no matter under which approach we define them.

The previous criterion can be derived using a simple approximation. We will follow the proof by Hastings (1982), who uses a discrete linear

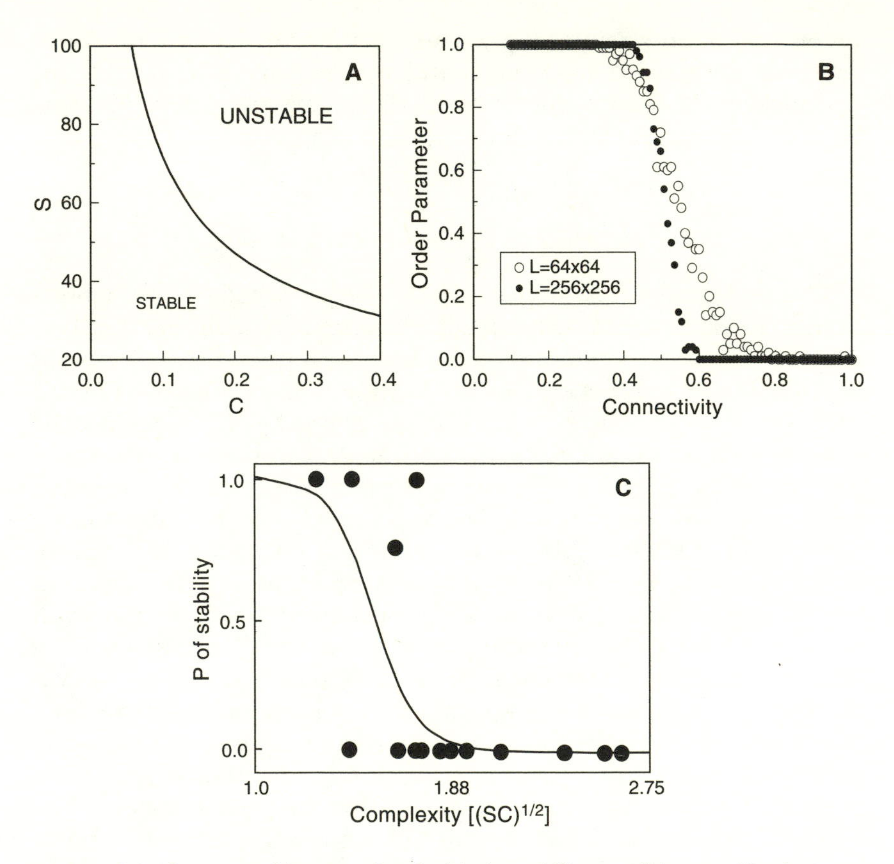

FIGURE 6.3. Phase transition associated with the stability condition used in the May-Wigner theorem. (A) Phase diagram, indicating the stable and unstable phases. (B) A numerical example of the phase transition phenomenon occurring close to the critical boundary. As the connectivity C increases, the probability shifts abruptly from one to zero close to $C = 0.5$. Two different ecosystem sizes ($S = 64$ and $S = 256$) have been used. For comparison, we also display the probability of having a stable (and feasible) web against complexity from experiments with microcosm communities (redrawn from Fox and McGrady-Steed, 2002).

model for the growth of perturbations $n_i(t+1) = \sum_j \alpha_{ij} n_j(t)$. Suppose that the perturbation from equilibrium abundances of our species verifies the constraint

$$\mathbf{n}^2 \equiv n_1^2 + n_2^2 + \ldots + n_S^2 = 1. \tag{6.8}$$

Considering that the entries of the $S \times S$ matrix are independent, and identically distributed stochastic variables, the average $\langle (A\mathbf{n})^2 \rangle$ can be written in this case as

$$\langle (A\mathbf{n})^2 \rangle = \langle (A\mathbf{n}_1)^2 + (A\mathbf{n}_2)^2 + \ldots + (A\mathbf{n}_S)^2 \rangle = \sum_{i=1}^{S} \langle (A\mathbf{n}_i)^2 \rangle \tag{6.9}$$

where $\mathbf{n}_i$ is a column vector whose components are $\delta_{ij} n_i$ with $j = 1, 2, \ldots, S$. Then each term can be obtained as

$$\langle (A\mathbf{n}_i)^2 \rangle = \left\langle \sum_{j=1}^{S} (A_{ji} n_i)^2 \right\rangle. \tag{6.10}$$

Now, remembering again that the matrix elements are independent, the crossed terms can be ignored and then

$$\langle (A\mathbf{n}_i)^2 \rangle = \left\langle \left(\sum_{j=1}^{S} A_{ji} n_i \right)^2 \right\rangle. \tag{6.11}$$

For large S, the last expression stands for the variance of a sum of SC independent stochastic variables, each following a distribution with zero mean and equal variance $(n_i)^2 \alpha^2$, then

$$\langle (A\mathbf{n}_i)^2 \rangle = SC\alpha^2 n_i^2. \tag{6.12}$$

Thus, using the constraint (6.8), the average $\langle (Ax)^2 \rangle$ can be written as

$$\langle (A\mathbf{n})^2 \rangle = \alpha^2 SC, \tag{6.13}$$

and we can see that if $\alpha^2 SC < 1$ then $\langle (A\mathbf{n})^2 \rangle < 1 (= \mathbf{n}^2)$, thus the system will be stable (for a detailed general analysis, see Hastings, 1982).

TOPOLOGICAL AND DYNAMIC CONSTRAINTS

One obvious limitation of random network models is the lack of a trophic structure. Ecosystems can be understood in terms of a layered pattern of species interactions in which a basal and a top level can be easily identified, together with one or more intermediate levels. Such a hierarchical component introduces important biases that have to be incorporated into any reasonable theory. As an example, an extension of May's linear stability analysis using structured models shows that larger stability domains can be reached (Hogg et al., 1989).

A different way of exploring the importance of network topology is to build small networks with a given, predefined arrangement of species interactions. This was done in particular by Pimm and Lawton, using LV models to assess the patterns of return times to equilibrium (Pimm and Lawton, 1977). Although this approach has some limitations (only four species are used, and the combined nature of the possible, larger

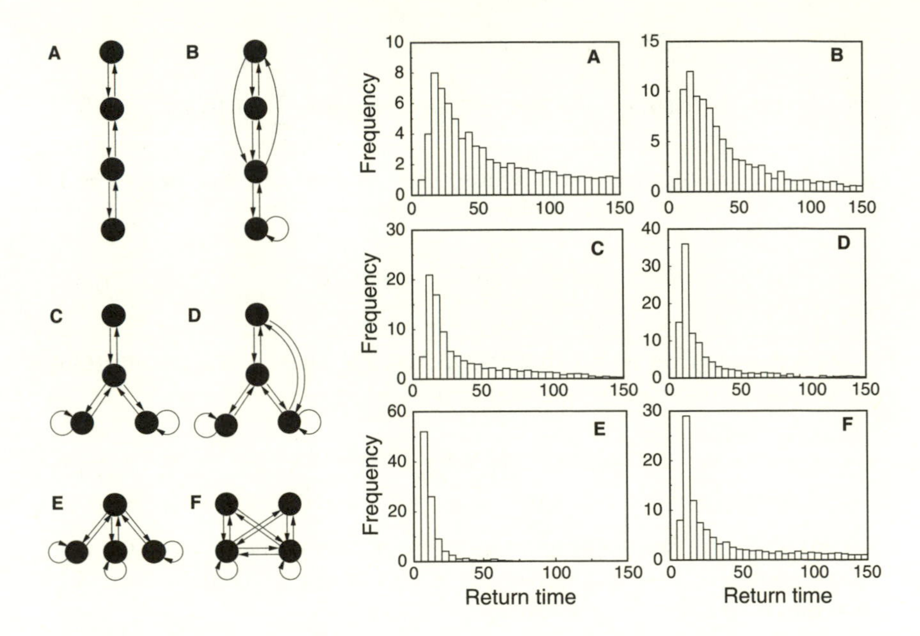

FIGURE 6.4. Left: (A–F) Six food web topologies considered in Pimm and Lawton's analysis of web stability and resilience. Right: Distributions of return times for the topological arrangements shown at left. Typically, longer food chains have greater frequencies of long return times.

systems, is not considered), it offers a good understanding of how large differences in ecosystem dynamics can emerge from basic topological constraints.

Under the previous general mathematical (LV) model defined for S interacting species, Pimm and Lawton considered six basic topological arrangements, shown in figure 6.4a–f. We can see that the four species are arranged in different ways from a four-level (a–b) linear food web to shorter (two-level) food chains (e–f). Using the Jacobi matrix A, they defined the return times to equilibrium after a disturbance as:

$$\tau_r = -\frac{1}{Re(\lambda_m)}, \tag{6.14}$$

where $Re(\lambda_m)$ is the real part of the largest eigenvalue, thus assuming that the system has a stable equilibrium point. The most interesting point raised by this study was the possible origin of the relatively short food webs observed in natural communities. Is the number of levels (N_L) in a food web simply proportional to productivity? If the length of the food chain were only dependent on the amount of energy flowing through the system, then one should expect increasingly larger food webs in systems with higher productivity. Different sources of evidence indicate, however, that the potential increase in N_L with productivity is counterbalanced by increasing (dynamic) instability (Post, 2002).

Experimental evidence for such conflict between productivity and stability comes mainly from small webs. An example are tree-hole food webs (Pimm and Kichting, 1987; Lawler, 1993) and involve up to four trophic levels: (1) detritus, (2) larval mosquitoes and chironomid midges; (3) larvae of a predatory midge; and (4) tadpoles. By using different amounts of litter (thus increasing or reducing productivity) it was shown that minimum levels of basal production are required to sustain coexistence among species but also that large levels of productivity lead to declines in the abundance of species at higher levels.

Higher nutrient levels have a destabilizing effect in both experimental (Morin, 1999) and model systems (Solé and Valls, 1992). Microcosm experiments clearly show that longer food webs are less stable (Lawler and Morin, 1993). Using model ecosystems, the differences are clear. In figure 6.4 (right) we can see the results of the numerical experiments by Pimm and Lawton. Here 2×10^3 webs are generated for each topology. The patterns of return times vary considerably depending on the topology of connections. In (a–c), the distribution of return times for webs without omnivores are shown, with a clear tendency toward increasing instability (i.e., larger return times) with larger N_L. These results seem to be in agreement with observations from communities of protists (Pimm and Kichting, 1987). On the other hand, omnivory can have a stabilizing effect (see d–f) at least for the simplest long linear chain but can also have a destabilizing effect when omnivores both consume and compete (an observation further confirmed by microcosm experiments; see Lawler, 1993). The role of omnivory in promoting stable dynamics in small food webs has been recently re-analyzed under the light of better data sets. Originally thought to be an exception, omnivory is now known to be a rule in ecological food webs (Morin, 2000). McCann and Hastings explored the stabilizing effects of omnivory using a simple three-species food chain (McCann and Hastings, 1997) showing that, typically, a bifurcation scenario from chaos to stable points is promoted by increasing levels of omnivory.

One conclusion seems clear from model and experience (at least when dealing with small systems): The number of trophic levels in ecological communities is only partially determined by energetic constraints, dynamical constraints being at least equally (if not more) important. This conclusion is supported by comparative studies of terrestrial and aquatic ecosystems (Hairston and Hairston, 1993). Measurements of the efficiency of energy transfer between trophic levels are consistent with the hypothesis that the fraction of energy consumed at each trophic level (and thus the efficiency of the process) is determined by competitive and predator-prey interactions. In other words, ecological efficiencies would not determine trophic structure; they would be a consequence of trophic interactions. In particular, it has been suggested that omnivory and intraguild predation truncate terrestrial food chains to three functional trophic levels. Due to the reduced abundance of herbivores under predators, plants would have more opportunities. Instead, aquatic ecosystems, under physical constraints derived from the medium, would restrict omnivory and intraguild predation, and four trophic levels would be observable. Although further research is needed, the evidence is compelling and might have important consequences in the long-term evolution of communities on land and sea, leading to differential responses during mass extinction events.

Let us conclude this section by mentioning the work of de Ruiter and coworkers on the importance of the patterning of interaction strengths on ecosystem stability (de Ruiter et al., 1995; see also Neutel et al., 2002, and Bascompte et al., 2005). These authors considered soil ecosystems (both native and cultivated), and seven food webs were studied. Here the links in the web were defined in terms of the energy flow through the system. From the food-web energetics, estimations of trophic interaction strengths were derived, following previous methods (May 1972; see also Hunt et al., 1987) and assuming a Lotka-Volterra dynamic, the impact on stability was estimated from the Jacobi matrix. The most relevant result of this study was the observation that the patterning of interactions (more than the specific values of energy flow at a given link) is extremely important. In this context, as other authors pointed out early on (Paine, 1988, 1992; Polis, 1994, 1998), interactions representing a relatively small rate of material flow can have a large impact on stability whereas interactions representing a large rate of material flow can have a small impact. These results thus support the view of complex ecologies as dynamical systems with higher-order properties arising to some extent from their intrinsic network structure.

INDIRECT EFFECTS

The intrinsic complexity of ecological food webs has an important consequence that illustrates perfectly well the relevance of the global view of ecosystems over a species-level analysis. It has to do with the fact that, together with direct effects among interacting species, there is another type of effect wherein two species affect each other through some intermediate species. This was already reported by Darwin in the *Origin of Species* (1859). He observed that bumblebees, which contribute to the pollination of the red clover (*Trifolium pratense*), had some

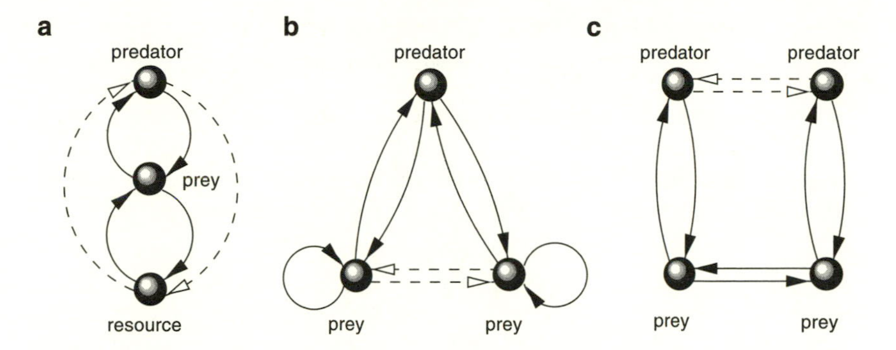

FIGURE 6.5. Indirect effects in food webs. Direct and indirect effects are indicated as solid and dashed lines, respectively. (a) An example of the simplest three-trophic-level model: the top and bottom levels interact indirectly in a positive way (see text). In (b) two preys common to a given predator led to a negative indirect effect. (c) An example of indirect mutualism, where the two consumers (predators) have a mutual (indirect) positive effect through their interaction with two competing species.

population dependence on field mice (which destroy their combs). But since the number of field mice depends on the abundance of cats (and other predators), Darwin concluded that "hence it is quite credible that the presence of feline animals in large numbers in a district might determine, through the intervention first of mice and then of bees, the frequency of certain flowers in that district."

Indirect effects propagate from one species to another via density fluctuations in one or more intermediary species in the food web and can lead to highly counterintuitive results (Yodzis, 1988; Vandermeer, 1990). The presence of such effects is easily seen by looking at the sign of direct interactions and then deducing the sign of higher-order interactions. This approach is generally known as qualitative stability and provides a powerful shortcut method when large communities are analyzed. Here the community matrix **A** is replaced by a new matrix **Q** (where $Q_{ij} \in \{+, -, 0\}$) in which only the sign of the links is indicated. Consider the simplest case, illustrated in figure 6.5a, of a three-trophic level system (Case, 2000). Here the qualitative matrix **Q** will be:

$$\mathbf{Q} = \begin{pmatrix} - & - & 0 \\ + & 0 & - \\ 0 & + & 0 \end{pmatrix}. \tag{6.15}$$

Since a chain connecting three species (in which the first and last are not directly connected) leads to an indirect influence through the intermediate species, the sign of the intermediate connections can be used to know the sign of the indirect effect. For this example, the indirect effect of species 1 on 3 (through species 2) is given by:

$$Q_{12}Q_{32} = (+)(+) = (+) \tag{6.16}$$

that is, species 1 indirectly benefits species 3 through its positive effect on species 2.

Another example is given by two prey that share the same predator although do not interact directly (fig. 6.5b). Now the matrix will be:

$$\mathbf{Q} = \begin{pmatrix} - & 0 & - \\ 0 & - & - \\ + & + & 0 \end{pmatrix} \tag{6.17}$$

and we can see that the two prey have a negative indirect effect through their common predator. This can be easily calulated from:

$$Q_{31}Q_{23} = (+)(-) = (-). \tag{6.18}$$

Thus, in spite of the fact that the two prey have no direct link, they effectively compete indirectly. This is called apparent competition (Holt, 1977). A third example is given in figure 6.5c and involves indirect mutualism. Here two predators prey on two strong competitors (the two prey species). If each predator is highly dependent on a different type of prey and competitive exclusion occurs in the absence of exploitation, then the two predators are effectively mutualists (see Menge, 1995, for other types of indirect interactions).

The previous examples and a large amount of field data show that, when put in the context of a community, interactions that would be of a competitive nature can lead, through indirect effects, to a mutualistic association (Stone and Roberts, 1991). Such a context-dependent scenario introduces an element of paradox in the classical theory of competitive equilibrium.

An obvious question emerges: How relevant are higher-order interactions in a complex community composed of many species? The question has many implications. One is the fact that if indirect effects are important, then a reductionist view of ecosystems in terms of basically separable species is simply useless. The second is that in such a network-dependent scenario no theory of long-term evolution at the species level can be extrapolated to larger, ecosystem scales. Evidence for the importance of indirect effects has been accumulating from field and microcosm studies (see Morin, 1999, and references therein). Actually, it might be the case that the effects of one species on another might be *more* important than direct, pairwise interactions (Yodzis, 1988).

A quantitative characterization of these effects can be obtained by using the inverse of the community matrix (Yodzis, 1988). Specifically, the total effect (i.e., both direct and indirect) of one species on another can be measured by determining how that species responds to additions of the other. For the general Lotka-Volterra model, we can consider the following modification:

$$\Phi_i(N_1, \ldots, N_S) = N_i \left(\epsilon_i - \sum_{j=1}^{S} \alpha_{ij} N_j(t) \right) + \delta_{ik} I_k \tag{6.19}$$

where $\delta_{ik} = 1$ for $i = k$ and zero otherwise. Here a continuous change I_k on species k is introduced. I_k is a very small perturbation (otherwise the linear stability approximation does not hold). This change can be positive or negative. As a consequence of this term, the equilibrium state N^* (which is assumed to exist here) will be modified. At the new equilibrium state $N^*(I_k)$ we have:

$$\epsilon_i - \sum_{j=1}^{S} \alpha_{ij} N_j^*(I_k) + \delta_{ik} I_k = 0. \tag{6.20}$$

Now we take the derivative of these equations with respect to I_k. This gives

$$\sum_{j=1}^{S} \alpha_{ij} \left(\frac{\partial N_j^*(I_k)}{\partial I_k} \right) + \delta_{ik} = 0. \tag{6.21}$$

By solving these equations at the equilibrium point N^* we get:

$$-E_k^T = A \partial_{I_k} [N^*(I_k)]^T \tag{6.22}$$

where

$$E_k = (0, \overset{k}{\ldots}, 1, \ldots, 0) \tag{6.23}$$

$$\partial_{I_k}[N^*(I_k)] \equiv \left(\frac{\partial N_1^*(I_k)}{\partial I_k}, \ldots, \frac{\partial N_S^*(I_k)}{\partial I_k} \right). \tag{6.24}$$

If matrix A is such that $\det |A| \neq 0$ (i.e., the inverse matrix exists), we have:

$$\frac{\partial N_i^*(I_k)}{\partial I_k} = -(\alpha_{ik})^{-1}. \tag{6.25}$$

These results are summarized as follows: The elements of the (sign-reversed) inverse of the community matrix give the total effect of one species on another. This includes direct and indirect terms.

A similar approach within multispecies communities of competitors was developped by Stone and Roberts (1991). These authors consider a set of S competitors using a model defined by a generalized LV system, and make use of a similar analysis but with changes in parameters instead of changes in population abundances as the source of perturbation. This example is especially interesting (in spite of the obvious limitations imposed by the type of community considered) because all direct interactions have a negative sign. Indirect effects, however, can also generate positive outcomes, which Stone and Roberts call advantageous in a community context or ACC. This term describes some general types of facilitation. Using randomly constructed models of competitive communities, and counting the fraction of negative elements in the inverse matrix, they found that a 20%–40% of interactions are expected to be ACC. In other words, a large number of positive interactions can emerge as a consequence of indirect effects in purely competitive assemblies.

This analysis can be extended to real food webs by using a number of assumptions (Yodzis, 1988). Starting with a given, real food web, one first assumes that the underlying dynamics can be described in terms of some given set of equations (such as those of the Lotka-Volterra type). The signs of the coefficients are known, but their strengths are seldom available except for a small number of webs. However, a number of well-known biological constraints help to provide rough estimate of their values. In this way, a community matrix can be constructed.

In figure 6.6a one of the webs studied by Yodzis is shown: the Narragansett Bay food web. The arrows indicate who eats whom, and the reverse links (which would have a negative sign) are not indicated. Now for each web, Yodzis generated 100 different matrices and computed their inverse. In this way one can actually obtain a different graph in which indirect effects and their sign are represented.[1] An instance of this for the Narangasset web is shown in figure 6.6b, where the largest effects of population changes in a given species on others are indicated by arrows. Here both direct and indirect effects were taken into account, using the inverse of the community matrix. Continuous and broken lines indicate positive and negative effects, respectively. One especially interesting result is that increases in some predators can lead to increases in the populations of some of their prey, as a consequence of indirect interactions. Another key result is that other realizations can result in very different graphs, and actually even the signs of some links can be reversed: These models reveal a strong sensitivity to parameters when looking at the effects that propagate through indirect pathways.

By averaging over many replicas, Yodzis defines the effect of a given species on another as directionally undetermined if less that 90% of the resultant elements $(\alpha_{ik})^{-1}$ had the same sign. In other words, this definition requires that a large percent of the links in the web maintain the same direction. What Yodzis found was actually surprising:

[1] Always under the constraint that we are perturbing equilibrium populations of one species and looking at the changes in the equilibrium densities of other species in the web.

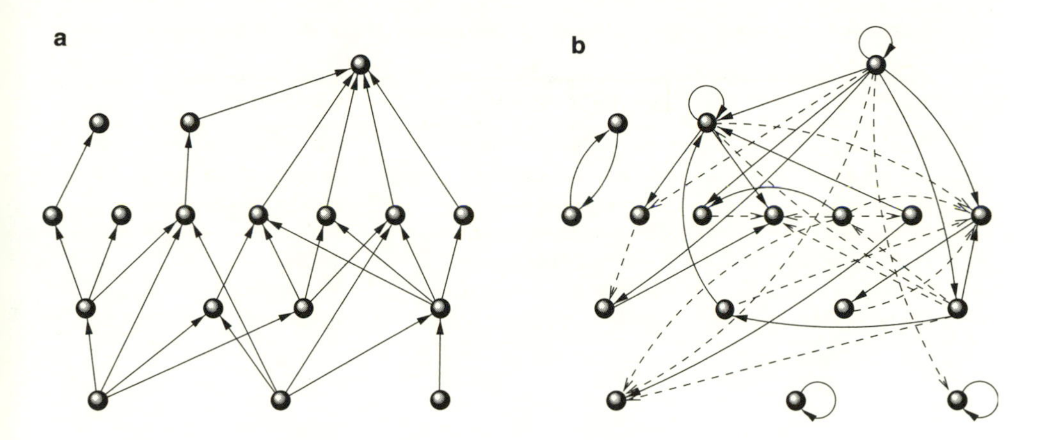

FIGURE 6.6. (a) Food web for Narragansett Bay. (b) Possible effects that small changes in equilibrium densities have on the equilibrium of other species (after Pimm, 1991). Here both direct and indirect effects are included. Continuous lines indicate positive effects (when increasing the abundance in one species promotes increases on another) and dotted lines stand for negative effects. A consequence of indirect effects is that increases in a given predator can result in increases in some of its prey.

Half of the effects derived from removing individuals of a given species lead to undetermined effects and in some cases where such effect was determined, the resulting effect is opposite of what we would expect (like prey populations decreasing after population decrease in one of its predators). He also found that more than 25% of competitors are actually (indirect) mutualists. The obvious consequence of Yodzis work is that, in general, it will be difficult to know if a given species will increase or decrease in population after the removal of another species in the web. There is indeed much evidence for indirect effects in removal experiments, although some experiments suggest that the outcome of the removal (at least the immediate one) might be more predictable (see Pimm, 1980). When interaction strengths are measurable, effects of disturbances can be predicted and related to some species-specific traits, such as connectivity, biomass, or relative abundance.

KEYSTONE SPECIES AND EVOLUTIONARY DYNAMICS

As discussed at the beginning of this chapter, a specially relevant feature of complex ecologies is the presence of particular species whose removal can cause large effects on the whole community. These are so-called keystone species (Paine, 1966; Bond, 1993), and their

TABLE 6.1. Keystone Species: Types, Including Mode of Action and Specific Examples

TYPE	MODE OF ACTION	EXAMPLE
Predators	Supress competitors	Otters, sea urchins
Herbivores	Supress competitors	Elephants, rabbits
Pathogens	Supress competitors	Myxomatosis virus
Mutualists	Effective reproduction	Pollinators, dispersors
Earth movers	Physical disturbance	Termites
System processors	Rates of nutrient transfer	Mycorrhiza
Social predators	Modify local succession	Army ants

identification as a common (perhaps inevitable) ingredient of ecosystem architecture has important implications for conservation issues. Indeed, their demise can lead, often through indirect effects, to the loss of many other species.

Keystone species can have different positions in the food web. Although top predators and basal species seem obvious candidates, intermediate species (particularly omnivores) are also as likely as others to be keystone species (see table 6.1). One significant problem, however, is the proper identification of possible candidate species in a given food web. Only the removal of a given node from the ecological graph gives a clear answer, and thus species deletion (either from a real or model ecosystem) can provide a direct, clear criterion. However, the effects might not be immediate but strongly delayed (as discussed in chapter 5 in relation to the extinction debt). Besides, in many cases the effects can propagate through the food web in a slow path. An example is the demise of jaguars and pumas in Central and South America (Terborgh, 1988; Wilson, 1992). Since these animals prey on a wide variety of small-sized species, their removal can cause further strong population changes. At the Barro Colorado Island, in Panama, their disappearance led to a tenfold increase in their prey populations, followed by an increased pressure on tree species with large seeds (on which they feed). This in turn lessens competition with other tree species with small seeds, eventually shifting forest composition in their favor. Quoting Wilson: "It seems inevitable that the animal species specialized to feed on them also prosper, the predators that attack these animals increase, the fungi and bacteria that parasitize the small-seed trees and associated animals spread, the microscopic animals growing on fungi and bacteria grow denser, the predators of these creatures increase and so the ecosystem reverberates from the removal of the keystone species" (Wilson, 1992).

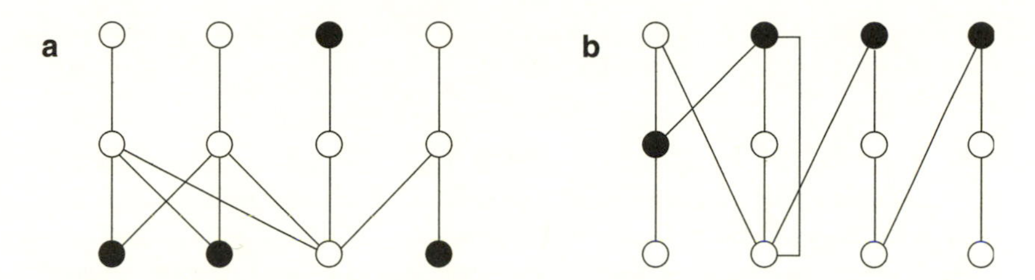

FIGURE 6.7. Examples of twelve-species three-trophic-level food webs with different topological arrangements. Here black and white circles indicate nodes whose removal leads to coextinction with a probability $P_c > 0.9$ and $P_c < 0.5$, respectively.

The effects of species removal, either by human-caused events or due to biotic and physical phenomena, is an essential ingredient of our understanding of community patterns. It ultimately defines in appropriate terms how relevant a given species is and its influence on ecosystem stability. The effects of species deletion are known to depend on the complexity of the food web and from which level they are removed (Pimm, 1991; see also Montoya et al., 2004, and references therein). How much can be learned from this?

Several theoretical studies have been developed in order to explore the effects of such deletion on food webs (Pimm, 1980). A simple approximation consists in using a Lotka-Volterra model with several trophic levels and different topological arrangements. Pimm considered a twelve-species model with three trophic levels. The model is defined as:

$$\frac{dx_i}{dt} = x_i \left(b_i + \sum_{j=1}^{S} \alpha_{ij} x_j \right) \tag{6.26}$$

with $i, j = 1, \ldots, S$. In order to properly define their place in different trophic levels, a number of simple assumptions must be made: For basal species, $b_i > 0$ and $\alpha_{ii} < 0$ (and $b_i < 0, \alpha_{ii} = 0$ otherwise). The signs of the $\alpha_{i \neq j}$ coefficients are defined depending on the specific interactions given by the web topology (see Pimm, 1979, 1980). By exploring different web topologies, one can identify the importance of each species in terms of the frequency of further species extinction after deletion.

The parameters in Pimm's study were generated at random, and those webs with $x_i > 0$ for all i (at the steady state) were used in the removal experiments. In figure 6.7a–b we plot two examples that illustrate the relevance of network topology. Here two types of nodes are indicated depending on their importance in triggering further extinction. Some nodes (black circles) have a very strong effect, leading to species loss with high probability (thus being candidates to keystone species) and some have smaller effects (white circles). This study revealed that:

1. Removal of basal species have typically small impact when polyphagous herbivores are present. A similar pattern was observable for the deletion of herbivores.
2. Extinction cascades due to the elimination of top species (predators) are more common when the species on which they feed are also polyphagous.

These observations show that the more complex the web, the more resistant to plant or herbivore species removals it will be. But such homeostasis seems counterbalanced by the effects associated to predator removal. These conflicting constraints are likely to be an important source of the complex responses associated to removal experiments. In this sense, predictions about which species will become extinct are expected to be difficult. This intrinsic unpredictability is enhanced by indirect effects (Yodzis, 1988) and reminds us once again of the importantance of network effects to the final fate of species survival. In this context, some studies indicate that the loss of particular species can be compensated by other similar species (Morgan Ernest and Brown, 2001; see also Lundberg et al., 2000) provided that they restore the energy flow by an appropriate use of available resources.

Emergence of Keystone Species

An important theoretical question is how common are keystone species. Are they expected to be present? It seems clear that all known, well-defined, and species-rich food webs have in common the presence of keystone species, indicating a possible general phenomenon shared by ecosystems in different conditions. One possible affirmative answer to the previous question comes from mathematical models within the context of autocatalysis in nonlinear networks (Jain and Krishna, 2001, 2002). These studies consider an abstract (but very general) model of species interactions. The model describes the concentrations of some chemical species x_i $(i = 1, \ldots, S)$ assuming normalization (i.e., $x_i \in [0, 1]$ and $\sum_i x_i = 1$). The time evolution is given by the replicator equations (Hofbauer and Sigmund, 1991)

$$\frac{dx_i}{dt} = \sum_{j=1}^{S} c_{ij} x_j - x_i \sum_{k,j=1}^{S} c_{kj} x_j \tag{6.27}$$

where $C = (C_{ij})$ is the matrix of interaction among species. This system evolves in time as follows. Starting from an initial configuration with S

nodes (species), each species is randomly wired to any other with some probability p. Let c_{ij} indicate the strength of the link between species i and j. The values of c_{ij} are randomly chosen from $[-1, +1]$ if $i \neq j$ and from $[-1, 0]$ otherwise. In other words, a link between two different species will be cooperative or inhibitory but self-connections are always negative. In this way, autocatalysis is not allowed and the persistence of a given species depends on cross-reaction with other species. The initial condition is obtained by first setting $x_i \in [0, 1]$ and then normalizing (such that $\sum_i x_i = 1$). The evolution of this net takes place in three steps:

1. Each population is allowed to evolve over a time interval T in such a way that x reaches a stationary state. At the end of this transient phase, we have $x \equiv \{x_i = x_i(T)\}$.
2. One of the least fit species is eliminated. Here fitness is defined just as the population size. One can use $P_i = x_i/[max(x_j)]$ as a probability of removal. All links c_{ij} are also removed.
3. A new node is added to the graph, together with its connections, which are set at random, as previously defined. The new population is rescaled again to $\sum_i x_i = 1$.

In spite of the simplicity of this model, it displays rather remarkable properties (see fig. 6.8). As time proceeds, the fraction of positive links becomes dominant, and cooperative structures thus develop. The complex ecology generated by the Jain-Krishna model is a consequence of spontaneous self-organization toward increasing cooperation. Together with this set, keystone species emerge and can be identified because their removal can have huge effects and even trigger large extinction events (fig. 6.8, bottom). In other words, network complexity can evolve toward highly organized, collectively stable structures but an ingredient of fragility (due to the presence of keystones) is always present. These two faces of complex ecosystems will be discussed below.

Keystone Species in Spatial Ecology

One important issue connecting the role of keystone species and spatial patterns is the likelihood of some of these species being particularly affected by habitat loss and fragmentation. This seems to be the case of many vertebrates (Wilson, 1992) and is well exemplified by army ants. Army ants (Holldobler and Wilson, 1990) have a huge impact on the ecological dynamics of many neotropical rain forests, and many species of vertebrates and invertebrates associated with them would face extinction if the army ants disappeared (Boswell et al., 1998). Their effects

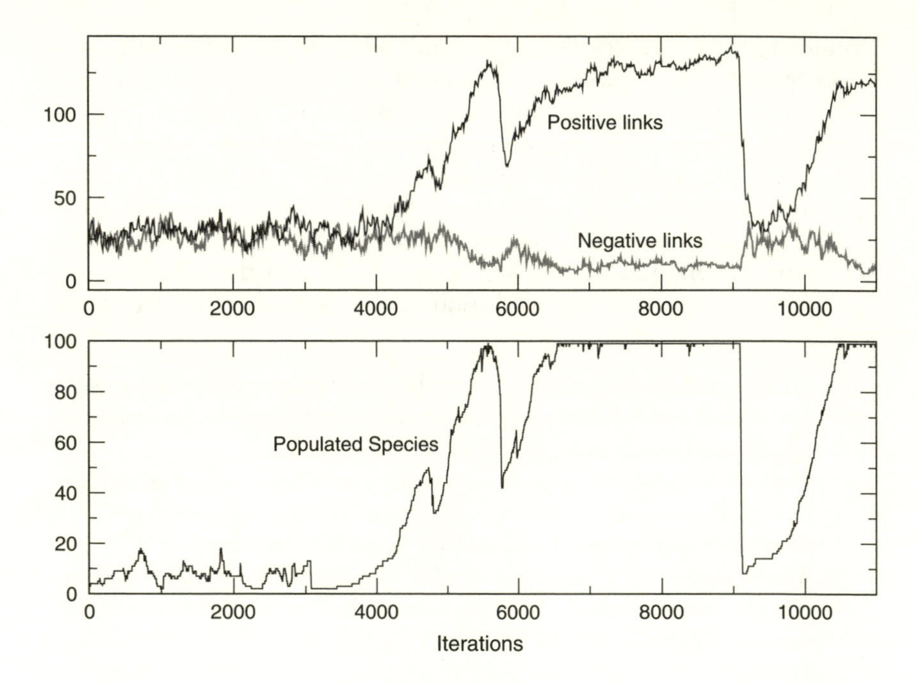

FIGURE 6.8. Top: Network dynamics in the Jain-Krishna model. As the autocatalytic set (ACS) becomes established, a rapid increase is observed in the number of positive links (which indicate the enhancement of cooperation) and the total population. Bottom: Rapid extinctions also punctuate the dynamics (a huge one occurs at $T = 900$) indicating the removal of a "keystone" species.

take place at different spatial and temporal scales. Their near-daily raids promote species diversity through creation of a spatial mosaic of habitat patches in different stages of ecological succession (Partridge et al., 1996). Besides, in Central and South America a large number of bird species are specialized in following the ant raids, feeding on insect prey that are flushed out from leaf litter by the raids (Willis and Oniki, 1978). The analysis of the effects of (random) habitat fragmentation, using realistic parameter values for the life cycle of army ant colonies, reveals that their populations would go extinct once percolation thresholds are reached (Boswell et al., 1998), thus confirming previous theoretical predictions (Bascompte and Solé, 1996; see also chapter 5). This opens the problem of what type of cascade of extinction might be triggered under these circumstances. In this context, current estimations of diversity loss based on species-area relations (Pimm and Raven, 2000) might actually be rather conservative. How food-web topology might alter these estimates is discussed in the next section.

COMPLEXITY AND FRAGILITY IN FOOD WEBS

One of the most successful approaches to the understanding of real food webs has to do with the graph patterns associated with the set of links between species. Specifically, and following the presentation in chapter 4, a food web can be described in terms of a graph $\Omega(V,E)$ consisting of a finite set V of vertices (species) and a finite set E of edges (i.e., feeding links) such that each edge e is associated with a pair of vertices v and w. The average degree will be indicated as $z = \bar{k}$.

In spite of the fact that ecological graphs are directed (i.e., links go from one species to another and the reverse connection will be typically different in strength and sign), most theoretical studies deal with nondirected graphs. In other words, two species appear connected if they share a common edge, irrespective of its particular properties. This approach has been rather successful, since many reported regularities from field data are fully recovered from nondirected graph arguments of a different nature (Cohen et al., 1990; Pimm, 1991; Warren, 1996; Williams and Martinez, 2000; see also Albert and Barabási, 2002). This observation is probably a consequence of fundamental laws constraining the basic topological arrangements of ecological networks that go beyond the specific rules of dynamical interaction.

The Niche Model

Although food webs are the result of population interactions and are thus nonstatic (Warren, 1989; Morin, 1999), several remarkable theoretical models lacking time evolution and explicit population structure have been able to match several relevant statistical features of real ecological networks (see Pimm et al., 1991). Beyond the random network, the so-called cascade model was particularly successful (Cohen et al., 1990). It assigned a random value $\eta \in [0, 1]$ to each species, and each one had a probability $P = 2CS/(S-1)$ of consuming only species j with $\eta_j < \eta_i$. This model provided a good estimate of species richness per trophic level. A powerful extension of this approach is the so-called niche model (Warren, 1996; Williams and Martinez, 2000; Camacho et al., 2002) where each species can consume all prey within one range of values whose randomly chosen center is less than the consumer's niche value. If we list the species as $s_1, s_2, \ldots, s_S$, the set of prey π_i of species S_i is

$$\pi_i \equiv \{s_j | \eta_j \in [r_i/2, \eta_j]\} \qquad (6.28)$$

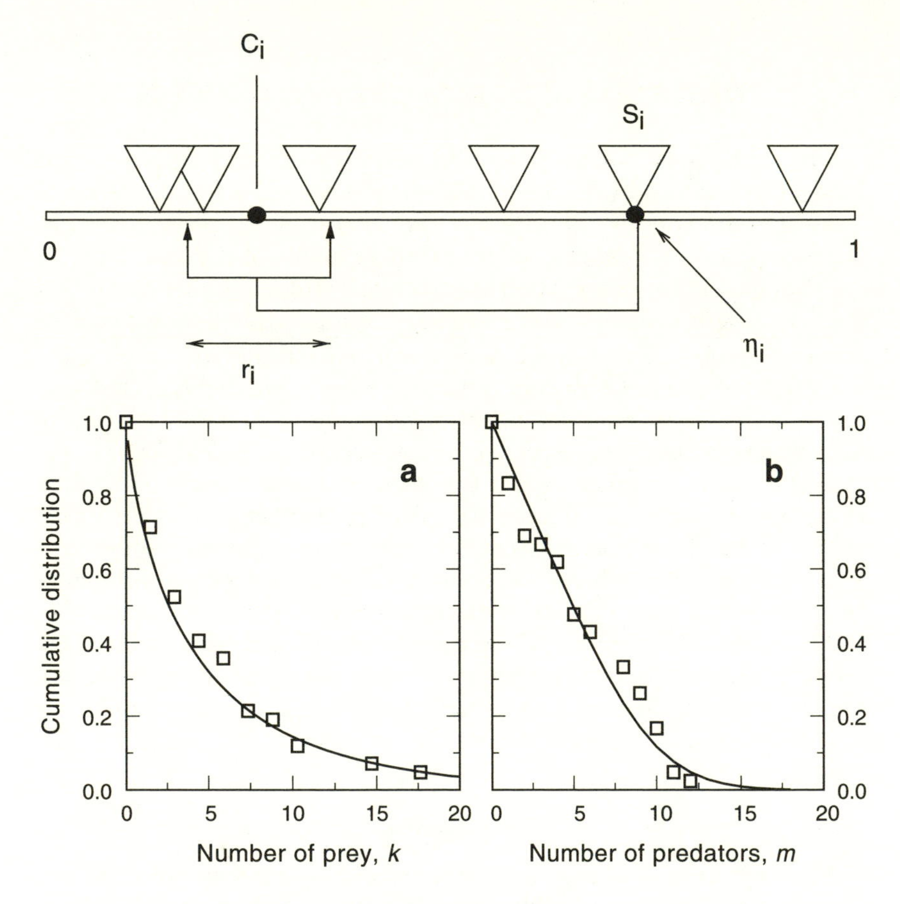

FIGURE 6.9. Top: Basic diagram for the (Williams-Martinez) niche model (see text). In (a–b) the cumulative distribution P_{prey} of number of prey k, and (b) P_{pred} of number of predators m for the St. Martin Island web are shown. The data agree well with the analytical predictions, indicated by the solid lines.

(see fig. 6.9, top). Through the previous rules it is clear that cannibalism occurs. Here $r_i = x\eta_i$ with $x \in [0, 1]$, a random beta-distributed variable, that is, with a density function

$$P_x(x) = \beta(1 - x)^{\beta - 1}. \tag{6.29}$$

Here β and S determine the average connectivity $\bar{k} = 2L/S$ (L is the total number of links). Numerical analysis of the WM model shows that several key quantities characteristic of empirical food webs (such as fractions of basal, top, and intermediate species; omnivory, mean chain length, or trophic similarity) are well reproduced (Williams and Martinez, 2000).

Analytical expressions for the distributions of prey for $S \gg 1$ and $L/S^2 \ll 1$ (i.e., large web sizes and sparse interaction matrices) can also be derived (Camacho et al., 2002). By using the (less noisy) cumulative distribution

$$P_{prey}(n) = \sum_{n' \geq n} p_{prey}(n') \tag{6.30}$$

it was shown that

$$P_{prey}(n) = \exp\left(-\frac{n}{2z}\right) - \frac{n}{2z} E_1\left(\frac{n}{2z}\right) \tag{6.31}$$

where

$$E_1(x) = \int_x^{\infty} t^{-1} \exp(-t) dt \tag{6.32}$$

is the exponential-integral function (Gradstheyn, 2000). The last functional form predicts an exponential decay in the number of prey for large n. Under similar conditions, the distribution of the number of predators of a given species, $P_{pred}(m)$, can also be derived (Camacho et al., 2002):

$$P_{pred}(m) = \frac{1}{2z} \sum_{m'=m}^{\infty} \gamma(m'+1, 2z) \tag{6.33}$$

where γ is the "incomplete gamma function" (Gradstheyn, 2000). For $m < z/2$ this function is approximately constant, while it decays with a Gaussian tail for $m \approx z$. An example of the good fit obtained from this approximation is shown in figure 6.9a–b for the St. Martin web. Good fits are also obtained for other food webs, although care must be taken in extending the conclusions of this study, since the webs chosen are actually not so sparse or highly species rich (with the exception of Little Rock) and are not highly resolved taxonomically (and parasites are typically not being considered). When dealing with sparse webs, such as those dominated by parasitoids and hyperparasitoids, scaling laws are observable (Solé and Montoya, 2001). Similarly, other no less relevant ecological webs, such as those defined among mutualists, display truncated power laws (see below).

A remarkable feature of the niche model is that it is a purely static approximation able to satisfactorily reproduce the most relevant quantitative patterns displayed by real food webs. Such remarkable agreement suggests that static approximations to food-web structure provide a solid framework for ecosystem analysis.

Small Worlds in Food Webs

As discussed in previous sections, food webs are not random. What type of network defines their architecture? The answer seems to be (Williams et al., 2002; Montoya and Solé, 2002) that food webs have a so-called small world structure (Watts and Strogatz, 1998).

Small-world (SW) networks lie somewhat between ordered (lattice-like) webs and Erdös-Renyi (Poissonian) graphs (see chapter 4). They involve rather remarkable properties that seem to be essential in understanding the organization of biological complexity (Strogatz, 2001). In order to properly define an SW web, let us consider two key properties of a graph Ω: the clustering coefficient and the average path length.

The clustering coefficient C measures the fraction of pairs of neighbors of a node that are also neighbors of each other. Let us indicate $s_i \in \Omega$ a given species ($i = 1, \ldots, S$). The set of nearest neighbors Γ_i will be obtained from $\Gamma_i = \{s_k \mid \xi_{ij} = 1\}$ now, defining

$$\mathcal{L}_i = \sum_{j=1}^{S} \xi_{ij} \left[\sum_{k \in \Gamma_i} \xi_{jk} \right]. \quad (6.34)$$

The clustering coefficient of the i-th species will be

$$c(i) = \frac{2\mathcal{L}_i}{k_i(k_i - 1)}, \quad (6.35)$$

where k_i is the connectivity of the i-th species. The clustering coefficient is defined as the average of $c(i)$ over all species, $C = \frac{1}{S} \sum_{i=1}^{S} c(i)$.

The average path length L is defined as follows: Given two species s_i, s_j, let $L_{min}(i,j)$ be the minimum path length connecting these two species in Ω. The average path length L will be:

$$L = \frac{2}{S(S-1)} \sum_{i=1}^{S} \sum_{j=1}^{S} L_{min}(i,j). \quad (6.36)$$

Two extreme situations can be considered in our context: random graphs with Poisson distribution and hierarchical graphs such as the Bethe lattice (chapter 4). It is interesting to see that in fact most static graph models of food webs involve features ranging from purely random nets (constant connectivity networks; see Martinez, 1991) to purely hierarchical models (the cascade model; see Cohen et al., 1990). Under the second class, a trophic species can only prey on a trophic species of lesser rank leading to a hierarchical structure sharing some features with some standard graph patterns such as Cayley trees (Watts, 2000). Random graphs will have very small clustering, and short path

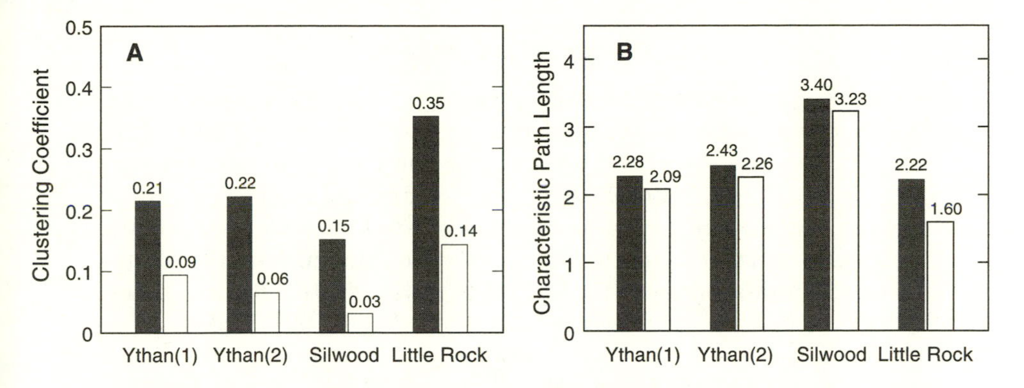

FIGURE 6.10. (A) Clustering coefficient for four food webs (see text). Here the dark bars are real webs and the white bars correspond to randomly generated webs with the same average number of links per species (averaged over 200 generated samples). We can see that in all cases the clustering is clearly larger than random. (B) Characteristic path length L with random values. Except for Little Rock, the difference between the random and real case is very slight (close to $L \approx 2$). For the two Ythan webs, obtained at different years with different resolutions, the improvement provides a better defined SW pattern.

lengths. It can be shown that for these graphs $C_v^{rand} \approx \bar{k}/S$ and $L^{rand} \approx log(S)/log(\bar{k})$. At the other extreme, a totally hierarchical graph will have $C_v^h = 0$ and L^h values of the order of the system size (Watts, 2000). These two characteristics are relevant here because they provide a measure of: (a) the number of links separating two species and thus the potential number of steps needed in order to propagate a disturbance from one to the other and (b) the presence of a local structure in trophic compartments, in particular the degree of omnivory or intraguild predation.

The analysis of taxonomically well-defined, species-rich food webs reveals their SW topology (Montoya and Solé, 2002; Camacho et al., 2002; see, however, Williams et al., 2002). The observed values of C and L can be compared with expected ones from the randomized webs with the same number of (total) links (fig. 6.10). As expected from an SW pattern, we see that all have very similar (and very short) distances L and a clustering coefficient from $C_v^{Yth1}/C_v^{rand} = 2.33$ to $C_v^{Silw}/C_v^{rand} = 5.0$ times larger than the random case. For the average path length, it was found that the values are close to $L \approx 2$, and thus we have short sequences of links propagating effects through food webs (i.e., following Williams et al., we have two degrees of separation). Such short L values are in agreement with empirical studies of cascade effects in both terrestrial and aquatic ecosystems (Pace et al., 1999; Williams et al., 2002).

Complexity and Fragility

Community fragility is far from understood. Some key issues, such as which species might be considered especially relevant to a given community, have led to a heated debate since Paine's definition of keystone species (Paine, 1966; Bond, 1993). It is commonly accepted that community fragility and persistence are related to how ecological communities are structured, specifically to how trophic links are distributed throughout the community (May, 1973; Pimm, 1991; McCann, 2000). But both the scarcity of high-quality data (Polis, 1991; Cohen et al., 1990; Williams and Martinez, 2000) and the lack of methods suitable for a detailed analysis of the complexity of food-web organization (Cohen et al., 1990) leads to the lack of a unified picture of community fragility and persistence. The fragile character of complex food webs is fairly well illustrated by the impact of invaders on community structure (Elton, 1958; Pimm, 1991). In this context, exotic predators can collapse food webs and promote the extinction of many species. As an example, the introduction of the Nile perch on Lake Victoria triggered a full ecosystem reorganization with widespread species loss. The lake had over four hundred species of fish (many of them endemisms) but today only three species appear in significant abundances (only one of them is native).

Networks exhibiting SW properties and heterogeneous distributions of connections present a characteristic response to the successive removal of their nodes, related to the way removals occur (Albert et al., 2000). When nodes are removed at random, the network exhibits high homeostasis. By contrast, if most-connected nodes are successively eliminated, the structure of the network reveals an intrinsic fragility that eventually leads to a breaking into many small subgraphs. The general nature of these results immediatly suggests their application to ecological networks. The main question is: How dependent is ecosystem fragility and persistence on graph architecture? In order to answer this question, the possible consequences for ecosystem stability against different types of species loss has been examined (Solé and Montoya, 2001; Dunne et al., 2002).

As it occurs with other types of complex networks, (species-rich) food webs display the robustness expected for long-tailed distributions of connections but also a high fragility against selective species removals in terms of (1) food-web fragmentation into disconnected subwebs, and (2) secondary extinctions (i.e., species that become extinct due to removals of other species) (Pimm, 1991).

Two types of species removals can be considered: random and selective. These correspond to removal of an arbitrary or most-connected node, respectively. To establish the number of connections for each species in each web, two kinds of ecological graphs must be considered: nondirected and prey directed. In the former, the global connection for each species i was measured, that is, the sum of inward links (i.e., the number of prey species consumed by i) and outward links (i.e., the number of predator species that feed on i). In prey-directed graphs, only the outward links for species i are considered. Thus, under selective removals, for the nondirected graph we delete successively the most globally connected species, whereas for the prey-directed graph, the order of deletion starts with the species with most outward connections and follows a decreasing sequence. Previous studies on nondirected and directed graphs have shown that the eventual effect of removal is network fragmentation (i.e., percolation, see chapter 4), which takes place in different ways depending on the type of removal used (Albert et al., 2000).

An example of the effects of directed and random removal on (static) ecological nets is shown in figure 6.11 for the Silwood Park food web. Here four snapshots (a–d) of the web are shown at different points through the directed removal process. The plot also shows the total fraction of extinct species against the fraction of selectively removed ones f. With random removal, the graph cannot be fragmented until extremely high species deletion has been introduced, and extinction rates remain at low values even for high f, so secondary extinctions are very small (almost nonexistent, [broken line]). However, what happens when most connected species are successively removed is entirely different. These webs are extremely vulnerable to this sort of removal. Their fragility can be seen in the large fraction of species that become coextinct at low values of species deletion, which reveals how fast secondary extinctions will occur (black circles).

The previous trend seems to be robust when considering different food webs, although the robustness of these webs (measured in terms of decreased coextinction levels) appears to heighten with increasing connection (Dunne et al., 2002). More specifically, increased connection delays the point at which the food web displays high sensitivity to removals. Having a unique, large food web with many species instead of many small subwebs with few species each reduces species risk of extinction. The reason is the so-called insurance effect (Naeem and Li, 1997; Levin, 1999; McCann, 2000). Expanding biodiversity increases the odds that an ecosystem will have (1) species that will respond differently under variable environmental conditions and perturbations, and (2) functional redundancy, that is, species that are capable of functionally replacing extinct species. The higher the level of biodiversity present in an unfragmented food web might support several

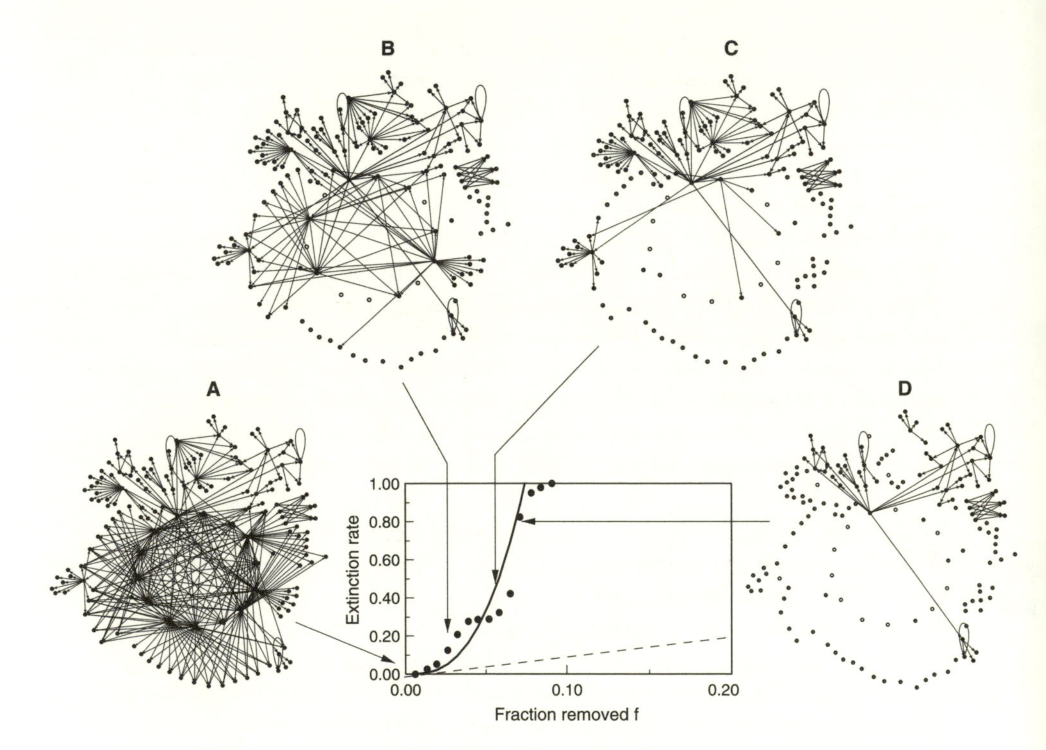

FIGURE 6.11. Effects of selective and nonselective removals in heterogeneous food webs. A sequence of selective removals is shown (A–D) for the Silwood Park food web. Isolated circles indicate removed and coextinct species, respectively. The fraction of extinct species (including coextinctions) is shown in the central plot against the fraction of selective removals. The dashed line indicates the expected extinction rate under random removal.

ecosystem functions that a fragmented web would not stand (Schuelze and Mooney, 1994; Levin, 1999). Nutrient cycling, carbon and water fluxes, and many other functions might be altered if fragmentation occurs. High biodiversity might also reduce the probability of secondary extinctions (Borvall et al., 2000), although no obvious relation between species numbers and the previous removal experiments seems to be present (Dunne et al., 2002).

Power Laws in Mutualistic Webs

We have seen that some ecological networks show heterogeneous patterns of connectivity, following well-defined broad statistical distributions. How common are these patterns? Are they present in other types of ecological webs, such as mutualistic networks? To address these questions, we need to look at a large data set and check for the frequency of different types of distributions. Jordano et al. (2003) have analyzed fifty-three webs of mutualistic interactions between plants and the animals that pollinate them or disperse their seeds. Interestingly, they found that the bulk of communities had connectivity distributions whose best fit was neither a power law nor an exponential distribution. The most frequent distribution (up to 65.6% of the networks studied) was a truncated power law (also called a broad-scale distribution, fig. 6.12):

$$P(k) \approx k^{-\gamma} e^{-k/k_x}, \tag{6.37}$$

where, as usual, $P(k)$ represents the cumulative distribution of the probability of interacting with k species, and k_x represents a critical connectivity. The exponential term is the cutoff. As soon as k approaches k_x, $P(k)$ decreases faster than described by the power law.

Jordano et al. (2003) concluded that the buildup process of these mutualistic networks is similar to the process shown by self-organizing systems based on the preferential attachment of nodes. However, there are constraints in the addition of new links beyond a certain limit that "filter the information" (Mossa et al., 2002), limiting the growth of scale-free topologies. There are forbidden links that chiefly arise from processes such as phenology or size uncoupling between pairs of species (Jordano et al., 2003). The consequent broad-scale distribution ensures that, in comparison to scale-free networks, these mutualistic nets are more robust to the loss of a keystone species (defined as highly connected).

Network Structure: Moving beyond Connectivity Distribution

Connectivity distribution is only a first approximation of structure. Other features, such as the presence of universal allometric scaling (Garlaschelli et al., 2003), must be considered in order to reach a complete picture of ecological webs. It should be noted that the degree distribution describes an average number of links, but it does not provide information on the patterns by which several groups of species are connected among them. For example, what is the relationship—if any—between the connectivity of a species (k) and the average connectivity of its nearest neighbors ($< k_n >$)? This relationship is called connectivity correlation (Krapivsky and Redner, 2001; Pastor-Satorras et al., 2001; Maslov and Sneppen, 2002) and provides an additional measure of structure uncovered by the distribution of connections. Protein networks have an inverse relationship between k and $< k_n >$ defined by a power law:

$$< k_n > \approx k^{-0.6}. \tag{6.38}$$

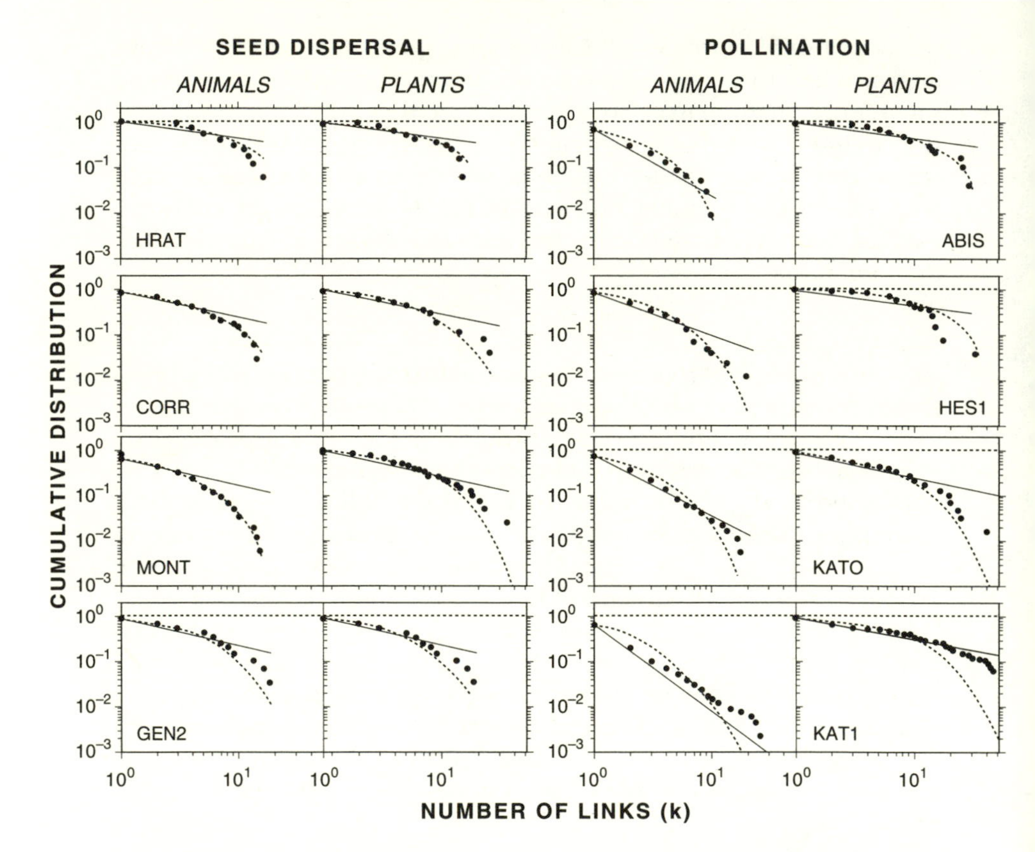

FIGURE 6.12. Connectivity distribution in mutualistic networks. The cumulative probability of interacting with k species is plotted in a log-log plot. Several communities of seed dispersal and pollination are shown. For each community, data is plotted separately for animals and plants. In each plot, dots represent data. The fit of data to a power law and to a truncated power law are represent as solid and dotted lines, respectively. As noted, most examples show connectivity distributions decaying as truncated power laws. From Jordano et al. (2003). Reproduced with permission of Blackwell Publishing.

That is, neighbors of highly connected proteins have low connectivity, and, similarly, low connected proteins tend to be connected to highly connected proteins. The resulting network is highly compartmentalized in subnetworks organized as a set of highly connected nodes, with few links among such subwebs (Maslov and Sneppen, 2002). As suggested by Maslov and Sneppen (2002), such structure would reduce the chances of the spread of deleterious mutations throughout the network.

Melián and Bascompte (2002b) analyzed the connectivity correlation patterns for food webs. They found a pattern that corresponds neither to random networks nor to protein networks. Food webs seem to have two characteristic domains with different assembly patterns. In the first domain, $< k_n >$ either decays very slowly or does not decay at all with k. In the second domain, after a threshold connectivity k_c, $< k_n >$ decays with k in a way similar to that in protein networks (fig. 6.13). Overall, the average connectivity of the nearest neighbors does not decay as fast with k as in protein networks. Whereas two highly connected proteins are unlikely to be connected with each other, the reverse happens in food webs. Food webs are more cohesive (i.e., less compartmentalized). While food webs would be less robust in relation to the spread of a contaminant, they would be more so in relation to fragmentation following species removals (Melián and Bascompte, 2002b). Similar descriptions of internal structure in complex networks have been recently provided by Ravasz et al. (2002) and Milo et al. (2002).

In relation to mutualistic networks, Bascompte et al. (2003) have looked for structural patterns in the relationship between sets of species. The key concept in their analysis is that of nestedness, a concept previously used in island biogeography when studying how a set of animals is distributed among a group of islands (Atmar and Patterson, 1993). To study this property, a matrix defining the interaction between the group of animals and the group of plants is used. Let us imagine that plants are represented in rows and animals in columns (fig. 6.14). In such a matrix, an element a_{ij} is 1 if plant i and animal j interact, and zero otherwise. Let us organize this matrix from the most generalist species to the most specialist. A matrix like this one is said to be perfectly nested (fig. 6.14a) if each species interacts only with the proper subsets of those species interacting with the more generalist species. An assembly pattern like this one is characterized by (i) a core of interactions (all generalist plants and all generalist animals interact among them), and (ii) highly asymmetrical interactions (a specialist animal [plant] interacts with the most generalist plant [animal]). Figure 6.14c shows a mutualistic network (a plant-pollinator system). As noted, this mutualistic network is highly nested (significantly more nested than a similar, randomly assembled network). Bascompte et al. (2003) showed that most of the mutualistic networks are highly nested, and that this pattern is the same for both plant-pollinator and plant-frugivore networks.

The previous result is interesting because ecologists have traditionally noted that the structure of ecological networks may affect their stability (May, 1972, 1973; DeAngelis, 1975; Pimm, 1979; Sugihara, 1982), with special emphasis on the idea of compartmentalization (Pimm and

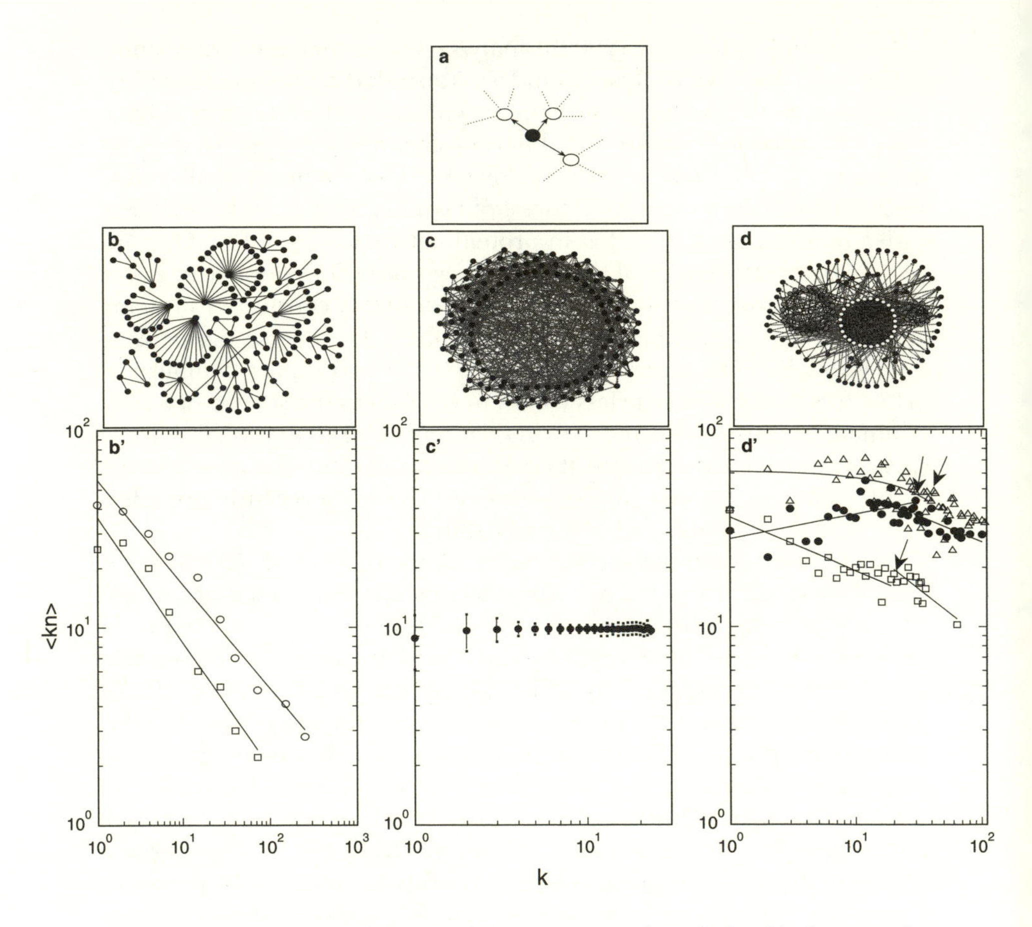

FIGURE 6.13. (a) Connectivity correlation measures the relationship between the connectivity k of a node like the black one, and the average connectivity of its nearest neighbors ($< k_n >$) here represented as white nodes. The figure compares the structure of three different types of networks: protein networks (b), random networks (c), and food webs (d). The connectivity correlation is represented for interaction (circles) and regulatory (squares) protein networks (b′); Maslov and Sneppen 2002), random webs (c′), and three highly resolved food webs (d′): Little Rock Lake (triangles; Martinez 1991), El Verde (circles; Reagan and Waide 1996), and Ythan Estuary (squares; Huxham et al., 1996). $< k_n >$ decreases with k as a power law for protein networks; random networks are uncorrelated; food webs show two domains with a significant shift in the relationship, represented by the arrows. Protein networks are more compartmentalized than food webs. From Melián and Bascompte (2002b). Reproduced with permission of Blackwell Publishing.

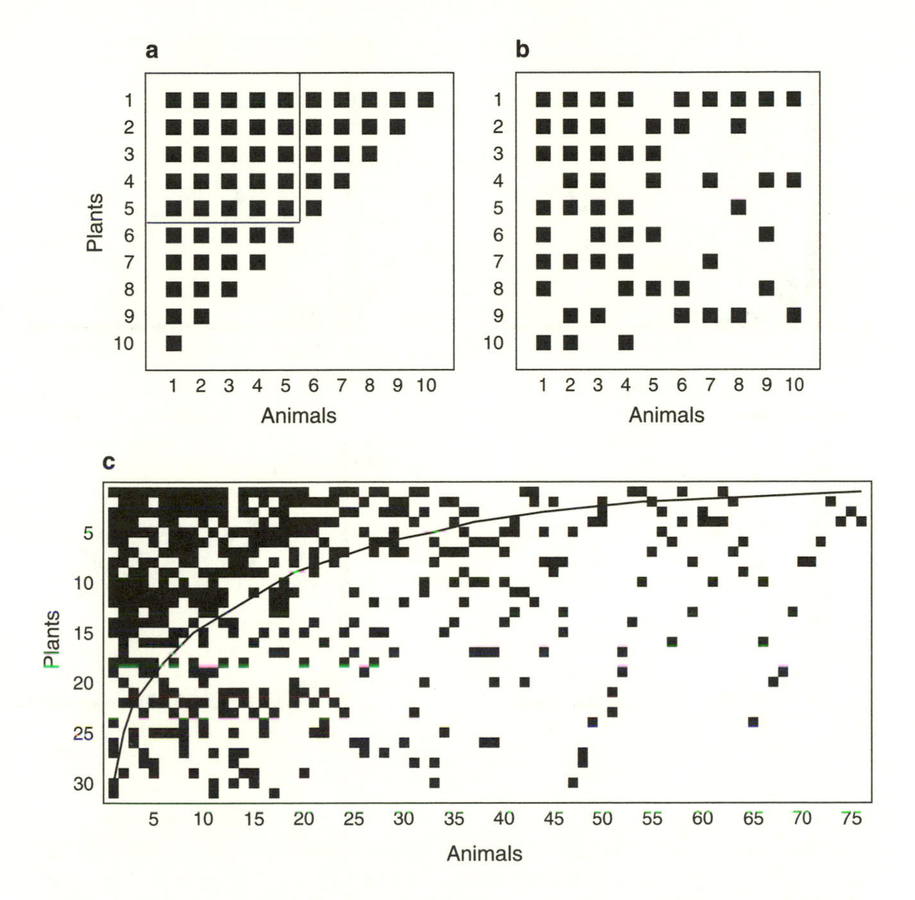

FIGURE 6.14. Plant-animal mutualistic matrices. A square indicates an interaction between plant i and animal j. a, b, and c represent a perfectly nested matrix, a random matrix, and the plant-pollinator network of Zackenberg, respectively. The box in (a) represents the central core of interactions. The isocline in (c) depicts the isocline of perfect nestedness. In a perfectly nested scenario, all the interactions would be located in the left-hand side of the isocline. The respective values of nestedness (an index ranging from zero to one) is 1 (a), 0.55 (b), and 0.742 (c). From Bascompte et al., 2003. Reproduced with permission of the National Academy of Science, WA.

Lawton, 1980; Raffaelli and Hall, 1992; Paine, 1992). Detailed measures, including interaction strength among communicating species, has shown that compartments are common and well defined (Krause et al., 2003). Using a different approach to mutualistic networks, Bascompte et al. (2003) have shown that mutualistic networks are neither randomly assembled nor compartmentalized, but highly cohesive.

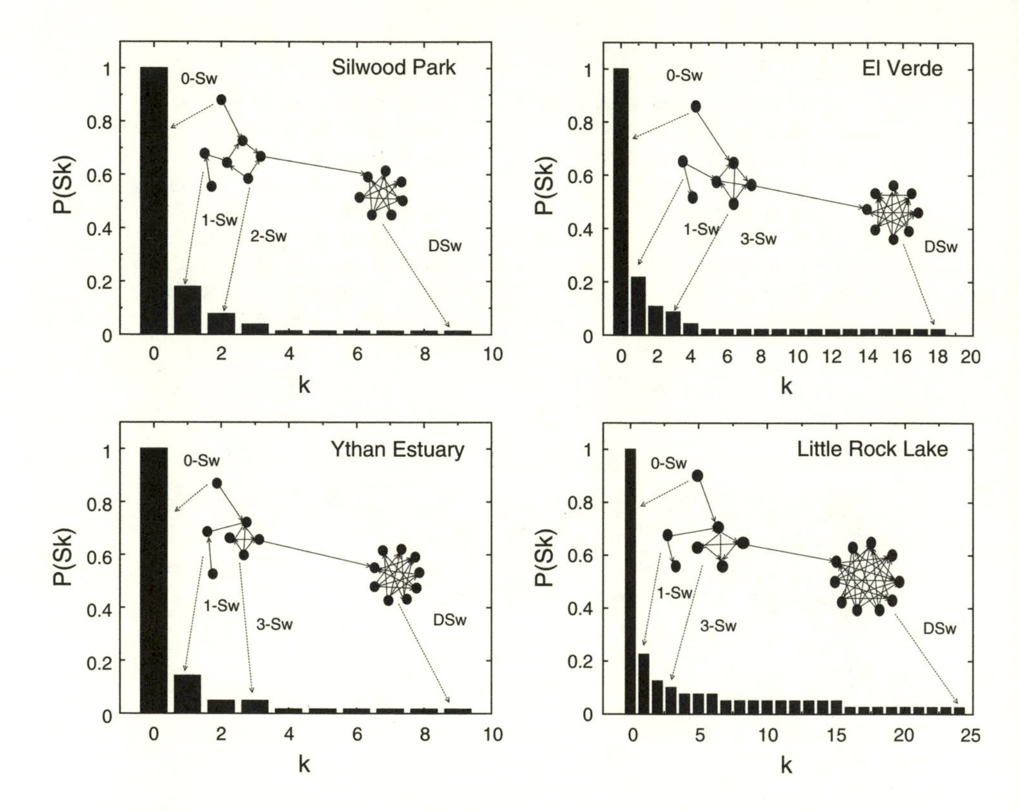

FIGURE 6.15. Cumulative subweb k-frequency distribution for the four most resolved community food webs. The inserts are intended to illustrate the different k-subwebs. As noted, the distribution is highly skewed; the best fit is in all cases a power law. This high heterogeneity depicts a case in which a single highly connected subweb provides cohesion to a large number of small disjointed subwebs. From Melián and Bascompte (2004).

Such pattern is observed when studying the heterogeneity of subwebs or modules within each food web. A k subweb represents a subnetwork formed by a group of species interacting with at least k other species of the same subweb (Melián and Bascompte, 2004). This is a useful representation for seeing how different modules are related to one another in food webs. This result holds when analyzing the best-resolved food webs. Interestingly, the k-frequency distribution is highly skewed (there is a large number of small k subwebs and only one most dense subweb; fig. 6.15). For the five most resolved food webs, the best fit in this distribution was provided by a power law, with an average $\pm SD$ of the critical exponent γ equal to 1.34 ± 0.57. This means that there is also a high level of structure in the organization of whole food webs into modules or subwebs. And this structure confers a high level of cohesion that, all things being equal, may increase the propagation of a contaminant but may also confer high robustness to fragmentation after the elimination of species, a result already seen when dealing with connectivity correlations.

COMMUNITY ASSEMBLY: THE IMPORTANCE OF HISTORY

Ecological systems are the result of a number of causal processes as well as historical contingencies. Simple models of two-species competition show that such path dependence can be easily interpreted under the language of nonlinear systems. Symmetry-breaking in particular provides the simplest illustration (chapter 2). For this particular case, when competitive exclusion takes place, the final winning population will be the one starting at the appropriate basin of attraction. In models where a single (globally stable) solution is present—such as when competitive coexistence is at work—no such path dependence is observed. However, once the dimensionality of the model grows, and more complex dynamical patterns are involved, multiple attractors are also available and contingency can play a leading role. An example of this multiplicity of attractors is shown in figure 6.16, obtained from numerical simulations of assembly in a five-species ecology (Law and Morton, 1993). The model used was a linear one and the topology is indicated in figure 6.16a. Different sequences of introductions lead to three possible "attractors" in the assembly process (fig. 6.16b). This is precisely what has been observed in experiments with aquatic microcosms (Drake, 1991). By using different orders of species introduction, different final compositions are obtained. Given the small size of these systems, one might conjecture that larger pools of species could lead to a larger diversity of final states.

The final organization of stable communities will have an important influence on their response to challenges of different types, such as human-driven habitat destruction or invasion by exotics. Despite a long history of study, ecologists still cannot predict the pathways or end points of community assembly with any degree of certainty. Much of what is known comes from a handful of easily studied systems, making it difficult to generalize from one community to another. It has been argued that in fact the final composition of a given ecosystem might derive from a largely historic process where competition or predation play a secondary role.

Community assembly refers to the development of complex ecosystems from a regional species pool, which depends on interactions among species availability, the physical environment, evolutionary his-

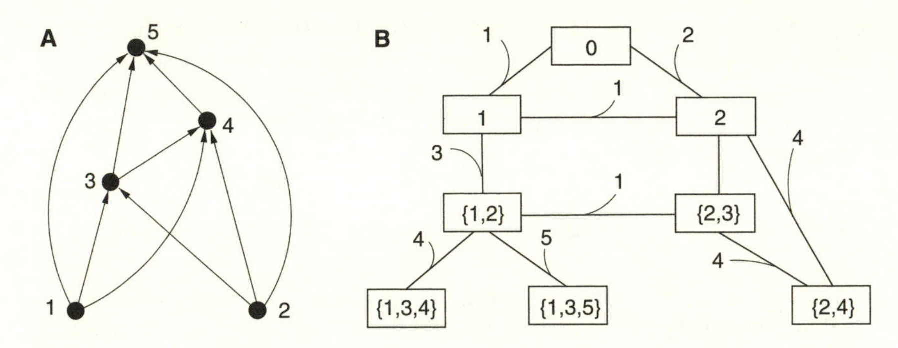

FIGURE 6.16. Multiple attractors in a model of community assembly. (A) Basic network used in the analysis, involving $S = 5$ species (see text). In (B) the possible pathways to community assembly are shown, involving all possible sequences of species interactions. Three final possible (stable) communities are obtained (bottom row).

tory, and the temporal sequence of assembly. We cannot easily predict which species are likely to invade or be lost from particular natural communities. We also know little about community responses to perturbations at different stages in development. Many attempts to create or restore communities fail, for reasons that often remain poorly understood.

In this context, mathematical models are among the most important tools in understanding community assembly. Along with theory, several early observations from field studies show that rich-species communities are more resistant to invasion by exotics than species-poor communities (Elton, 1958). This is one particularly important observation. On the one hand, it points to a nontrivial property of complex communities that requires an explanation. On the other, processes that decrease species diversity can allow invasion events that might eventually disrupt community stability in a catastrophic way.

The simplest scenario of multispecies community assembly is provided by a community of competitors (Levins, 1968; Case, 1990, 1991). If N_i is the population size of species i (here $i = 1, \ldots, S$, then the LV model will be defined as:

$$\frac{dN_i}{dt} = \frac{r_i N_i}{K_i} \left(K_i - \sum_{j \neq i}^{S} \alpha_{ijN_j} \right) \tag{6.39}$$

where $\alpha_{ii} = 1$, and r_i and K_i stand for growth rates and carrying capacities of the i-th species, respectively. This is a multiparametric model, but its complexity can be strongly reduced by means of appropriate transformation. The simplified equations are:

$$\frac{dN_i}{dt} = \frac{r_i}{r_m}\left(1 - \sum_{j \neq i}^{S} \alpha_{ij} N_j\right) \tag{6.40}$$

where $r_m = max\{r_i; i = 1, \ldots, S\}$. New species invade the system with a small initial population $\theta \ll 1$. The new species is randomly chosen from the S-species pool. Introductions occur at each τ step, and the outcome of the invasion process (as it occurs in experimental microcosms) is community dependent. A simple, approximate criterion for invasion success can be established as follows. Let k be the new invader species that enters the system at $t = \tau$. Success after invasion is given by $(dN_k/dt)_{t>\tau} > 0$ or, in other words,

$$\sum_{j \neq k} \alpha_{kj} N_j < 1 - \theta. \tag{6.41}$$

This condition is interesting, since it reflects two relevant features. One, that a good competitor can fail to invade the resident community if the last inequality does not hold. Two, that the previous result suggests that a multiplicity of communities can fulfill the inequality, and thus multiple outcomes of the assembly process are possible.

The latter is a crude condition of success, since it only provides a criterion a short time after invasion. In this context, four different scenarios have been defined by Case (1991):

1. Community augmentation: Invasion is successful and the invader population remains positive in time.
2. Rejection failure: The invader is rejected and no further changes are observed in the resident community structure.
3. Indirect failure: The invader initially grows but it is eliminated later in the dynamic. As a consequence of the invasion, some residents become extinct.
4. Replacement: The invader succeeds at the expense of one or more residents.

Indirect failure is a rare event but clearly illustrates the unexpected features exhibited even by these simple communities (Case, 1991). An example of the numerical results obtained in these experiments, using a uniform distribution of competition coefficients over the unit interval (i.e., $\alpha_{ij} \in [0, 1]$) is shown in figure 6.17. Here averages are obtained over two hundred runs, and the three main pathways of invasion have

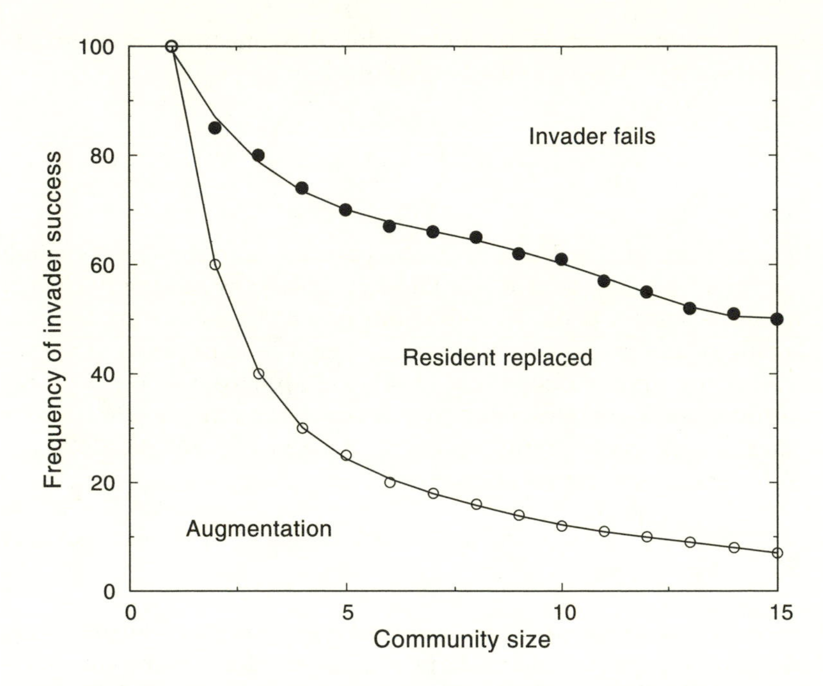

FIGURE 6.17. Frequency of invader success, augmentation and replacement events as a function of system size (number of species) of the core community being invaded.

been considered, with frequencies drawn against community size. The plot shows that augmentation rapidly declines as a larger community is built, indicating that invasion success becomes less and less common. Both failure and replacement frequency increase in an asymptotic fashion. Similar results are obtained using more detailed models of competition.

SCALING IN ECOSYSTEMS: A STOCHASTIC QUASI-NEUTRAL MODEL

As discussed at the beginning of this chapter, an important feature displayed by rich-species ecosystems is the presence of long-tailed species-abundance distributions (Pielou, 1969; May, 1975; Bell, 2000, 2001). These distributions provide a statistical view of the whole community in terms of the relative abundances of the different species. Species-abundance distributions involve the number of species $S(n)$ represented by n individuals. The total number of species S and the number of individuals N is obtained from $S(n)$ through

$$S = \int_0^\infty S(n)dn \,, \tag{6.42}$$

$$N = \int_0^\infty nS(n)dn \,, \tag{6.43}$$

where we have assumed that $S(n)$ is given by a continuous distribution.

Two distributions are ubiquitous: the lognormal and the power laws. Both share a trait: They display long tails indicating that a few species are rather common and most are rare. The long-tailed distributions displayed by real ecologies have been interpreted in terms of stochastic multiplicative processes governed by the conjunction of a variable number of independent factors (May, 1975). Power laws can be considered a limiting case of lognormal behavior and are observed in a number of ecological scenarios, from rain forests to marine ecosystems (Caswell, 1976; Hubbell, 1997, 2001; Solé and Alonso, 2000). These distributions are commonly found in a number of situations, including, for example, multiplicative processes (May, 1975), coagulation-fragmentation models (Camacho and Solé, 2000), coherent noise (Newman and Sneppen, 1996), and systems with interacting units having complex internal structure (Amaral et al., 1998). In physics, they are known to be common close to critical points, as discussed in chapter 4 (Binney et al., 1992; Solé et al., 1996b; Stanley, 1971; Solé et al., 1999). An additional scenario that might account for the presence of power laws suggests that the systems have self-organized into a critical state (Bak et al., 1987). Self-organized critical (SOC) systems require a number of strong constraints in order to operate (see chapter 4), and it is not clear that such constraints will be present in a broad range of situations (Solé et al., 1999; Solé and Goodwin, 2001). Perhaps more important than validating the possible scenarios suggested to explain these distributions is providing a theoretical framework where not only $S(n)$, but also the other regularities presented in this chapter are accommodated in a closed way.

Partially inspired by SOC ideas, a mechanism of ecosystem organization has been suggested in which the observed patterns result from the spontaneous driving of complex ecologies toward the marginal instability boundary defined by the species-connectivity (S-C) relation (Solé et al., 2002a, b; McKane et al., 2000). The basic model is a mean-field (spatially implicit) interacting particle system (see Durrett and Levin, 1994; Durrett 1999). Here a system involving N individuals is considered, together with a pool $\Sigma(S)$ of S possible species. This setup can be understood in terms of a given area with a finite number of sites that is invaded by individuals from an outside pool. Interactions among individuals are introduced through a random matrix Ω. This matrix is fixed and has a predefined connectivity C. Two basic rules exist:

1. Immigration: A randomly chosen site occupied by species B is replaced by a species randomly chosen from the species pool, that is, $A \in \Sigma(S)$:

$$B \xrightarrow{\mu} A \tag{6.44}$$

   This occurs with probability (of immigration) μ.
2. Interaction: Two randomly chosen individuals belonging to species $A, B \in \Sigma(S) - \{0\}$ will interact if $\Omega_{AB} \neq \Omega_{BA}$ (in particular this includes the case of no-interaction). The result of the interaction will be:

$$A + B \xrightarrow{\Omega} 2A \tag{6.45}$$

   if $\Omega_{AB} > \Omega_{BA}$ (and $A + B \rightarrow 2B$ otherwise).

This model does not take into account the local character of ecological interactions. Besides, the way one species replaces another is defined in terms of an abstract type of interaction event (which might be competitive replacement or predation), but extensions to these basic rules have been developed and provide similar results (Solé et al., 2002a, b). By simulating it, we can see that populations can fluctuate in complex ways (fig. 6.18b). An example of the fluctuations displayed by the model is shown in figure 6.18b for a system with $\mu = 0.01, C = 0.1, N = 4 \times 10^3$ and a pool involving $S = 400$ species. Five species are represented, and we can appreciate a high variability in their populations through time. For small C or large immigrations, the type of fluctuation becomes similar to a random walk, as expected. It has been shown that for small S values and high C deterministic chaos can also be observed (Solé et al., 2002a, b).

In figure 6.18a the corresponding evolution of the rank distribution for these parameters is shown. After a transient time, a stationary state is obtained (even though species turnover is operating all the time). When compared with figure 6.1, we can see close similarities in the way the model and real ecological succession behave.

In order to obtain the stationary distribution of species abundance, we can consider the one-step birth and death process (a particular class of Markovian process) where only single-unit changes are allowed to occur. Let us start with the master equation for this model. If $P(n,t)$ is the probability (for any species) of having n individuals at time t, the

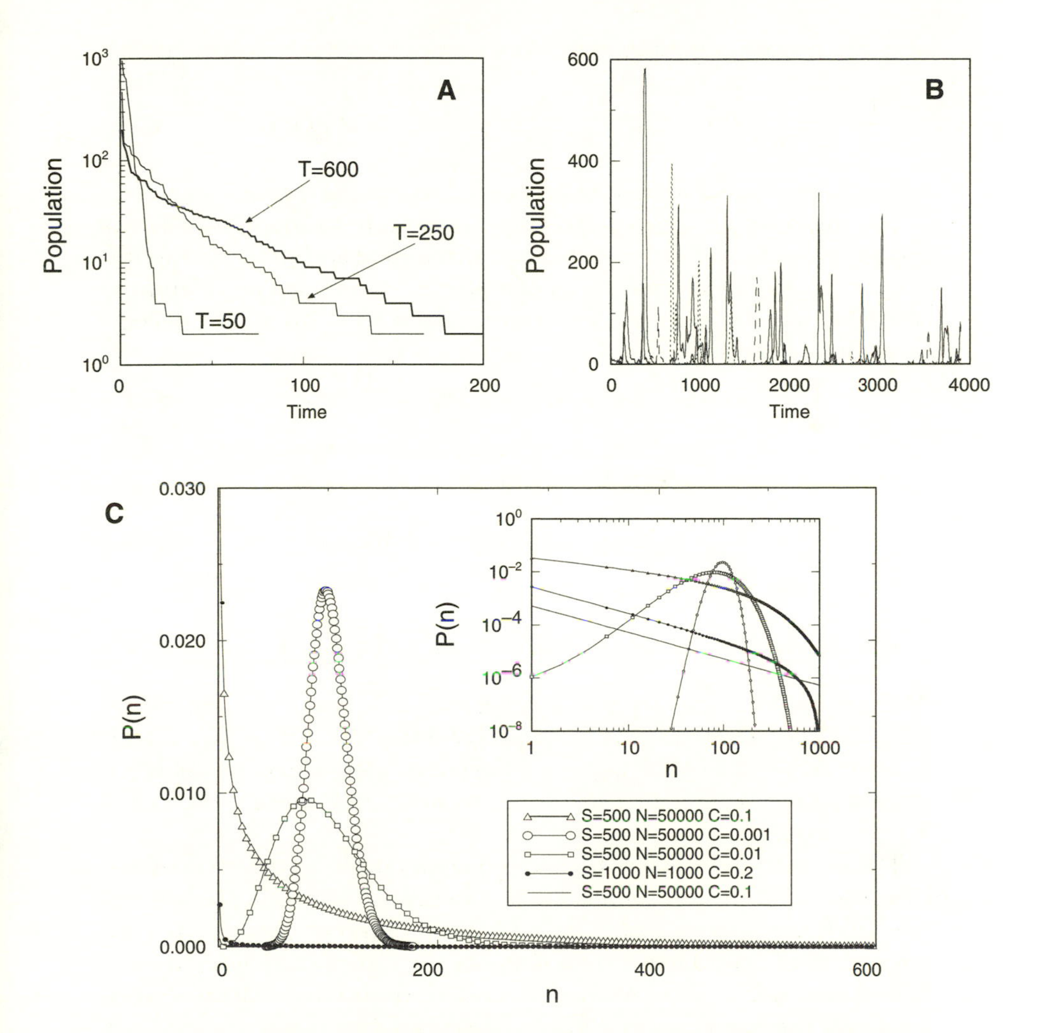

FIGURE 6.18. (A) Time evolution of the species-rank distributions for the stochastic model with $\mu = 0.01$, $C = 0.1$, $N = 4 \times 10^3$ and $S = 400$. We start with an initial condition where only two species are present. (B) Time fluctuations in population abundances for five species, under the previous conditions. (C) Species-abundance distributions obtained from the stochastic model using different values of the parameters S, N, and C. As the immigration rate decreases, long-tailed distributions become more common. In this plot we can see: Gaussian (white circles), Log-normal (white squares) and power-law (black circles) distributions. In the inset, the same distributions are shown in log-log scale. The scaling exponent for the low-immigration regime is indicated. It agrees with the theoretical prediction of $\gamma = 1$, obtained from the mean-field model.

one-step process is described by (Van Kampen, 1981; Renshaw, 1995):

$$\frac{dP(n,t)}{dt} = r_{n+1}P(n+1,t) + g_{n-1}P(n-1,t) - (r_n + g_n)P(n,t) \tag{6.46}$$

where $r_n \equiv W(n-1|n)$ and $g_n \equiv W(n+1|n)$ are the transition rates.

Two different processes will contribute to each transition rate in our model: those linked with interactions through Ω and those due to immigration. As an example, let us consider r_n. Asuming a population of N individuals, S species and a matrix connectivity C_π, r_n will be decomposed in two terms:

$$r_n \equiv W(n-1|n) = W_\Omega(n-1|n) + W_\mu(n-1|n) \tag{6.47}$$

where W_Ω and W_μ indicate interaction and immigration-dependent transitions, respectively. It is not difficult to show that:

$$W_\mu(n-1|n) = \mu\left(1 - \frac{1}{S}\right)\frac{n}{N} \tag{6.48}$$

$$W_\Omega(n-1|n) = 1 - (1-C_\pi)^2\frac{n}{N}\left(\frac{N-n}{N-1}\right) \tag{6.49}$$

The first of them is easily understandable: We choose one individual from the species considered (with probability n/N) and replace it with another individual belonging to a different species (with probability μ). In order to guarantee that it belongs to some *other* species, the factor $1 - 1/S$ is used.

The second term can also be derived from simple arguments. In this case, we take, with probability $1-\mu$, two individuals from the system, one from our species and another from a different species. If they interact (with probability C^*) then one of them wins, as defined by our rules (two possible pairs are available and the probability that the second wins over the first is just one half). The probability C is given by

$$C^* = 1 - (1-C_\pi)^2, \tag{6.50}$$

which is easily understandable: Since the elements of Ω are chosen at random, the probability of no-interaction is $P[\Omega_{ij}, \Omega_{ji} = 0] = (1-C_\pi)^2$, and interaction will occur otherwise.

The final one-step transition rates are given by:

$$r_n = C^*(1-\mu)\frac{n}{N}\frac{N-n}{N-1} + \frac{\mu}{S}(S-1)\frac{n}{N} \tag{6.51}$$

$$g_n = C^*(1-\mu)\frac{n}{N}\frac{N-n}{N-1} + \frac{\mu}{S}\left(1 - \frac{n}{N}\right) \tag{6.52}$$

These one-step processes occur whenever the stochastic process consists of the birth and death of individuals. The name does not imply that it is not possible for n to jump by two or more units at a time Δt, but only that the probability for this to happen is $\mathcal{O}(\Delta t^2)$. The previous transition probabilities are completed by the natural boundary conditions of minimum zero population and maximum N population (i.e., we have $r_0 = 0$ and can define $g_{-1} = 0$ and $g_N = 0$ and define $r_{N+1} = 0$).

From the previous definitions, see that $r_0 = 0$ and $g_N = 0$, and we define $g_{-1} = 0$ and $r_{N+1} = 0$. The stationary distribution $P_s(n)$ is obtained from $dP(n)/dt = 0$ using standard methods (Van Kampen, 1981), namely by writing

$$P_s(n) = \frac{g_{n-1}\,g_{n-2}\cdots g_0}{r_n\,r_{n-1}\cdots r_1} P_s(0)\ , \tag{6.53}$$

and using the normalization condition $\sum_{n=0}^{N} P_s(n) = P_s(0) + \sum_{n>0} P_s(n) = 1$, to show that

$$(P_s(0))^{-1} = \sum_{n=0}^{N} \binom{N}{n} (-1)^n \frac{\Gamma(n+\lambda^*)}{\Gamma(\lambda^*)} \frac{\Gamma(1-\nu^*)}{\Gamma(n+1-\nu^*)}\ . \tag{6.54}$$

Here $\lambda^* = \mu^*(N-1)$ and $\nu^* = N + \mu^*(N-1)(S-1)$, where $\mu^* = \mu/[(1-\mu)SC^*]$.

This sum takes the form of a Jacobi polynomial $P_N^{(\alpha,\beta)}(x)$ (Abramowitz and Stegun, 1965) with $\alpha = -\nu^*$, $\beta = \lambda^* + \nu^* - (N+1)$ and $x = -1$, which can itself be expressed in terms of gamma functions for this value of x. Therefore using $\beta(p,q) = \Gamma(p)\Gamma(q)/\Gamma(p+q)$, the stationary, normalized solution can be written in the form

$$P_s(n) = \binom{N}{n} \frac{\beta(n+\lambda^*, \nu^* - n)}{\beta(\lambda^*, \nu^* - N)}\ . \tag{6.55}$$

As the connectivity (or the immigration rate) increases, the distributions change from Gaussian to lognormal and to power laws, as shown in figure 6.18c. This is expected since very small connectivities will make interactions irrelevant, with fluctuations dominated by random events. We can simplify (6.55) in the region of interest: N and S large, and $N \gg n_{max}$ where $n = 1, \ldots, n_{max}$. If in addition $\lambda \ll 1$, which corresponds to μ small: $\mu \ll SC^*/N$, we get a scaling relation (Solé et al., 2002a, b):

$$P_s(n) = \mathcal{K} n^{-1} \exp(-n\mu^* S)\ , \tag{6.56}$$

where $\mathcal{K} = (\mu^* S)^{\lambda^*}/\Gamma(\lambda^*)$ in agreement with simulations and field data.

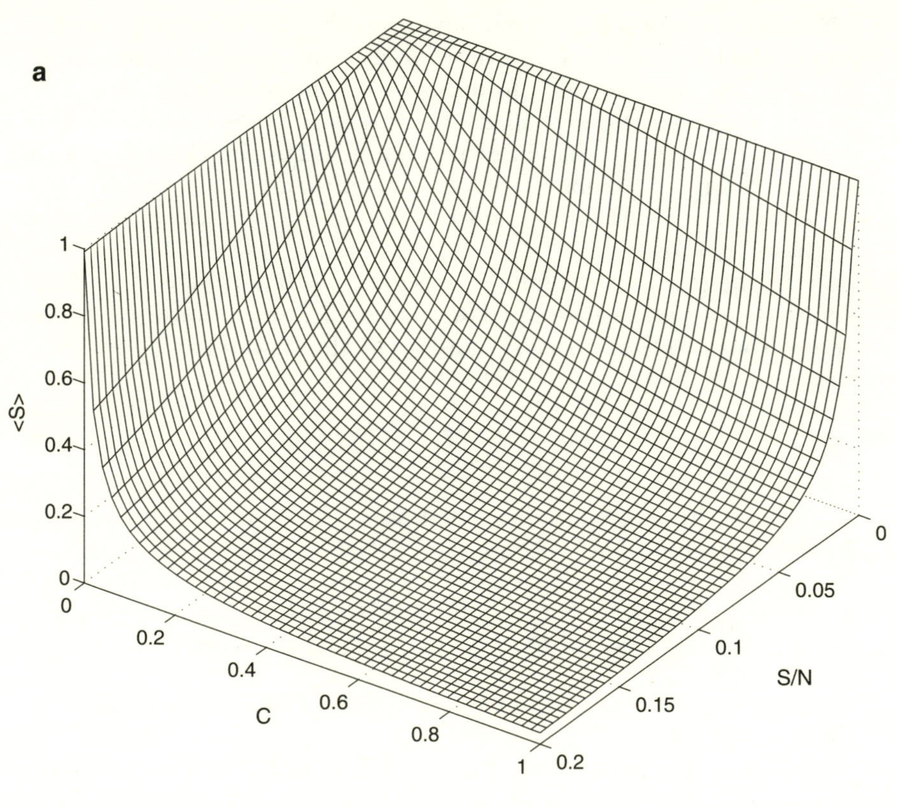

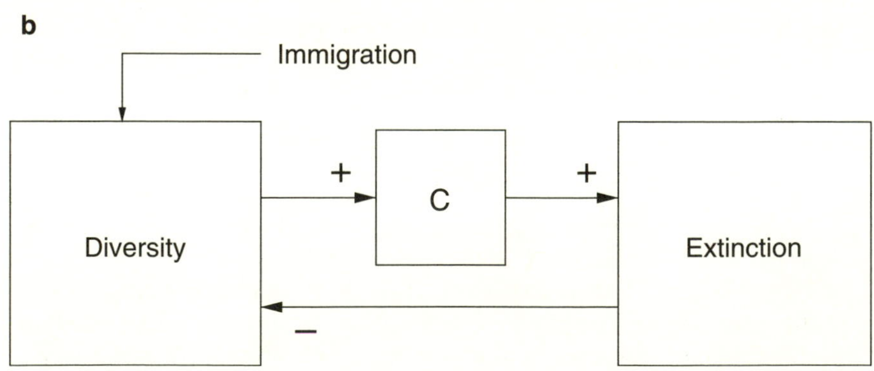

FIGURE 6.19. (a) Fraction of remaining species at steady state as a function of S/N and connectivity (here $N = 5 \times 10^4$, $\mu = 10^{-3}$). For this parameter range, hyperbolic-like behavior, i.e., $S \sim C^{-1}$ is observed at high values of S/N. (b) Feedback loops leading to instability. For an ecosystem, the constant addition of new species by immigration increases the likelihood of interactions. Such interactions limit diversity and trigger extinctions thus inhibiting further increases of diversity.

TABLE 6.2. Comparison between Field Data and the Stochastic Model. HF: highly fluctuating; (+): A lower bound is provided by an estimate from the Hawaiian data of Keitt and Marquet (1996), with $\theta \approx 1$.

PROPERTY	FIELD DATA	MODEL
Species-Connectivity	$S \approx kC^{-1+\epsilon}$, $0 \leq \epsilon \leq 0.5$	$S = kC^{-1+\epsilon(\mu)}$
Scaling in diversity	$P(n) \sim n^{-\gamma}, \gamma \approx 1$	$P(n) \sim n^{-\gamma}, \gamma \approx 1$
Lifetimes	$N(T) \approx T^{-\theta}; \theta \approx 1.6$ (+)	$N(T) \approx T^{-\theta}; (1 < \theta < 3/2)$
Population Dynamics	HF, ($\lambda_L \approx 0$)	HF, Marginal
Gaussian distribution	Not observed	$C \to 0, \mu \to 1/2$
Lognormal distribution	Common	$C \to 0.5, \mu \approx 0.1$
Power law distribution	Observed	$C \to 1, \mu \to 0$.
Species-Area relations	$S \sim A^z, 0 < z < 1$	$S \sim A^z, 0 < z < 1$

From this solution we can compute the stationary number of species $< S >= (1 - P(0))S$ and find the relation between this quantity and the probability of having two species connected, C^*. Under the same conditions that led to (6.56), we find that $P_s(0) = (\mu^* S)^{\lambda^*}$, which in this parameter range gives $< S >= S\mu^* N \ln(1/(\mu^* S)$. Substituting in the expression for μ^* gives

$$< S >= \frac{\mu N}{(1-\mu)C^*}\left[\ln\left(\frac{1-\mu}{\mu}\right) + \ln C^*\right] \approx \mathcal{A}(C^*)^{-1+\epsilon(\mu)} , \qquad (6.57)$$

where ϵ is given by $\epsilon^{-1} = \ln((1-\mu)/\mu) \approx \ln \mu^{-1}$ and $\mathcal{A}$ is a constant equal to $N\epsilon^{-1}\exp(-\epsilon^{-1})$. We have thus recovered the observed scaling relation reported from field studies. The relation between $< S >$ and C is shown in figure 6.19a for different S/N values and $\mu = 0.001$.

This model has been shown to reproduce other key features exhibited by species-rich ecosystems, such as the scaling in the lifetime distributions (Keitt and Stanley, 1998) or—when defined on a lattice—the species-area relations found in rain forest plots (Hubbell et al., 1999). Some of the basic results are summarized in table 6.2.

A straightforward extension of this model (which keeps the previous simplicity but introduces some additional elements of realism) allows us to reproduce a wide range of quantitative properties and actually includes most of the relevant models of species interactions described in this book (Solé et al., 2002a, b). These sub-cases include standard predator-prey models, the contact process, the forest fire model, or Hubbell's neutral model, among others (see appendix 3).

The patterns displayed by the model are a consequence of the spontaneous tendency of the system toward increasing diversity and thus increasing likelihood of interactions among species. Since increasing diversity is essentially inevitable (either by immigration or, on a long time scale, through speciation), the system is always pushed toward high S. At the beginning, interactions might be unlikely to occur (especially at low C values) but as S grows (here a species will interact, on average, with $\approx C < S >$ species), the system creates a negative feedback that generates extinctions. Eventually an equilibrium is reached between both tendencies. Here we can identify immigration as a driving force (as defined in chapter 3), and the extinction (species turnover) plays the role of the order parameter (fig. 6.19b). It has been suggested that a new concept, self-organized instability, can be a useful alternative way of defining the process described here (Solé et al., 2002b). Two reasons to not call it SOC are that interactions among species are non-uniform (are weighted by the interaction matrix) and that the driving is not slow: Immigration occurs all the time without complete system relaxation, as SOC requires. As the system moves toward the critical boundary defined by the S-C line, the likelihood of invaders will decrease, and population fluctuations will increase in size. Since this is a marginally stable state, it would explain the dominance of near-zero Lyapunov exponents observed in field data (chapter 2). The large fluctuations will provide the source of long tails in the observed distributions and are ultimately responsible for rarity. Three basic properties (time fluctuations, species-abundance distributions, and the S-C scaling) are thus simultaneously explained by this approach. The implications of this scenario for evolutionary ecology have yet to be explored, although it has been suggested that it might help to understand key events of large-scale evolution, such as megafaunal extinctions (Forster, 2003). The relationship between web structure and its evolution produces a large number of questions, and some models have provided insight in this area (Caldarelli et al., 1998; Bastolla et al., 2001). Some of these ideas will be explored in the next chapter.

CHAPTER SEVEN

Complexity in Macroevolution

EXTINCTION AND DIVERSIFICATION

Current estimates indicate that 99% of the total species that sometimes inhabited our planet are extinct. Extinction, thus, is eventually the fate of all species and shows unequivocally that no biota is infinitely resilient (Chaloner and Hallam, 1989; Elliott, 1986). How biotic groups vanish and how biotas recover from large extinction events are important problems with immediate implications for current ecologies (Erwin, 1993). In this context, we are facing a human-driven episode of biodiversity loss comparable with previous mass-extinction events and yet with unknown consequences. Understanding the future of our biosphere requires a knowledge of our past. The fossil record of life, even if incomplete and fuzzy, provides a laboratory of evolutionary experiments, sometimes richly replicated (Jablonski, 1999).

When looking at the history of life on Earth over the last 540 million years (the so-called Phanerozoic eon), we see a pattern of increasing diversity interrupted by extinction events of a variable magnitude. This is shown in figure 7.1 for marine animals. The plot has been generated by indicating the number of families becoming extinct per stage.[1] The Phanerozoic starts with the so-called Cambrian explosion (CE) event, involving the (geologically) sudden apearance of most metazoan body plans (Conway Morris, 1998). This event took place about 530–520 million years ago. Although multicellular organisms were present before this event, available evidence indicates that the abundance and variety of both fossils and fossil traces rapidly rose close to the CE horizon. An interesting observation from the CE is that modern ecologies are observable right at the Cambrian boundary: The Burgess Shale fauna (fig. 7.1b) in particular, looks remarkably modern, with well-defined groups of suspension feeders, deposit feeders, and carnivores, all linked through a complex food web (Conway Morris, 1986). In fact, this is inferred from the species-rank distribution, which appears to be the characteristic one in stable communities (chapter 6).

Together with well-defined regularities, such as the presence of a community structure that seems to be universal, the history of life on Earth is not free (at all) of contingencies (Gould, 1989; Conway Morris,

[1] Stages are uneven intervals of time with a typical length of 7 million years, and recognizable through specific geological and paleobiological markers.

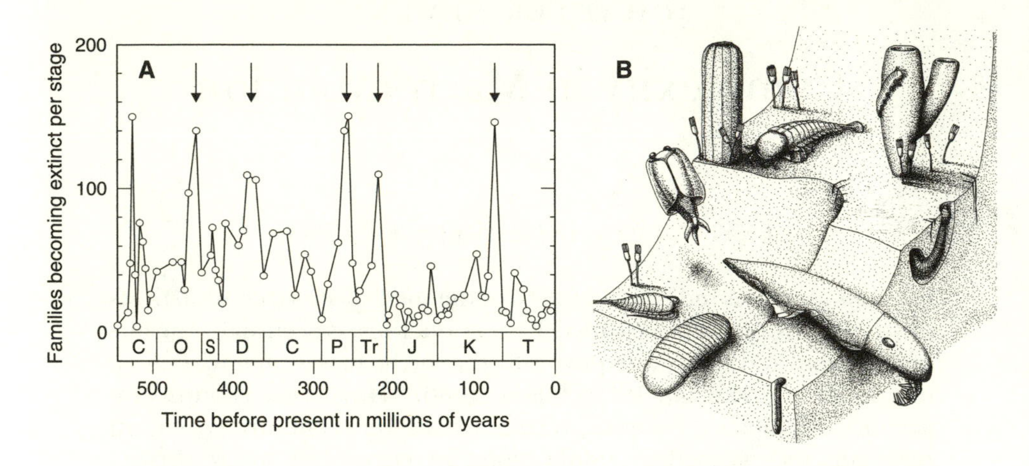

FIGURE 7.1. (A) Time series of extinctions for marine animal families per stratigraphic stage. The arrows indicate the position of the so-called big five mass extinction events. Here the sequence of stages is (from left to right): Cambrian, Ordovician, Silurian, Devonian, Carboniferous, Permian, Triassic, Jurassic, Cretaceous, and Tertiary. (B) A reconstructed paleoecology of a Burgess Shale–like fauna (drawing by R. V. Solé).

1998). Some groups of organisms have experienced wide fluctuations in their diversity, and in some cases today's well developed groups faced an almost complete extinction at some point in their evolutionary history. This is the case of echinoids at the Permian boundary: The *Miocidaris* clade was the only group to survive (Erwin, 1993, 1998a). This was a highly opportunistic group with a worldwide distribution. The scarce group of survivors gave rise to today's extremely diverse and ecologically important modern echinoids. The story thus contains ingredients that require some understanding of the biotic scenario but also some components that introduce chance into the play. Both must be considered together.

INTERNAL AND EXTERNAL FACTORS

How relevant are internal, biotic effects, to the overall dynamics of the biosphere? To some authors, there is no way to decouple the biotic processes from the physical/geochemical ones. This corresponds, for example, to Lovelock's geophysiology where biogeochemical cycles must be taken into account as an essential part of the whole picture (Lovelock, 1988; Lenton, 1998). In this context, some paleontologists have suggested that a self-contained model of large-scale evolution should include indirect biogeochemical factors (Plotnick and McKinney, 1993). To others, biotic responses are less important: The overall pattern would be dominated by the relationship between the number of extinct species and the size of asteroid impact (Raup, 1986, 1993). The idea that impacts are the leading cause of biotic change is appealing and supported by several sources. However, there is no clear correlation between some large-impact craters and the expected extinction event (Poag, 1997). Though these craters (~100 km diameter) were large enough to have produced considerable large-scale environmental stress, no extinction seems to be associated to them. Careful analysis of some well-known groups does not correlate with this picture. An example is provided by the analysis of North American mammals (Alroy, 2003). Although this group experienced a major mass extinction at the Cretaceous/Tertiary boundary (strongly tied to an impact event) no such correlation is found afterward. Specifically, Cenozoic impact events (two of them very large) do not correlate with pulses of biotic change.

But some well-defined observations give support to the relevance of a biotic scenario where ecologies change and evolve in a coordinated fashion (Maynard Smith, 1989; Vermeij, 1987; see also Jablonski et al., 1996). The fact that a surprisingly constant diversity is observable within particular groups of organisms, in spite of continuous extinction and diversification, is an example of such observations. Such an equilibrium between speciation and extinction reveals an important prevalence of biotic factors. In a related context, it is interesting to see that community composition can display long periods of stasis followed by periods of rapid change (Benton, 1995a). An example is demonstrated in Caribbean shallow-water ecosystems (Jackson et al., 1996; Jackson 1997): After many millions of years of evolutionary calm, more than half of this coral and mollusk fauna turned over within less than one million years. A similar situation is found in many terrestrial communities throughout the Quaternary (Vrba, 1985). The rapid changes experienced by these communities (often under minor climate changes) have been interpreted within the context of metapopulation dynamics (Jackson et al., 1996) and the threshold effects associated with them (chapter 5).

Cascade effects have also been reported in the paleobiology literature, dealing with the propagation (through food webs) of initial perturbations involving a limited number of species or guilds. This seems to be the case in the collapse of marine food chains with the end-Cretaceous phytoplankton crisis (Arthur et al., 1997; Jablonski, 1998). For this example, it is known that marine primary productivity decreased rapidly at the K/T boundary resulting in the selective extinction of filter and suspension feeders (those directly dependent upon the influx of organic matter). The extinction of these groups was followed by the concurrent disapparance of certain organisms in higher trophic levels. Although this and other examples are directly linked to the bolide impact hypothesis (which caused the collapse at lower levels in trophic chains due to anoxia), the ability of systems to return to their initial states (in terms of diversity) does not seem to scale directly with the size of the perturbation: Some mass extinctions seem to have more profound effects than others (Droser et al., 1997).

Another important aspect is the evolutionary dynamic resulting from so-called evolutionary arms races and how they affect extinction and diversification (Vermeij, 1987). A good example of arms races is provided by the ammonoidea (Ward, 1992). This group of organisms was very successful and is present in the fossil record over a period of 320 million years. It went nearly extinct in several moments of its evolutionary history and displayed an enormous amount of variation in shapes and sizes. Starting from small, smooth spiral shells, organisms in this group evolved large, thick shells with fractal patterns (in their suture lines) and very complicated modifications of their basic structure. A large body of evidence shows how these changes evolved as a response to predators. And although some authors have linked extinction events in this group to changes in sea level, the relation, if present, is highly nonlinear (Solé, 2001). Although the ammonites disappeared at the end of the Cretaceous period (when the large K-T extinction event took place) they were already declining.

Time Series: Random or Correlated?

Two extreme positions provide the boundaries for a wide range of views of the Phanerozoic pattern of extinction and diversification. The first essentially establishes that the history of life on Earth is the result of microscale biotic constraints plus a dominant, large-scale dynamic driven by external perturbations of a different nature. The second says that the ecological scenario plays a leading role in shaping the long-term changes in the dynamics of the biosphere. As it happens in most cases in evolution, the final answers to the questions come from a somewhat intermediate situation, where the world biota must be considered "interactive, coherent and possessing its own dynamic, while not immune to abiological factors" (Conway Morris, 1998). As shown below, the two components contribute in a significant way to our understanding of extinction and diversification through the Phanerozoic.

A first approximation to the overall pattern of extinction can be gathered from the study of paleobiological time series. Early work in this area indicated the presence of an unexpected structure: a long-term ($T \approx 26$ Myr) periodic component in the power spectrum (Raup and Sepkoski, 1984, 1986; Sepkoski, 1989). The only explanation that can be posed for such long-term periodicity is that mass extinctions have an ultimately extraterrestrial cause. Astronomical forces seem to be the only ones to act with sufficient precision to explain the rather exact schedule of mass extinction events incorporated into the periodic theory. The regular timing of the cosmos could possibly inflict mass extinction on the earth via climatic changes or periodic bolide impact events. Although no clear candidate accounting for this periodicity has been confirmed, and some re-analysis questioned its importance, it is interesting to note that a periodic component remains detectable under different statistical approaches (Plotnick and Sepkoski, 2001; Dimri and Prakash, 2001).

Beyond the presence of some particular trends in the fossil record, it seems necessary to explore null models based on stochastic dynamics in order to determine if random processes can account for the regularities displayed by the fossil record. The simplest approach to modeling the evolutionary pattern of speciation and extinction is to consider clade evolution in terms of branching processes (BP, chapter 4). In this context, a given (monophyletic) group would owe its existence to the interplay of lineage branching (speciation, at a rate λ) and lineage termination (extinction, at a rate μ). As discussed in chapter 4, if $\mu < \lambda$, the BP proceeds by expanding the number of branches and thus the clade typically persists and grows. But if $\mu > \lambda$, extinction will occur (Raup et al., 1973; Raup, 1985). Of course this is a simple approximation of cladogenesis since the BP is homogeneous (i.e., the rates of extinction and speciation are constant). However, it offers a null model that is a useful reference point for the analysis of macroevolutionary dynamics.

By using small modifications of the homogeneous BP, simulated phylogenies can be generated. In figure 7.2 an example is shown (Raup et al., 1973). Here each spindle gives the changing diversity of a given taxon (the width of the spindle at a given time is proportional to diversity) and four runs of the algorithm are shown. Sometimes, stochastic simulations of this type give phylogenies a look (qualitatively) similar to those observed in the fossil record. On the other hand, if we consider the resulting time series of clade fluctuations under the assumption of a BP, subcritical ($\mu > \lambda$) and supercritical ($\mu < \lambda$) scenarios will display a constant decay or increase of clade numbers in time. Only close to the critical point will complex fluctuations be observed. Since the long-term average branching rates estimated from available data indicate that branching rate is just slightly greater than extinction rate

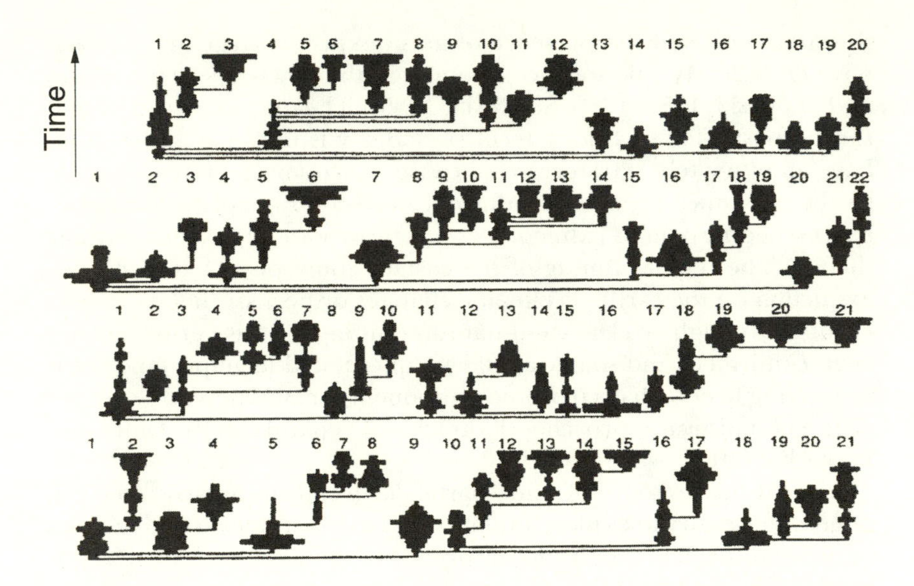

FIGURE 7.2. Four examples of simulated phylogenies (Raup et al., 1973). In this computer simulation phylogenies are represented in terms of trees of spindles, simulating higher taxa. Adapted from Raup et al., 1973.

(Gilinsky, 1991) some diversity-dependent mechanism must be at work. Jack Sepkoski suggested that these rates should be influenced by the diversity of clades D (Sepkoski, 1978, 1979, 1984, 1996): For small D, competition among clades would be small and thus $\lambda > \mu$ but once D approaches some "equilibrium diversity" D^m, origination rates would decay and extinction rates increase until some equilibrium was reached.

If a simple stochastic process involving uncorrelated asteroid impacts were at work, with no further response by the biosphere (except at very short time scales), we should expect a random time series from the fossil record. This is what we obtain from stochastic models but it is not obvious that it is the situation here. A first analysis of family-level databases (the Fossil Record-II data base; see Benton, 1993, 1995a) suggested that different extinction and diversification metrics were consistent with 1/f-dynamics (Solé et al., 1996a). These correlations were indicated by both the power spectra and the estimation of Hurst exponents (chapter 4). Some further re-analyses with other methods provided support for this observation (Amaral and Meyer, 1999; Solé et al., 1998), but other independent studies suggested that either the correlations were too weak (Kirchner and Weil, 1998) or were due to the decline of extinction rates

(Newman and Eble, 1999), but more consistent with a f^{-2} spectrum at large frequencies. The latter result seems appealing, since there is reasonable evidence from such decline, which would provide a source of correlations, although the nature of the decline itself has not been explained yet. Since the disparate results obtained from different studies might be related to the presence of trends and the uneven and finite nature of the time series, some authors have carefully measured the power spectrum by means of the so-called Lomb method (Dimri and Pakrash, 2001). This analysis gave new support to the $1/f$ scenario and suggested that long-range correlations are at work.

Some other authors have suggested that (as happens with many other natural systems) the correct description of these data sets should be provided by multifractal (instead of the fractal) formalism (Plotnick and Sepkoski, 2001; see also, in a similar context, Peters and Foote, 2002). One possible way of detecting long-range correlations is provided by lacunarity analysis (Allain and Cloitre, 1991; Plotnick et al., 1996). Lacunarity provides a scale-dependent measure of the homogeneity of an object, whether or not it is fractal. Starting from a time series of size T, a sliding window of size r is used and for each r the average number $M_1(r)$ of extinctions (or originations) as well as the corresponding variance $M_2(r)$ are calculated. The lacunarity $L(r)$ is given by $L(r) = M_2(r)/(M_1(r))^2$ and it typically displays scaling (i.e., $L(r) \sim r^{-\chi}$) for self-similar objects (both fractal and multifractal), although different slopes might be observed if multiple scales are involved. This is the case of generic-level biodiversity patterns (Plotnick and Sepkoski, 2001) as shown in figure 7.3. An interesting feature of the scaling in extinctions is the deviation that occurs at small r (shown by means of an arrow). Interestingly, it is close to the time scale associated with the $\approx$ 26 million-year periodicity.

Plotnick and the late Sepkoski (2001) suggest that one possible source of scaling in lacunarity is the presence of multiple scales that would interact in a multiplicative fashion. Specifically, their cascade model is defined as follows. A given interval of discrete steps $\{t_1, \ldots, t_N\}$ is partitioned into k pieces ($k = 2, 4, \ldots, N$) of the same length N/k. Each partition P_k is then used in order to generate a set $L_k = \{\xi_k(t_1), \ldots, \xi_k(t_N)\}$ where $\xi_k(t_i)$ takes identical values within each subset of the partition. For example, for L_1, we have $\xi_k(t_i) = p_{11}$ for all $i \in \{1, N/2\}$ and $\xi_k(t_i) = p_{12}$ otherwise. For L_2, four consecutive values (p_{21}, p_{22}, p_{23}, p_{24}) will be available, and so on. The final time series $\{X(t_i)\}$ is obtained in a multiplicative fashion:

$$X(t_i) = \prod_{k=2}^{N} \xi_k(t_i) \tag{7.1}$$

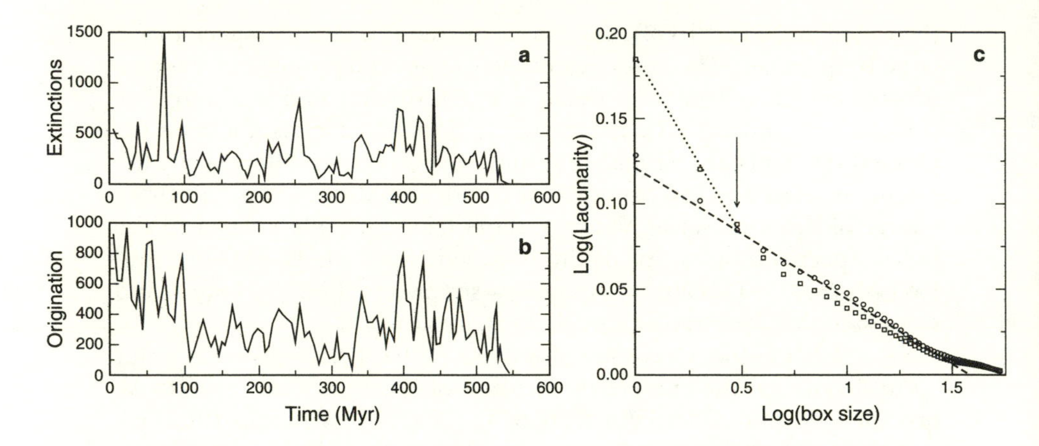

FIGURE 7.3. Fractal structure of paleobiological time series from lacunarity analysis. In (a) and (b) generic-level biodiversity patterns are shown (Plotnick and Sepkoski, 2001). In (c) the results of the lacunarity analysis are shown, with a decay in origination that is consistent with a self-similar time dynamic.

The values taken at each subinterval in each partition (i.e., p_{ki}) are taken from a Gaussian distribution with mean 0.5 and variance 0.05. This process (explicitly self-similar) gives a times series that displays scaling in the power spectrum and high values of the Hurst exponent, and is also consistent with long-range correlations.

SCALING IN THE FOSSIL RECORD

The features exhibited by the time evolution of extinction and diversification suggest that biosphere dynamics is far from random. The possible presence of long-range order needs to be both incorporated into theoretical approaches and explained in a consistent way (Gisiger, 2001). In this sense, although the spectral properties of the fossil record time series are suggestive, a complete theory can be obtained only by considering different statistical properties. Coincidence in one given statistical pattern can be due to chance. Simultaneous matching of different regularities should be a requirement to any theoretical approximation.

It is interesting to see that several scaling laws emerge when looking at the Phanerozoic patterns of extinction and diversification. Not all regularities are scaling functions (see next section) but some key data sets seem to exhibit this property. The most remarkable of these are summarized below.

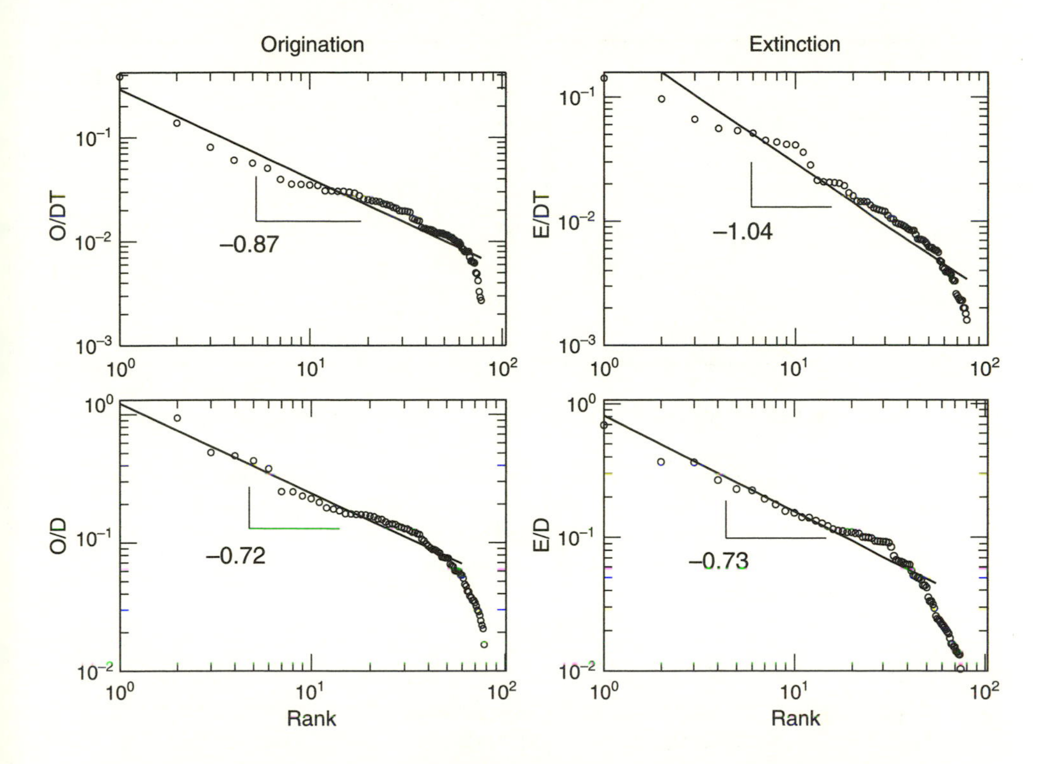

FIGURE 7.4. Scaling in the origination-extinction data from the Fossil Record II. Here the data set has been represented against rank, using: (a–b) origination and extinction per stage and length of interval (here indicated as O/DT and E/DT, respectively) and (b–c) origination and extinction per stage (O/D and E/D, respectively). A scaling exponent ξ close to one is found (the fit is calculated over $1 \leq r \leq 50$).

Extinction Sizes

The distribution of extinctions $N(s)$ of size s follows a power-law decay with $N(s) \propto s^{-\alpha}$ with $\alpha \approx 2.0 - 2.5$ (Solé and Bascompte, 1996; Newman, 1997). A detailed analysis of the frequency histograms of both origination and extinction (using generic level biodiversity) suggests that log-normal distributions, instead of pure scaling laws, are the characteristic feature (Plotnick and Sepkoski, 2001), and interpolating from the tail of this distribution some authors suggest a somewhat larger exponent $\alpha \approx 2.5$ (Sibani et al., 1998).

Given the scarcity of data, a different way of estimating the scaling exponent is to make use of the rank distribution $N(r)$ (see chapter 6). Here $N(r)$ gives the magnitude of the event with rank order r. In figure 7.4 four different plots are presented, including both extinction

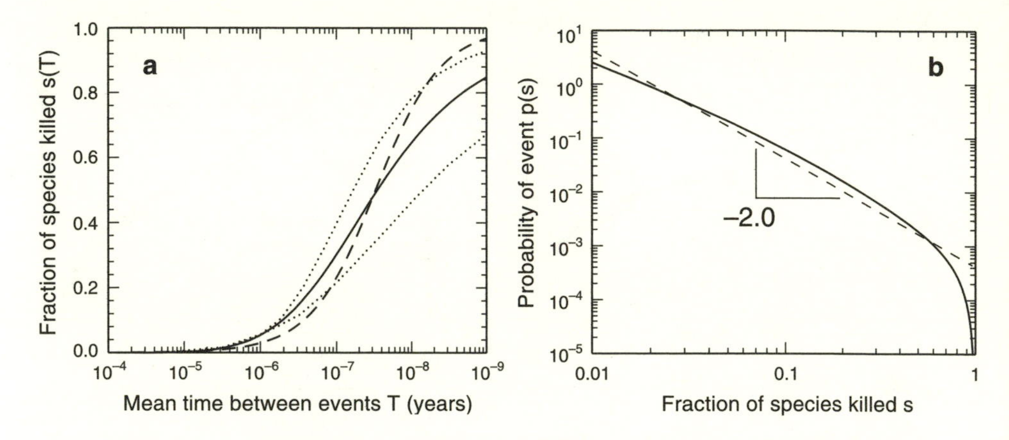

FIGURE 7.5. (a) Kill curves for Phanerozoic marine species. The solid line is the curve given by Raup (1991), and the dotted lines are the associated errors. The broken line is the best-fit kill curve obtained by assuming a power law in extinction sizes. (b) The extinction distribution corresponding to Raup's kill curve (redrawn from Newman, 1996).

and origination from the Fossil Record II compilation (Benton, 1993). By using the rank distribution, we take advantage of the whole data set. It can be shown that the scaling exponent α of the frequency distribution relates to the scaling in the rank distribution

$$N(r) \sim r^{-\xi} \tag{7.2}$$

through the following relationship:

$$\alpha = 1 + \frac{1}{\xi}. \tag{7.3}$$

From figure 7.4 we can see that $\xi \approx 1$ and thus this exponent is compatible with the $\alpha \approx 2$ value, although a significant deviation is observable toward larger values of α (which would fluctuate between two and three).

An additional support for a power law in extinction is provided by Newman's analysis of Raup's kill curve (Newman, 1996). Raup (1991) introduced the concept of kill curve that is closely related to the distribution of extinctions. The curve (fig. 7.5a) is actually a cumulative frequency distribution of extinctions, where the fraction of species s killed is plotted against the (mean) waiting time T. This curve is assumed to have a functional form

$$s(T) = \frac{[log\, T]^a}{e^b + [log\, T]^a}, \tag{7.4}$$

with a best fit (to data from Phanerozoic marine families) given by $a = 5$ and $b = 10.5$. The connection between $p(s)$ and $s(T)$ is obtained by considering the number of extinctions $P(s)$ of size *greater* than s per unit time (Newman, 1996):

$$P(s) = \int_s^1 p(s')s'. \tag{7.5}$$

The mean time T between events of size s or greater is just $1/P(s)$, and thus we establish a connection between $s(T)$ and $p(s)$. The resulting distribution is shown in figure 7.5b, and is fully compatible with a power law with an exponent $\alpha = 1.9 \pm 0.4$. In a similar way, one can estimate the expected kill curve under the assumption of a scaling in extinction events following an exponent α. The kill curve has a functional form:

$$s(T) = \left[\frac{T_0}{T} + 1\right]^{\frac{1}{1-\alpha}}, \tag{7.6}$$

and estimation of the parameters α and T_0 (a characteristic time scale) gives a kill curve that is shown in figure 7.5a (broken line) for $\alpha = 2.0 \pm 0.2$.

Lifetime Distributions

Another statistical distribution that seems to fit the scaling behavior is the lifetime distribution of genera (Sneppen et al., 1995; Newman and Palmer, 2003). As noted by Raup (1991), the distribution (fig. 7.6a) is skewed, with an average of $\langle T \rangle \approx 20$ million years. Specifically, the lifetime distribution $N(T)$ follows a power-law decay $N(T) \propto T^{-\kappa}$ where $\kappa \approx 2$ (Sneppen et al., 1995). Some estimates provide a smaller value (Newman and Palmer, 2003), but the estimated scaling exponent seems to be reasonably well determined.

A power-law distribution of taxon lifetimes is interesting for several reasons. It indicates that the emergence and extinction of groups not only follows highly skewed laws, but it also suggests complex dynamical patterns. Skewed lifetime distributions appear close to criticality in branching processes and other complex systems (chapter 4). As

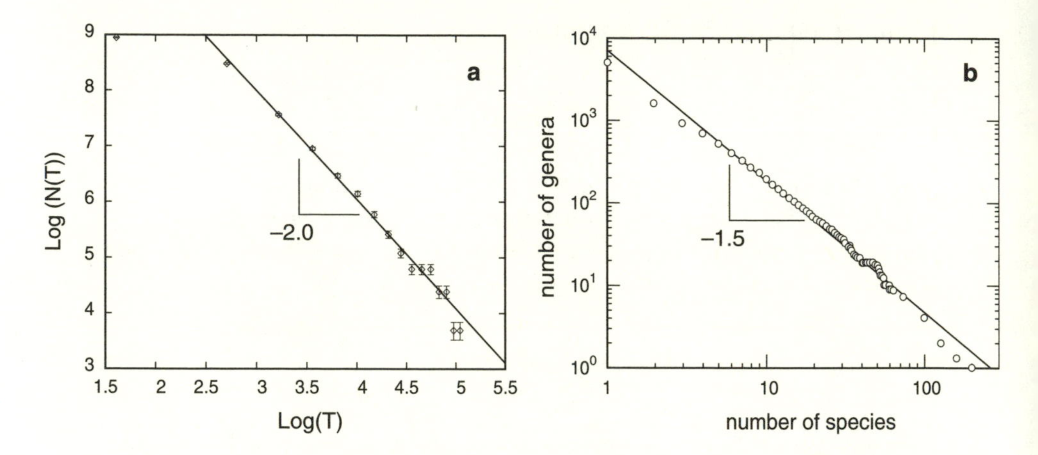

FIGURE 7.6. (a) Lifetime distribution for genera, with a power-law fitting that is consistent with a scaling law $N(T) \sim T^{-2}$. (b) Histogram of the number of species per genus of flowering plants. The solid line is the best power-law fit to the data (after Willis, 1922).

an example, the branching process at criticality follows a scaling law $N(T) \sim T^{-1}$. In this sense, we see that although a critical branching process (CBP) would be a tentative explanation for the observed scaling behavior (since an equilibrium between origination and extinction is known to be present), the scaling exponents are very different.

Fractal Taxonomy

A different source of scaling is observable at the level of taxonomic organization in the fossil record. Scaling laws in phylogenetic patterns have been discussed in different contexts (Willis, 1922; Williams, 1944; Burlando, 1990, 1993). Early work by Willis and coworkers revealed that the number of genera N_g involving S species follows a power-law distribution with $N_g(S) \propto S^{-\alpha_b}$. The analysis of well-known groups of plants (see, for example, fig. 7.6b) gives an exponent $\alpha_b \approx 3/2$. This result reminds us of the exponent obtained for the distribution of tree sizes in a branching process at criticality. Since a given tree can be understood in terms of the whole set of species generated by a given ancestor, it can be easily mapped into a genus. However, other analyses that are somewhat closer to our scales of interest give a different characteristic value $\alpha_b \approx 2$ that seems to be a general one relating taxa and subtaxa beyond the species-genus scale (Burlando, 1990, 1993). Specifically, by examining size-frequency distributions of taxa with different numbers of subtaxa (and for different geologic periods), it was shown that the exponent was consistently close to two for most data. This result indicates that the overall organization of taxonomy is self-similar and provides a structural source of scaling.

Why worry about these scaling laws? These are rather intriguing observations. They suggest that there is no obvious separation between small and large events (as should be expected if two different mechanisms were operating for small and large extinctions). This occurs in spite of the fact that in principle species duration and other biologically relevant properties have well-defined characteristic scales. Scaling, as it has been discussed in other chapters, can originate from a number of mechanisms. All these mechanisms are, at one scale or another, present throughout the history of life. Which of them are more relevant? Most models in the paleontological literature deal with qualitative approximations to different questions concerning the persistence and evolution of taxa. But due to their qualitative character, it is difficult to validate them against real data. Nevertheless, the fact that it is too often assumed that models *must* involve a set of parameters to be tuned, makes the output of them dependent on a number of arbitrary choices. Scaling laws, with their intrinsic simplicity (together with the observed features of time series and the law of constant extinction) provide the best platform from which to develop and test models against reality. In the spirit of chapter 4, we might hope that the complexity of the biosphere follows some fundamental laws of organization that might be describable through simple models. In this context, parameter-free models able to reproduce *several* of the observed trends displayed by the fossil record must be developed as potential theoretical explanations of how macroevolution has happened.

Understanding the previous patterns and their consequences requires a detailed record of life's history at different levels but also the development of an appropriate theory (Maurer, 1999). In the following sections we discuss some of the ecological-based models proposed in order to explain and interpret the temporal patterns of macroevolution at different levels. Although null models are necessary in defining the simplest hypotheses, given their intrinsic (and desirable) simplicity, some relevant features of the fossil record cannot be compared with them. By introducing some additional sophistication (but maintaining the simplicity), models can profit from a range of different observable patterns that can be compared with theoretical predictions.

CHAPTER SEVEN

COMPETITION AND THE FOSSIL RECORD

During the history of our biosphere, many once-dominant groups of animals and plants have waxed and waned. Faunal and floral replacements, so-called biotic replacements are a characteristic feature of macroevolutionary patterns. The origins of such replacements have been controversial (Benton, 1996). Are they a consequence of competitive interactions among clades? As discussed in the previous section, the closeness between extinction and origination rates suggests a biotic control. In such a case, competition among clades should be expected and models of large-scale dynamics, including competitive interactions, should be considered.

One of the best-known models of large-scale evolution in terms of competition was presented by Jack Sepkoski in the late 1970s. We have explored competitive interactions and how competition can explain some simple laboratory experiments, their relative importance and how spatial degrees of freedom can decrease the role of competitive exclusion in some circumstances. Actually, two alternative theories must be considered and tested when looking at the outcome of interactions in paleoecologies. One is competitive exclusion (competitive replacement) due to ecological superiority. The second (known as independent replacement) says that one group declined and became extinct due to some cause unrelated to the presence of a second group. In the latter scenario the second group displays a radiation after the first group has been cleared away (Benton, 1996; Miller, 2000).

Testing between the two alternatives is not an easy task. External events can favor one of the species and harm the second in a way that looks like competitive replacement. But if one group does become extinct clearly before the rise of the second, no competitive exclusion would be (in principle) involved. However, if the two groups are ecologically related and their rise or decline is correlated in time, competition is likely to be involved. Some examples are suggestive but not conclusive: This is the case of the replacement of mammal-like reptiles by the dinosaurs or the replacement of perisodactyl mammals (such as horses) by artiodactyls (such as deer). Two other examples are even better candidates. The first is the replacement of conifers and ferns by flowering plants (angiosperms) as a dominant group. The latter's colonization superiority under a wide range of conditions made them rapidly increase at the expense of gymnosperms (Niklas, 1997). Another example is the replacement of brachiopods by bivalves throughout the Permo-Triassic (PT) mass extinction, which will be analyzed in some detail.

During the Paleozoic (i.e., the 25 million years before the PT event) the brachiopods dominated. These are filter-feeding invertebrates, still present today but largely replaced by bivalve molluscs. A careful analysis of the record shows that bivalves were steadily increasing (in terms of number of genera) through the Permian and that they suffered less than the brachiopods at the Permian extinction. In the aftermath, bivalves outperformed brachiopods. Since these two groups have roughly similar morphologies as well as habitat and feeding requirements, competitive interactions would be expected. However, some authors have argued against this view (Gould and Calloway, 1980). It has been suggested that the Permian extinction simply reset the initial diversities and that extrapolation of microevolutionary theory to large-scale evolution might be misleading. These are certainly open questions, but it is worth looking at an example of explicit extrapolation of competition theory in macroevolution (the Sepkoski's analysis of brachiopod-bivalve replacement) and exploring its possible strengths and pitfalls.

The brachiopod-bivalve replacement can be explored by means of a two-clade competition model (Sepkoski, 1978, 1979). Sepkoski's model involves, in its simplest form, two coupled equations. The variables involved are not species abundances, but diversities of given clades D_i, where $i = 1, 2$ is the clade number. If a given clade grows at a rate r_i (the rate of diversification) and has an equilibrium diversity D_i^*, then Sepkoski's choice is the following two-dimensional coupled Lotka-Volterra model:

$$\frac{dD_1}{dt} = r_1 D_1 \left(1 - \frac{D_1 + D_2}{D_1^m}\right) = \phi_1(D_1, D_2) \tag{7.7}$$

$$\frac{dD_2}{dt} = r_2 D_2 \left(1 - \frac{D_1 + D_2}{D_2^m}\right) = \phi_2(D_1, D_2) \tag{7.8}$$

where D_i^m are the carrying capacities of each clade.

This is a particular case of the L-V model where the competition coefficients are set to one. For this particular case, no coexistence point is allowed, and only exclusion can occur. The Jacobi matrix for this system is (chapter 2):

$$L(D_1, D_2) = \begin{pmatrix} \dfrac{\partial \phi_1}{\partial D_1} & \dfrac{\partial \phi_1}{\partial D_2} \\ \dfrac{\partial \phi_2}{\partial D_1} & \dfrac{\partial \phi_2}{\partial D_2} \end{pmatrix} \tag{7.9}$$

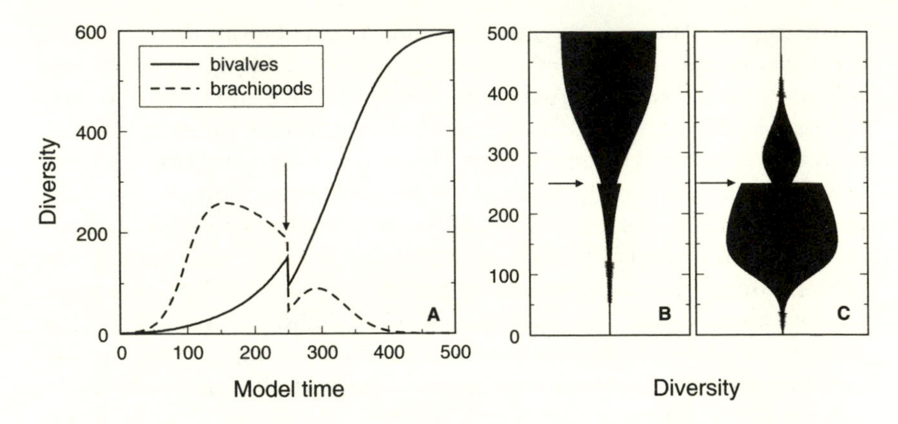

FIGURE 7.7. Competition between clades: (A) Numerical simulation of Sepkoski's equations for the bivalve-brachiopod replacement. An extinction event is introduced at $t = 250$ (arbitrary units). (B–C) Spindle diagrams corresponding to the diversification of both groups.

that is to say,

$$L(D_1, D_2) = \begin{pmatrix} r_1\left(1 - \dfrac{D_1 + D_2}{D_1^m}\right) - r_1\dfrac{D_1}{D_1^m} & -r_1\dfrac{D_1}{D_1^m} \\ -r_2\dfrac{D_2}{D_2^m} & r_2\left(1 - \dfrac{D_1 + D_2}{D_2^m}\right) - r_2\dfrac{D_2}{D_2^m} \end{pmatrix}. \tag{7.10}$$

For the exclusion point $(D_1^m, 0)$, we get two eigenvalues $\lambda_1 = -r_1$ and $\lambda_2 = r_2(1 - D_1^m/D_2^m)$, which will be negative if $D_1^m > D_2^m$. For that case the first clade is stable and the second goes extinct (these are the assumptions of Sepkoski's model, together with $r_1 < r_2$). In figure 7.7 the dynamics of the model is shown, for $D_1^m = 300, D_2^m = 300, r_1 = 0.03$ and $r_2 = 0.06$. The initial condition is $D_1(0) = D_2(0) = 1$. These parameters were estimated from the available series of bivalves and brachiopods (Sepkoski, 1996). The extinction introduced at $t = 250$ mimics the PT event, and is given by decays of 38% and 75%, respectively. In figure 7.7a an arrow indicates the Permian crisis. The curve given by the broken line (brachiopods) shows a rapid increase (due to the condition $r_2 > r_1$), but it also shows a steady decay before the extinction event. This is consistent with the observation that this group was declining before the end-Permian crisis. In figure 7.7b–c we plot the so called spindle diagram, commonly used in paleobiology. We can appreciate the changes in diversity with time and the effects of the external stress. Sepkoski concluded that the nonlinearities in the diversity histories of these two groups are consistent with the hypothesis that brachiopods have been displaced through competition with species of bivalves. This theoretical model also suggests that the end-Permian event was not a fundamental discontinuity in the histories of the two groups.

How reliable are these models of competition to the history of life at such long time scales? One possible flaw of this approximation is the assumption that the parameters describing ecological interactions are fixed through time. This is a strong point, but it might happen that, due to evolutionary constraints (perhaps derived from morphological or developmental factors), the average effects among ecologically related groups can be considered roughly constant over time.

RED QUEEN DYNAMICS

A classical theoretical approach to evolution in multispecies communities was proposed by Leigh van Valen in 1973 and is known as the Red Queen (RQ) hypothesis. This model was introduced as a theoretical explanation for the observation that the probability that a species becomes extinct is approximately independent of its length of existence (Van Valen, 1973; see also Benton, 1995b). In other words, the fossil record suggests that a species might disappear at any time, irrespective of how long it has already existed.

The so-called law of constant extinction can be formulated as follows: Given an initial number of S_0 species (or genera or families), the number of surviving species $S(t)$ at a given time $t > 0$ is:

$$S(t) = S_0 e^{-\mu t}$$

where μ is the probability of extinction of a species (per Myr). Actually, although this law is essentially correct, on average the fine-scale pattern is much more episodic. This is shown in figure 7.8 (Raup, 1986). Here we use sets of families present in the fossil record at a point in time (the so-called pseudocohorts). The number of survivor units of each group is then plotted through time. We can clearly appreciate a roughly exponential decay together with sharp drops in the survivorship associated to mass extinctions.

We would have intuited certain species within any group to become longer lived. Take mammals, for example: Careful analysis of their extinction rates shows that species of modern mammals are just as likely to become extinct as were their ancestors living 200 million years ago (Benton, 1995b). But if evolution leads to improvement through

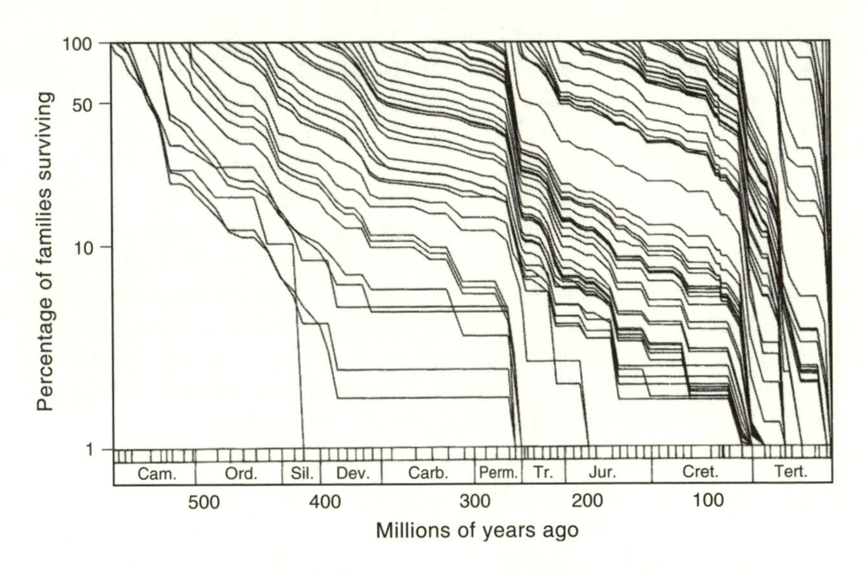

FIGURE 7.8. Successive cohort survivorship curves for 2,316 extinct families during the Phanerozoic (redrawn after Raup, 1991). Together with an average tendency to decline in an exponential fashion (which would give straight lines in this linear-log plot), we observe discrete events that punctuate the dynamics of survivorship.

adaptation, why is it that modern mammals have the same extinction probabilities as their ancestors? Van Valen's interpretation is that species do not evolve to become any better at avoiding extinction. If adaptation improves species progressively through time, a decreasing probability of extinction should be expected: Older species would last longer. Van Valen suggested that constant extinction probability would be the result of an always changing biotic community, with species continually adapting to one another's changes. The name for this conjecture alludes to the Red Queen's remark in Lewis Carroll's *Alice through the Looking Glass*: "Here, you see, it takes all the running you can do, to keep in the same place." Van Valen's view of evolution is that species change just to remain in the evolutionary game. As a consequence of this theory, extinctions occur when no further changes are possible: If genetic variability is not enough, the player is removed from the ecological game.

A model was developed by Maynard Smith and coworkers (see Stenseth and Maynard Smith, 1984, and references cited) in order to test the plausibility of the previous picture, namely, the continuous (co-)evolution of species even in a constant environment. Their model considers a fixed number S of interacting species. It is assumed that some fitness measure ϕ can be defined, and a maximum fitness ϕ_i^* is supposed to exist for every species in a given fixed, external biotic environment. At a given time, the fitness ϕ_i and the maximum ϕ_i^* take different values, and each species tries to reduce the so-called lag load, defined as

$$L_i = \frac{\phi_i - \phi_i^*}{\phi_i}; \quad i = 1, \ldots, S. \tag{7.11}$$

If β_{ij} is the change in the lag load L_i due to a change in L_j, then a mean-field equation for the average lag load $\langle L \rangle = \sum_i (L_i)/S$ can be derived. This is done by first separating, for each species, changes due to "microevolution of coexisting species" from those linked with their own microevolution (Stenseth and Maynard Smith, 1984). The whole equation for the lag load variation in a given species is

$$\delta L_i = \delta_c L_i - \delta_g L_i \,, \tag{7.12}$$

which simply says that it typically increases due to changes in the other species and decreases due to microevolutionary changes in the species under consideration. This can be written in the following way,

$$\delta L_i = \sum_{j=1}^{S} \beta_{ij} \delta_g L_j - \delta_g L_i \,, \tag{7.13}$$

where β_{ij} (with $\beta_{ii} = 0$) is the increase in L_i due to a (unit) change in L_j. If we assume that most species are close to their adaptive peaks, any evolutionary change in one species will have a deleterious effect on the rest of the species. The time-continuous equivalent formulation of this model is

$$\frac{dL_i}{dt} = \sum_{j=1}^{S} \beta_{ij} k_j L_j - k_i L_i. \tag{7.14}$$

By taking the average on both sides of the previous equation, we obtain the following expression for the evolution of the average lag load:

$$\frac{d\langle L \rangle}{dt} = \frac{1}{S} \sum_{i=1}^{S} \left\{ \sum_{j=1}^{S} \beta_{ij} k_j L_j - k_i L_i \right\}. \tag{7.15}$$

Assuming now that $k_i = k$ for all $i = 1, \ldots, S$, the average lag load equation can be written as

$$\frac{d\langle L \rangle}{dt} = \frac{k}{S} \sum_{j=1}^{S} (\Psi_j - 1) L_j, \tag{7.16}$$

and it has a steady state solution if $\Psi_j = 1$ for all $j = 1, \ldots, S$. In other words, if

$$\Gamma \equiv \sum_{i=1}^{S} \beta_{ij} = 1; \quad \forall j. \tag{7.17}$$

Otherwise, it can be shown that $\langle L \rangle$ will decrease (increase) for $\Gamma < 1$ ($\Gamma > 1$). The previous identity is telling us that the equilibrium state of this system is reached through a balance between the reduction of the individual lag load of each species and the increases due to coevolutionary changes in the remaining partners. And the main result of this model is that the Red Queen picture, in which evolution of species proceeds at an approximately steady rate, is indeed feasible even in the absence of changes in the physical environment.

There is a deep connection between this result and the presence of a dynamic instability outlined at the end of chapter 6. As species coevolve, their interactions with other species change. Additionally, immigration from outside the ecosystem will occur, thus introducing further changes in diversity and connectivity. These connections can simply change in time or some new ones can appear or disappear, as new ways of exploiting other species emerge or as new specializations arise. A given tree species in the rain forest can become resistant to some plague, or more able to disperse its seeds. A consequence of this evolutionary adaptation will be an increase in population size. But in a rich ecosystem, larger populations offer greater opportunities for parasites to profit. Larger populations thus are only transient phenomena, as emerging parasites or competitors will find a way to exploit them. Increasing numbers of connections will be inevitable since diversity is usually linked with the emergence of a web of interactions. But after the threshold is reached, increasing numbers of links or changes in their strengths will lead to instability. The system is always forced to change, and species turnover will be the rule (McKinney and Drake, 1998).

EVOLUTION ON FITNESS LANDSCAPES

The Red Queen model, as defined in the previous section, lacks a true evolvable set of parameters. In that sense, the dynamics of a complex coevolving ecology with changing interactions is not explicitly excluded. One of the first attempts to understand large-scale evolution in terms of a complex adaptive system with simple interactions among different species was made by Kauffman and Johnsen, who used previous theoretical ideas on fitness landscapes (Kauffman, 1993; Adami,

2000; see also Gavrilets, 1997; Gavrilets and Gravner, 1997). Kauffman himself elaborated extensively this appealing intuitive concept of the great geneticist Sewall Wright (who used the word "adaptive" instead of "fitness" landscape). The basic idea is that single species can be characterized in terms of a string of genes defining the genotype. Strings have an associated real number. This number is the fitness of the string in terms of the phenotype it produces, and the distribution of fitness values over the space of genotypes defines the fitness landscape. Adaptation is then thought of as a process of "hill climbing" toward higher, nearest peaks. We can imagine a simple situation (see fig. 7.9a) where the fitness only requires the specification of two traits whose characteristics can be measured (the X and Y axis of the plot). Imagine that these characteristics describe the shape of the organism and that different combinations are allowed. The fitness landscape gives us an idea of how optimal these combinations are, and for a given fixed environment a number of peaks corresponding to best-fit combinations are expected to be present.

Depending on the distribution of the fitness values, the fitness landscape can be more or less rugged. The ruggedness of the landscape is a crucial property, strongly constraining the dynamics and leading to universal phenomena. But if macroevolution has to be modeled, then many different, coupled species have to be taken into account. Each species is characterized by the number of genes N and by another parameter K, which is in fact related to the degree of ruggedness. The parameter K indicates how many other genes influence a given gene, that is, the richness of epistatic interactions among the components of the system. In this way, changes in a given gene depend on K other genes. These ingredients define a so-called NK fitness landscape (Kauffman and Levin, 1987: Kauffman, 1993).

Imagine, for simplicity, that we use a binary description of each trait. Then each possible allowed genotype will be a string of bits and we can define some fitness value for it. The fitness landscape is now restricted to an N-dimensional hypercube $\mathcal{H}_N$. For a very simple case with $N = 3$, a three-dimensional cube $\mathcal{H}_3$ is enough (fig. 7.9c). The fitness of a given string is obtained by means of a table of values, as the one shown in figure 7.9b. Here a $N = 3, K = 2$ system is shown, together with the corresponding landscape, here just a simple three-dimensional cube. We can see that, for each particular combination of zeros and ones, a given contribution to the fitness is associate with each unit. The fitness of the organism is then obtained as the average fitness (i.e., $w = (w_1 + w_2 + w_3)/3$, fig. 7.9b, last column in the table).

The fitness $\phi(\mathbf{S}_i)$ of a given string $\mathbf{S}_i = (S_i^1, \dots, S_i^N) \in \mathcal{H}_N$, together with the fitness of the local neighbors of $\mathbf{S}$ (i.e., the set $\Gamma_i = \{\mathbf{S}_k\}$) will be

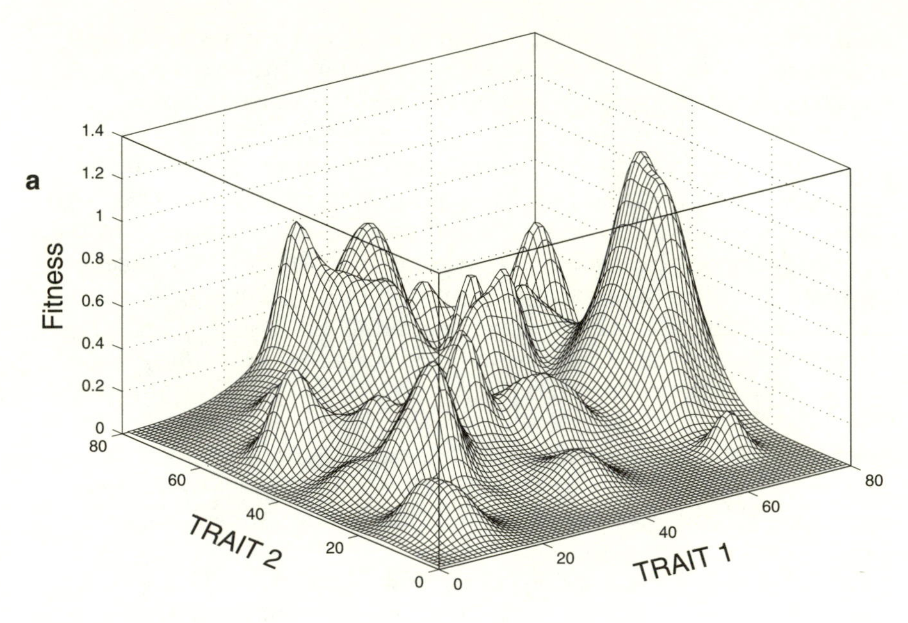

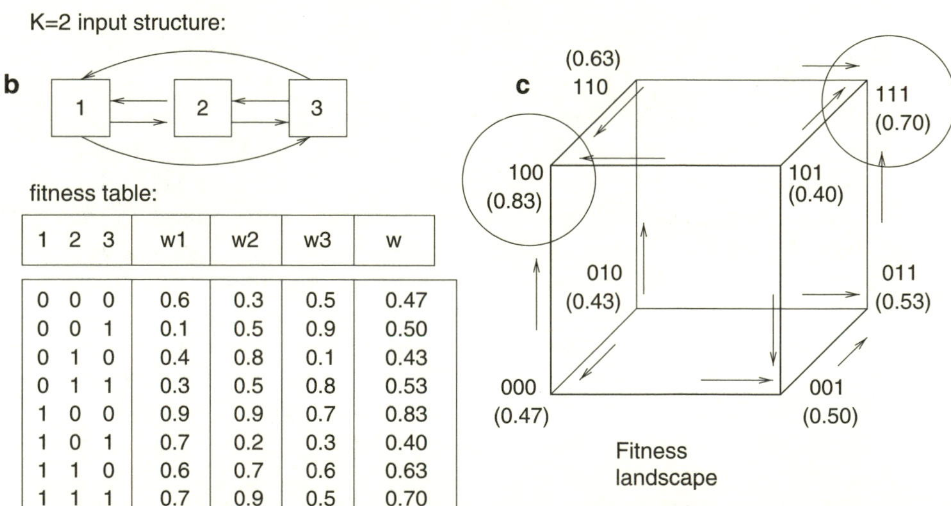

1 2 3	w1	w2	w3	w
0 0 0	0.6	0.3	0.5	0.47
0 0 1	0.1	0.5	0.9	0.50
0 1 0	0.4	0.8	0.1	0.43
0 1 1	0.3	0.5	0.8	0.53
1 0 0	0.9	0.9	0.7	0.83
1 0 1	0.7	0.2	0.3	0.40
1 1 0	0.6	0.7	0.6	0.63
1 1 1	0.7	0.9	0.5	0.70

FIGURE 7.9. (a) The fitness landscape, as depicted by Sewall Wright: Here a simplified view of it is shown as a surface on a two-parameter space. In (b–c) a Boolean description of states is used in describing the genotype of a given organism with only three genes ($N = 3$), and two epistatic interactions (i.e., $K = 2$) are introduced (b). The fitness of each string is randomly defined by assigning a random value $w_i \in [0, 1]$ for each gene in each string. The total fitness is computed as the average $w = \sum_i w_i/N$. The resulting landscape is indicated in figure (c). The end points of the adaptive walks that can be performed on this cube are indicated by means of circles.

used in order to define the evolutionary dynamics on the NK landscape. The dynamic proceeds by means of so-called adaptive walks. Here for a single species we choose a given trait and flip a coin (i.e., mutate) a single element s_i^j. Then we look at the fitness table, and if the average fitness of the new configuration is larger than the previous one, an adaptive walk occurs and so a movement in the fitness landscape. If not, no walk is allowed to occur. This simple procedure leads to a hill climbing in the landscape until a local peak is reached. A local fitness peak will occur at a given $\mathbf{S}_L \in \mathcal{H}_N$ if $\phi(\mathbf{S}_L) > \phi(\mathbf{S}_j)$ for all $\mathbf{S}_j \in \Gamma_L$. Once a local optimum is reached, the walk stops.

This *NK* model has been widely explored, and many relevant results concerning its statistical properties have been derived (Kauffman and Levin, 1987; Kauffman, 1993; Weinberger, 1989, 1990, 1991). In particular, since the fitness of a given gene is epistatically affected by K other genes, it can be shown that increasing K leads to a rapid increase in the number of local maxima. This is actually a universal feature of a wide range of models from statistical physics, such as spin glasses (Mezard et al., 1999). For $K = 0$ (the so-called Fujiyama landscape), no interactions among different traits are present and a smooth landscape is obtained, with a single global maximum and an expected number of walks $L_w = N/2$ to reach the optimum. This is a highly correlated, simple landscape. At the other extreme, when $K = N - 1$, the landscape is fully random. As a rule, the more interconnected the genes/traits are, the more rugged the landscape. These constraints produce the multipeaked nature of the landscape and lead to a large number of suboptimal solutions. A statistical study of this landscape is easy to perform (Kauffman and Levin, 1987). Several interesting properties have been reported, among others: (a) the number of local fitness optima is maximum; (b) the expected number of fitter one-mutant variants drops by half at each improvement step; (c) the length of adaptive walks to optima are short, with a characteristic value $L_w \approx log(N)$.

The last property is actually susceptible to being tested from available data. In this context, Kauffman suggested that the asymmetry observed between the Cambrian event, where all basic phyla became established, and other events of wide diversification (such as the Permo-Triassic crisis), where no new phyla or classes emerged, might result from the generic features of adaptive evolution on rugged landscapes (Kauffman, 1989; Solé et al., 1999). Instead of an explanation based on ecological opportunity or developmental canalization between the Cambrian and the Permian, Kauffman suggested that the Cambrian event is an example of hill climbing on a rugged landscape. Specifically, the first multicellular organisms would have started from low-fit states and rapidly explored a large diversity of alternative morphologies, thereby establishing phyla. This rate would rapidly slow down and new variants be very hard to reach, even under the presence of great ecological opportunity. One test of this hypothesis was performed by Gunter Eble (1998), who analyzed the increase in innovation (in terms of new body plans) against the number of tries (measured in terms of cumulative genera) and found a logarithmic increase, as predicted.

Adaptive Walks in Morphospaces

A particularly neat example of the applicability of fitness landscapes to evolution is the work of Karl Niklas on the patterns of morphological diversification in plants (Niklas, 1994, 1997). This work involves a comparative analysis of fossil plant phenotypes with simulated forms resulting from adaptive walks on a fitness landscape. The available set of forms is called a morphospace, the domain of all conceivable phenotypes (McGhee, 1999). Each point in this (multidimensional) space has some assigned fitness.

To define a realistic morphospace is not easy. Any choice will contain some number of arbitrary assumptions. A morphospace is a context-dependent structure: Relative fitness depends on environmental constraints. But there are a number of fundamental constraints relating the physical laws and processes governing the exchange of mass and energy between plants and their environment (Nobel, 1983). The successful scaling theories in biology (Brown and West, 2000) fully support the view that, in spite of the overwhelming variety of life, universal principles of morphological organization pervade most context-dependent scenarios.

The phenotypes defined by Niklas models introduce a small number of highly relevant features, such as mechanical properties or the ability to capture light and nutrients. These properties lead to epistatic contributions to fitness: Mechanical constraints, for example, will alter the pattern of ramification and thus the capacity to gather light. As predicted by NK models, an increasing number of functional obligations leads to an increasing number of phenotypes with similar relative fitness (and thus to a larger number of adaptive peaks/attractors). The morphospace used by Niklas is three-dimensional. It involves three tasks: light harvesting, mechanical stability, and reproduction (Niklas, 1997; Niklas and Kerchner, 1984; McGhee, 1999). The scalar parameters are: (a) p, the probability of branching termination (which is a measure of bifurcation frequency); (b) γ: the rotation angle of the branch and (c) ϕ the branching angle. For each set (p, γ, ϕ), a fitness function $F(p, \gamma, \phi)$ is defined. In Niklas analysis, the three functional obligations (growth, survival, and reproduction) can be quantified by means of closed-form equations derived from biophysics and biomechanics.

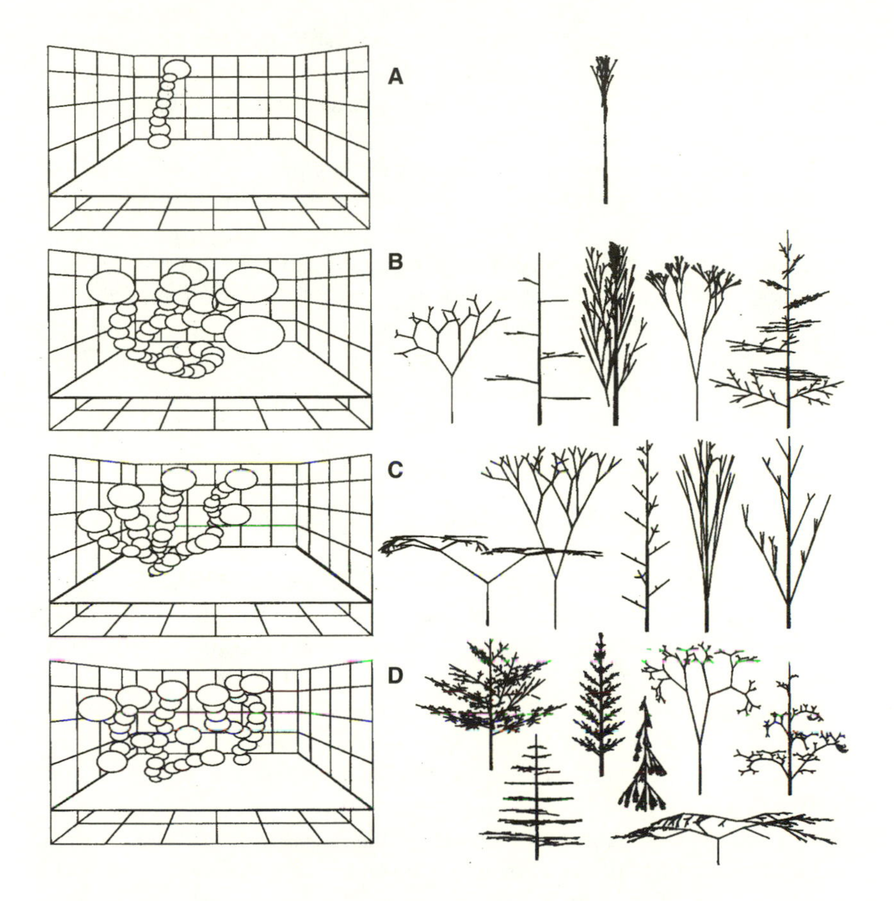

FIGURE 7.10. Adaptive walks through plant morphospace of early vascular land plants (adapted from Niklas, 1997). Here the morphospace is schematically represented at left and the final structures that are obtained (at local optima) are shown at right for each case. The walks are shown as "bubbles" that indicate one step in the walk. Their size scales with the search domain that had to be explored to locate the next most-fit plant morphology. From top to bottom, the number of constraints to be simultaneously optimized (mechanical stability, reproductive success, and light interception) combines and/or increases in number (see text).

As it is illustrated in figure 7.10, the final result of many adaptive walks on this landscape will depend on the number of tasks considered. If only mechanical resistance is chosen (fig. 7.10a), then a single peak is available and thus only one morphology is reached. If two tasks are considered simultaneously, then several peaks are accessible through the walks, and an increasing diversity of plant forms is obtained (fig. 7.10b–d). If the three constraints are introduced simultaneously, the whole spectrum of vascular land plant morphologies is obtained. In spite of the limitations implicit in this approximation (Niklas, 1997), the outcome of the multi-task scenario is spectacular. On the other hand, early evolution of plants on land took place on a basically uninhabited landscape and thus was presumably driven by physical laws. And perhaps more importantly, this work shows that "organic complexity may not impose the severe limits on evolution that are sometimes envisioned" (Niklas, 1997). Instead, it suggests that the evolution of complex organisms (and indirectly that of ecosystems) may lessen the burden of climbing the adaptive peaks of a fitness landscape, by means of the multiplicity of accessible, low-fit peaks.

Coevolution on Rugged Fitness Landscapes

Now the problem is how to obtain a more complete picture of an evolving system formed by many species in interaction. This can be done by using the so-called NKC model (Kauffman and Johnsen, 1991). The new parameter C introduces the number of couplings between different species and allows one to obtain an evolutionary link with the stability-complexity debate raised in chapter 6. Once again, each species is represented by just a string. But now the previous table must be extended in order to take into account that each trait receives input from C other traits belonging to different species. These traits are chosen at random between the S species involved.

The NKC model thus involves three basic parameters describing: (a) the number of traits required to characterize a given species (N), (b) the number of so-called epistatic interactions among genes in the same species[2] (K) and (c) the number of interactions among traits of different species (C). In figure 7.11a we show a simplified view of the landscapes of three coupled species. Here $N = 3$, and the local peaks are indicated by black circles. Some value of K is assumed to be defined. The specific set of traits displayed by each species is indicated as a string. The connections between traits of different species are indicated by means of arrows. We can see that species 2, 3, and 4 are already at their local peaks, but species 1 is not (A, white circle). It is close to a local peak (B) and thus in the next step it will perform an adaptive walk toward the peak at B. In figure 7.11b, a possible result of the adaptive walk is shown: Species 1 is now at the local peak as is species 2. But now species 3 and 4 are not located in fitness peaks: Their landscapes have been modified by the adaptive move of species 1. This is what occurs

[2] In this model it is assumed that all individuals can be considered equivalent and thus only one genotype (instead of a population) represents the whole population.

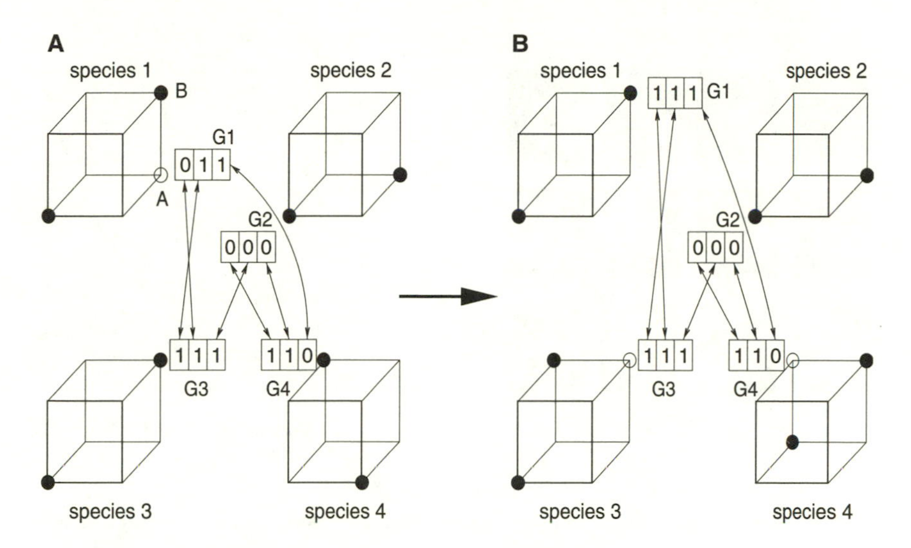

FIGURE 7.11. Evolutionary dynamics in the NKC model. Here four species are considered, each one described by $N = 3$ traits and thus represented as a three-dimensional fitness cube. Local optima are indicated by means of black circles, and current states by white circles. Only the interactions among traits influenced by other species are shown. The movement of species 1 from A to B (an adaptive walk) is indicated in (A). As a result of this movement in fitness space, the landscapes of the other species becomes modified (B); new local optima are at work and further coevolution will take place.

at the chaotic regime: Changes in a given species propagate through the system. There is however another regime (characteristic of small connectivities): the frozen, ordered regime where species reach their local peaks and remain there.

Using extensive computer simulations, Kauffman and Johnsen demonstrated the presence of a wide variety of dynamical behaviors as the parameters were tuned. One particular property is the presence of two well-defined regimes. These regimes are the chaotic phase, where changes in the ecosystem are always taking place (i.e., the system does not settle down in a number of local optima, and one might talk about a chaotic Red Queen phase), and the ordered (or frozen) phase where local optima are reached by all species. These steady states roughly correspond to Nash equilibria (Kauffman, 1993; Hofbauer and Sigmund, 1991). An example of a particular case where species are located on a two-dimensional lattice is shown in figure 7.12a, where twelve successive steps are shown. Frozen and changing species are indicated as black and white squares, respectively. We can see that the changing

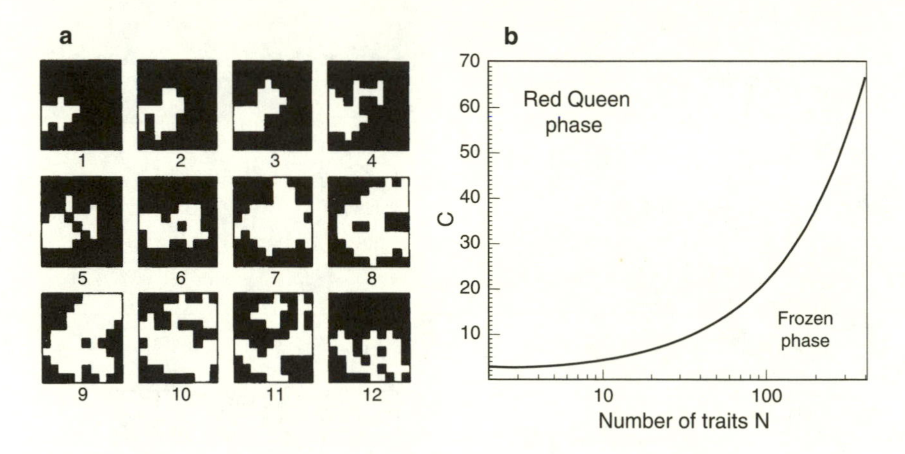

FIGURE 7.12. (a) An example of the coevolutionary dynamics in a simple ecosystem with square lattice topology (here a 10×10 system is used, so that $S = 100$ species are present). Each species is linked only to one of its four nearest neighbors ($C = 1$) and $K = 10$. After some transience, the system reaches a Nash equilibrium (adapted from Kauffman, 1993). (b) The phase space of the NKC model for $K = N - 1$, defined as the number of interactions among traits of different species (c) versus the total number of traits (N). The critical line separates the frozen phase from the chaotic (Red Queen) phase.

(Red Queen) domains define a cluster that shrinks and expands through time as a consequence of coevolutionary responses.

Just at the boundary between both regimes, species in a finite system reach local peaks but any small perturbation generates a coevolutionary avalanche of changes through the system. The distribution of these avalanches is a power law, as expected for a critical state. Kauffman and Johnsen mapped these avalanches into extinction events, suggesting that the number of changes in species is proportional to the extinction of less-fit variants. If this analogy is used, then the obtained scaling relation for the number of avalanches $N(s)$ involving s changes is $N(s) \propto s^{-1}$, which does not agree with the value reported from the fossil record. However, a recent variation of this model (see Newman and Palmer, 2003) allowing changes in the parameters (i.e., to the coevolution of connections among species) has been shown to self-organize to the critical state with avalanches following the correct $\alpha = -2$ exponent. This model shows that, as species tune their own landscapes (by readjusting their parameters and so the landscape ruggedness), the whole system drifts toward the instability boundary.

The presence of two phases can be derived through simple theoretical arguments for $K = N - 1$ (Bak et al., 1992). The basic idea is the following: We know that for this landscape the number of walks L_w

required for a given species to reach a local fitness peak is on average $L_w \approx log(N)$. Let us assume that all species are located at local peaks and that a perturbation is performed: One of the species (say species 1) is moved to a random position in its fitness landscape. This species will start to climb toward some local peak. If interactions among species are rare, the other species will remain unaffected by the adaptive walks of species one. But if C is large enough, then the other species can see their landscapes modified and start to change too.

Each adaptive walk of species 1 involves a change in a given trait. But in fact the fitness of other species depends, through C, on the values taken by the genes/traits of species 1. If any of these genes/traits are among the ones that changed through the walk, the affected species will be set back in evolution. The question is, what is the critical condition that defines the combination of N and C that can trigger a "chain reaction" able to percolate through the system? The critical condition is easily obtained. The probability that a given trait in a random species depends on species 1 is C/N. The critical condition is that at least one change in a species occurs. This means:

$$L_w \frac{C_{crit}}{N} = 1, \tag{7.18}$$

that is, when, on the average, one out of C randomly chosen genes is among the L_w changed genes. This gives the critical line in the (N, C) space:

$$C_{crit} = \frac{N}{\ln(N)}, \tag{7.19}$$

which is shown in figure 7.12b. When the number of traits (the species complexity) is such that $C > C_{crit}$, the interactions between different genotypes constantly modify the underlying fitness landscapes, and coevolutionary avalanches occur.

The suggestion that large-scale evolution might be related to critical dynamics triggered some further theoretical efforts that culminated in the formulation of the so-called Bak-Sneppen (BS) model (Bak and Sneppen, 1993; Sneppen et al., 1995; Bak, 1997). The model, which has been extensively explored under different approximations, provided the simplest approach to evolution through self-organized critical dynamics (chapter 4). The BS model allows us to understand some of the key features of coevolutionary dynamics under self-organized criticality: Regardless of the state in which they start, the BS ecosystems always tune themselves to a critical point. As with the NKC model, coevolutionary avalanches are triggered by small events, and a toy representation of a fitness landscape is used. As with the NKC model, avalanches of activity are defined, and scaling laws are obtained, although the exponents, again, fail to reproduce the real ones. In summary, the previous models provide a powerful picture of large-scale evolutionary dynamics, but the comparison between the model and the fossil-record data is difficult. In this context, although co-evolutionary avalanches might be related to true extinction events, it is desirable to make use of models that explicitly introduce extinction and diversification (Maddox, 1994). Beyond the approaches dealing with coevolutionary avalanches, two possible scenarios can be introduced: models considering externally driven dynamics and models in which ecological responses play a leading role. Each will be considered in the following sections.

EXTINCTIONS AND COHERENT NOISE

The simplest extinction model involving external stresses was presented by Newman (1996, 1997). It is actually a direct translation of a previous model of coherent noise in large physical systems (Newman and Sneppen, 1996). The model has a fixed number N of species, which in the simplest case do not interact. The predictions of the model are not greatly changed if one introduces interactions, and the noninteracting version makes a good starting point because of its extreme simplicity. The absence of interactions between species also means that critical fluctuations cannot arise, so any power laws produced by the model are definitely of noncritical origin.

The key assumption of Newman's model is that the extinction of a species is caused by a change in its environment, which may take place in many different forms. During its lifetime, a species will in general experience a number of these stresses, and for any given stress, each species will have a certain tolerance (or lack of it). This tolerance is defined by a species fitness measure denoted by x.

When the stress occurs, species with higher values of x are less likely to become extinct than those with lower values. The ability to withstand stress could depend on many factors, such as ability to adapt to a new environment ("generalists" versus "specialists"), or possession of particular attributes. For convenience, x is assumed to take values between zero and one, though this is not a necessary condition. On the other hand, the level of the environmental stress will be represented by a single number η, which is chosen independently at random from some distribution $p_{stress}(\eta)$ at each time-step. In this way, a true extinction takes place and its size is directly determined from the number of removed species. In order to maintain a constant number of species, the system is repopulated after every time-step with as many new species as have just become extinct. The extinction thresholds x_i for the new species can either be inherited from surviving species, or can be chosen at random from some distribution $p_{thresh}(x)$. In the following, a uniform distribution for x_i will be assumed.

In its simplest form, the model defines an ecosystem consisting of some large number N of species, where each species has a fitness x_i that can take values between zero and one. The dynamics of these fitnesses follows the next two rules, which are applied repeatedly:

1. At each time-step, a small fraction f of the species, selected at random, evolves, and their fitnesses x_i are changed to new values chosen at random in the range $0 \leq x_i < 1$.
2. We choose a stress level η randomly from some distribution $p_{stress}(\eta)$, and all species whose fitnesses x_i lie below that value become extinct and are replaced by new species whose fitnesses are also chosen at random in the range $0 \leq x_i < 1$.

The random selection of η at each step can be thought of as imposing external stresses that coherently (i.e., simultaneously) influence species with fitness lower than η. In order to complete the model, $p_{stress}(\eta)$ needs to be specified. Several choices are possible. If asteroid impacts are the source of stress, skewed distributions of stress are the reasonable choice. Using different distributions, Newman found that, in a fashion reminiscent of the self-organized critical models, many of its predictions are robust against variations in the form of $p_{stress}(\eta)$, within certain limits. In spite of its simplicity, the outcome of coherent noise dynamics is remarkable (Newman and Sneppen, 1996; Newman, 1996, 1997). In particular, scaling behavior is obtained despite the absence of critical dynamics: power-law-event size distributions are typical even in the absence of any interaction between different species, and the observed extinction size distributions are statistically similar to those observed from the fossil record.

In figure 7.13a, the distribution of extinction events is shown using a Gaussian form:

$$p_{stress}(\eta) \sim e^{-\eta^2/2\sigma^2} \tag{7.20}$$

with $\sigma = 0.1$ and zero mean. A perfect power law is obtained (except for a plateau at very small extinction sizes) with an exponent 2.02 ± 0.02, in agreement with fossil data. Although the exponent varies somewhat as the stress distribution is changed, it is always close to two. An analytic derivation of this result was presented by Newman and Sneppen (1996), and although further refinements can be introduced, such as

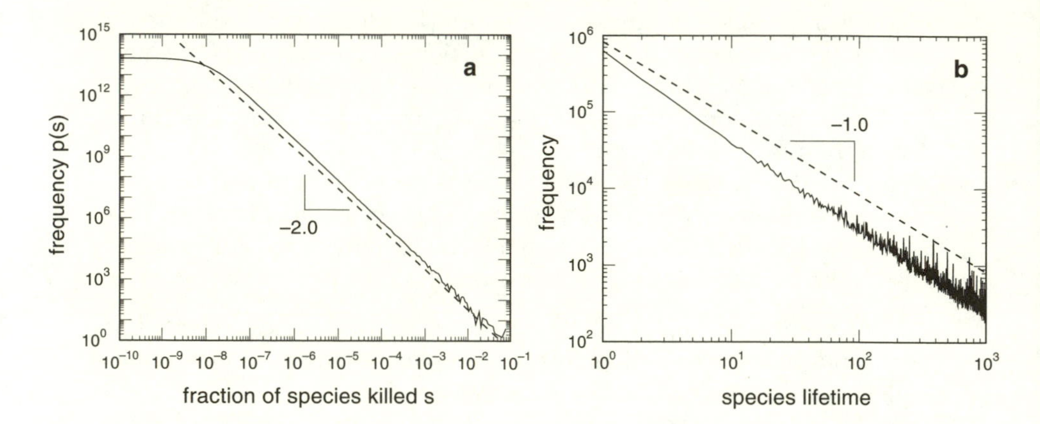

FIGURE 7.13. (a) Distribution of the sizes of extinction events taking place in the Newman's model (1996). The distribution is power law in form with an exponent of $\tau = 2.02 \pm 0.02$ except for extinctions of very small size, where it becomes flat. (b) The distribution of the lifetimes of species. It follows a power law with an exponent close to one.

inheritance of the x_i values from the survivors, the previous result is robust.

Other key quantities are easily obtained. In figure 7.13b, the predicted distribution of species lifetimes is shown (see also Adami, 1995). Again, a scaling law is obtained but, as noted by Newman, fossil data concentrate on higher taxa, where more reliable data are available. Such higher level information can be obtained by properly defining genera. Although the model does not (explicitly) include taxonomic organization, genera can be defined and their lifetimes calculated. By following the phylogenetic trees derived from a given founding species, the lifetime of these trees (that define a genus) provide the distribution $N(T)$ for the model. The result is a power law with an exponent $\tau = 1.0 \pm 0.1$ (fig. 7.14a). Additionally, the structure of the taxonomic trees agrees with the observation of scaling in the genera-species distributions. An example is shown in figure 7.14b: The scaling exponent is close to 3/2. Although the scaling in the genera lifetimes strongly deviate from the observed values, the one for the taxonomy agrees with Willis's observations.

Many variations of the basic rules defined by this model have been explored (Newman, 1997), and they confirm the previous regularities and provide some predictions, such as the presence of aftershock extinctions.

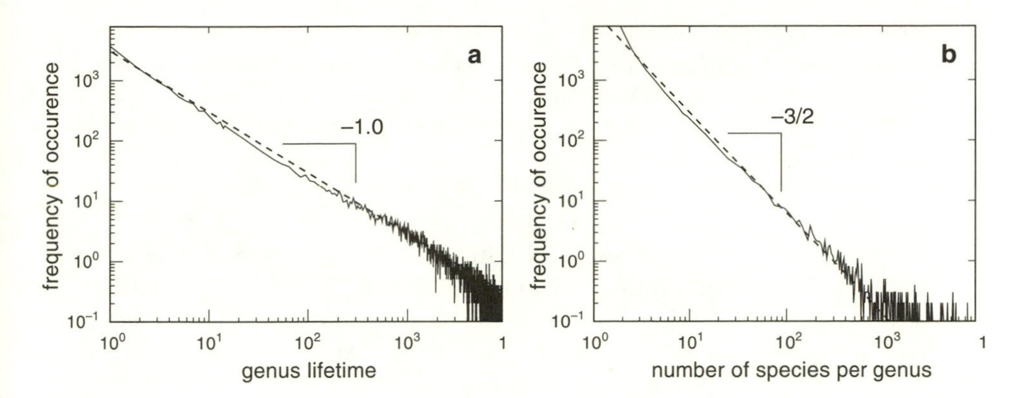

FIGURE 7.14. Genera-level results from Newman's model. (a) The distribution of genus lifetimes measured in simulations of the model. A scaling law is obtained, with an exponent $\tau = 1.0 \pm 0.1$. (b) Histogram of the number of species per genus. The distribution has an exponent of 1.6 ± 0.1.

NETWORK MODELS OF MACROEVOLUTION

How relevant are nonlinear phenomena in large-scale evolution (Rand et al., 1994; Solé and Bascompte, 1996)? The previous model shows that, at least to some extent, several key features of the fossil record can be accounted for through an externally driven mechanism. This approach is very different from the critical dynamics considered in the NKC or the Bak-Sneppen models. Nevertheless, the question of scale plays a particularly important role. As Simon Levin suggests, critical states are likely to be relevant in terrestrial ecosystems where immigration/speciation events are especially important, but clearly external events have played an important role in shaping our biosphere and it is far from clear that critical phenomena might be operating at all scales (Levin, 1999). Additionally, it is becoming increasingly clear that atmosphere-biosphere interactions play a key role in biodiversity dynamics and thus the relationships between biotic and abiotic factors can be bidirectional. Multiple attractors are often involved (chapter 2) and thus rapid shifts can take place associated with relatively mild changes in climatic variables. The coupling between climate and ecosystem structure and function produces complex behavior (Higgins et al., 2002; Lenton and van Oijen, 2002, and references cited).

In previous chapters several aspects concerning the nonlinear behavior of complex ecologies were presented. We discussed different problems relating connectivity, stability, and diversity, and some field data seemed to support the idea that current ecologies might display some features characteristic of marginally stable states (chapter 6). Can a similar situation take place at the evolutionary scale? The presence of such phenomena at two very different scales is significant, and we might ask if an ecological-based network theory of species changes can provide some understanding of large-scale evolution.

A simple model of large-scale evolution involving a set of N interacting species can be easily defined (Solé, 1996; Solé and Manrubia, 1996, 1997; Solé et al., 1996a). In this model, species interactions are introduced by means of a $N \times N$ connectivity matrix $\mathbf{W} = (W_{ij})$, and evolution is represented through changes in its elements. The "state" of each species is described by a binary variable $S_i \in \{0, 1\}$ ($i = 1, 2, \ldots, N$), for the i-th species. Here $S_i = 1$ or $S_i = 0$ if the species is alive or extinct, respectively. So the whole ecosystem is described in terms of a simple, directed graph in which the connections are initially set to random values. Each species receives input from some others. A given input can be positive or negative. These signs indicate that the given species is favored or harmed by the species that sent the input. The first case would correspond to a prey or mutualist and the second to a predator or parasite, for example. In this model, species are in fact described by their sets of connections.

A small number of simple rules is introduced in a number of ways (Solé, 1996; Solé et al., 1996a; Solé and Manrubia, 1996); in its simplest form, as indicated in figure 7.15a–d, these are:

(i) Random changes in the connectivity matrix. At each time step we pick up one input connection for each species and assign to it a new, random value without regard for the previous state of the connection. This rule introduces changes into the web. These changes can be due to evolutionary responses or to environmental changes of some sort. In other words, changes derived from coevolution among two species, innovation at the species level, and/or environmental-driven changes are lumped together within this rule.

(ii) Extinction. Changes in the connectivity will eventually lead to extinctions. In this model, we compute the total sum of the input for each species, and this sum defines the condition for extintion: If it is negative, the species is extinct ($S_i = 0$) and all its connections are removed. Otherwise, nothing happens ($S_i = 1$). In other words, the state of the i-th species is updated following

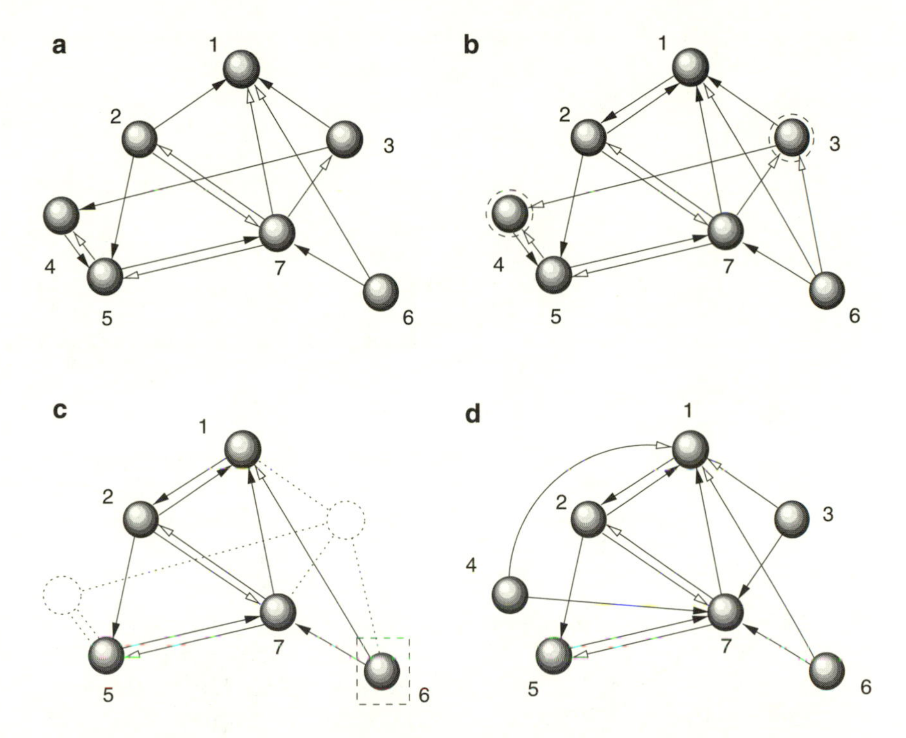

FIGURE 7.15. An example of the rules used in the evolution model. Here an $N = 7$ network is shown, with a given initial connectivity (a). Here black and white arrows indicate negative and positive interactions, respectively. The first step is to introduce changes in the links. In (b) some links have been changed (even in sign) as a consequence of the first rule. Different pairs of interactions are observed, including competition, predation or parasitism, and mutualism. In this example, two species (4 and 3) become nonviable and will die in the next step. In (c) the extinction rule has been applied and the two species and their links removed. Species 6 is selected (among the survivors) and copied into the two empty sites. This defines the third rule, including diversification. The copies carry the links of their parent species, which will be modified once the first rule is applied again.

the dynamical equation:

$$S_i(t+1) = \Phi\left[\sum_{j=1}^{N} W(i,j)\right] \tag{7.21}$$

where $\Phi(z) = 1$ for positive z and zero otherwise.

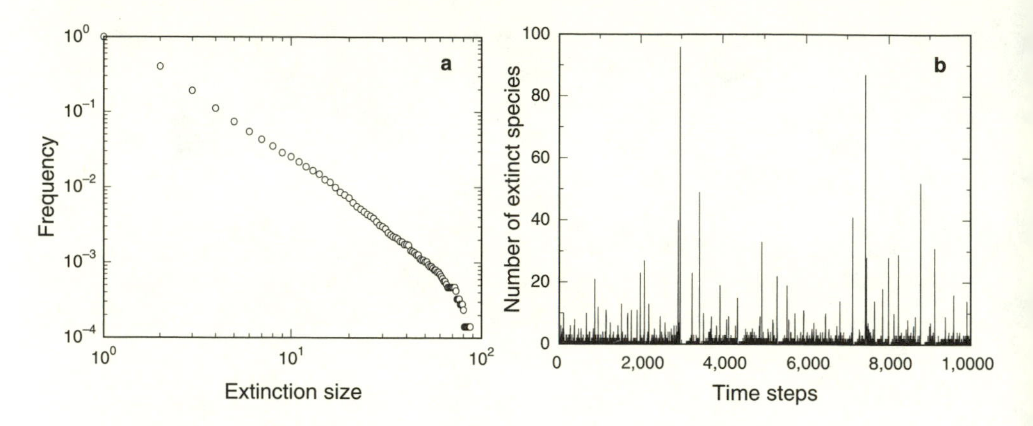

FIGURE 7.16. (a) Power-law (cumulative) distribution of extinction events in the network model. The corresponding time series for this case is shown in (b), displaying the first 10^4 steps in the simulation. Here $N = 150$ species are being used.

(iii) Diversification. A number of species can disappear as a consequence of the previous rule. These empty sites will be refilled by diversification. Each extinct species is replaced by a randomly chosen survivor. The replacement is made by simply copying the connections of the survivor into the empty site.

This model and its variants have been analyzed in detail (Solé, 1996; Solé and Manrubia, 1996; Manrubia and Paczuski, 1998; Solé et al., 1998). In particular, this model shows a strongly nonlinear behavior with avalanches of extinction as well as the correct power-law distribution of extinction sizes (although with an exponent typically close but higher than $\alpha = 2$; see also Drossel, 2001). An example is shown in figure 7.16a–b, where the temporal dynamics of extinctions is represented together with the statistics of extinction sizes. We can see that both small and very large events are generated by the same dynamical rules. Most of the time, the extinction of a given species has no consequences for the other species. But from time to time, a given (keystone) species with positive input to others might disappear. The removal of this species can have a destabilizing effect on others. This effect can further propagate, leading to a mass extinction event.

What is the origin of the extinction pattern? We first must see how a given species can shift toward a negative sum of inputs. The reason is easy to understand from rule (i). Since changes of links among species are random, and the new values are chosen from a uniform distribution, an expected consequence is that, in the long run, the sum of inputs will decay to zero. If we look at the sign of the links, so that $P(W^+) = P[W_{ij} > 0]$ and $P(W^-) = P[W_{ij} < 0]$, the time evolution of the positive links can be described in terms of a master equation (chapter 4):

$$\frac{dP(W^+,t)}{dt} = w(W^- \to W^+)P(W^-) - w(W^+ \to W^-)P(W^+) \qquad (7.22)$$

where $P(W^+) + P(W^-) = 1$, starting for example from an initial condition $P(W^+, 0) = P_0$. Since, from the first rule, we have $w(W^- \to W^+) = w(W^+ \to W^-) = 1/2N$, the master equation reads:

$$\frac{dP(W^+,t)}{dt} = \frac{1}{2N}\left[1 - P(W^+)\right] \qquad (7.23)$$

and an exponential decay is obtained:

$$P(W^+,t) = \frac{1}{2}\left[1 + (2P_0 - 1)e^{-t/N}\right]. \qquad (7.24)$$

As a consequence, the sum of inputs $\mathcal{F}_i = \sum_j W_{ij}$ will also decay as: $\mathcal{F}_i(t) \sim e^{-t/N}$, predicting an exponential decay in the probabilities of survival, as expected from the Red Queen hypothesis. This rule actually introduces the basic ingredients of van Valen's theory. All species in the system continue changing through time (either due to biological or environmental causes), eventually reaching extinction. The ultimate fate of all species is to go extinct, and so an exponential decay in the survival probability will be observed. Here, however, there is no intrinsic, species-level variability (in terms of a genotype), and the fate of a given species will be dominated by network responses and chance. Ecological-driven phenomena are the key forces in the long run, although small-scale events take place all the time. The nature of the decay is exponential on average but episodic when looking at pseudocohorts (see fig. 7.17), consistent with Raup's analysis (compare with fig. 7.8).

Mean Field Model

Detailed analytic study of the previous model is a difficult task, mainly due to the complexity arising from the random matrix structure of the interactions. A further simplification can be obtained by mapping the previous set of rules into a linear model (Manrubia and Paczuski, 1998), which is actually a mean field approach to the network model. The starting point is again a set of N species, now characterized by a single integer quantity ϕ_i ($i = 1, 2, \ldots, N$). This quantity will play the

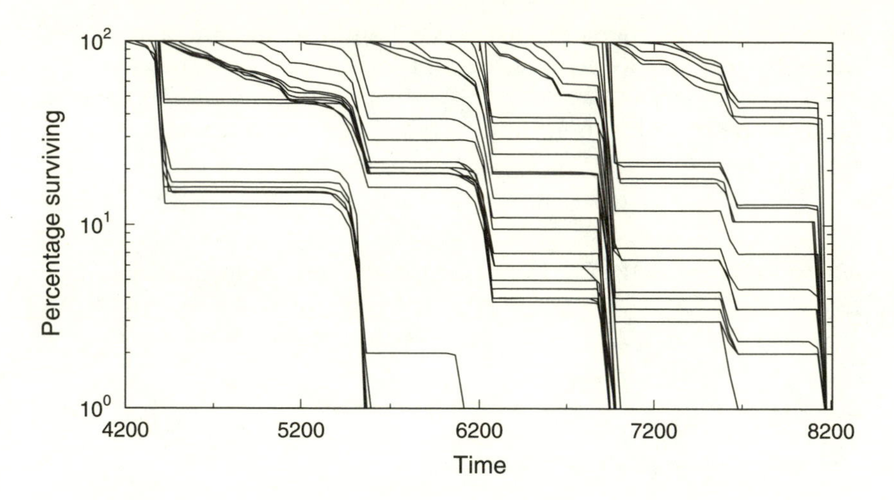

FIGURE 7.17. Pattern of species extinction in the network model of macroevolution: On average, species go extinct in an exponential way, but this average pattern is punctuated by coextinction associated with mass extinction events (to be compared with figure 7.8).

role of the internal field. Each species is now represented by this single (integer) number $\phi_i \in \{-N, -N+1, \ldots, -1, 0, 1, \ldots, N-1, N\}$ that represents the sum of inputs from other species. So in this case the whole set of connections defining each species is replaced by a single number.

The dynamics consists of three steps: (a) with probability $P = 1/2$, $\phi_i \to \phi_i - 1$, otherwise no change occurs (this is equivalent to the randomization rule in the network model); (b) all species with $\phi_i < \phi_c$ (below a given threshold) are extinct. Here we use $\phi_c = 0$ but other choices give the same results. The number of extinct species, $0 < E < N$, gives the size of the extinction event. All E extinct species are replaced by survivors. Specifically, for each extinct site (i.e., when $\phi_j < \phi_c$), we choose one of the $N - E$ survivors ϕ_k and update ϕ_j to: $\phi_j = \phi_k$; (c) after an extinction event, a wide reorganization of the web structure occurs (Solé and Manrubia, 1997). In this simplified model, this is introduced as a coherent shock. Each of the survivors is updated as $\phi_k = \phi_k + q(E)$, where $q(E)$ is a random integer between $-E$ and $+E$.

The previous rules reproduce the most relevant features of the observed dynamics in the fossil record, including the presence of well-defined transient trends (Solé et al., 1998). This mean-field approximation allows us to obtain some analytic results. The discrete dynamic is given by the following three-step process:

$$N(\phi,t+1/3)=\frac{1}{2}N(\phi,t)+\frac{1}{2}N(\phi+1,t) \tag{7.25}$$

$$N(\phi,t+2/3)=N(\phi,t+1/3)+N(\phi,t+1/3)\sum_m \frac{m}{N-m}P(m) \tag{7.26}$$

if $\phi > 0$ and zero otherwise. Finally:

$$N(\phi,t+1)=N(\phi,t+2/3)-N(\phi,t+1/3)+\sum_{q>-\phi} N(\phi+q,t+1/3)P(q). \tag{7.27}$$

From these equations, the full master equation for the dynamic reads:

$$N(\phi,t+1)=\frac{1}{2}\sum_{q=-\infty}^{+\infty}\sum_m \frac{P(m)}{2m+1}\theta(m-|q|)[N(\phi+q,t)-N(\phi+q+1,t)]$$
$$+\frac{1}{2}[N(\phi,t)+N(\phi+1,t)]\sum_m \frac{mP(m)}{N-m}, \tag{7.28}$$

where two basic statistical distributions, which are self-consistently related, have been used. These are:

$$P^*(q)=\sum_m \frac{P_e(m)}{2m+1}\theta(m-|q|), \tag{7.29}$$

which is an exact equation giving the probability of having a shock of size q. The second is $P_e(m)$, the extinction probability for an event of size m. We have a mean-field approximation relating both distributions:

$$P_e(m)=\sum_q P^*(q)\delta\left[\sum_{\phi=1}^{q-1} N(\phi)-m\right]. \tag{7.30}$$

Equation (7.30) introduces the average profile $N(\phi)$, that is, the time-averaged distribution of ϕ-values. For the mesoscopic regime $1 \gg q \gg N$, by Taylor-expanding the master equation,

$$N(\phi)=\frac{1}{2}\sum_{q=-\infty}^{+\infty}\sum_m \frac{P(m)}{2m+1}\left\{2N(\phi+q)+\left.\frac{\partial N}{\partial \phi}\right|_{\phi+q}+\frac{1}{2}\left.\frac{\partial^2 N}{\partial \phi^2}\right|_{\phi+q}+\cdots\right\}$$
$$+\frac{1}{2}\left\{2N(\phi)+\left.\frac{\partial N}{\partial \phi}\right|_{\phi}+\frac{1}{2}\left.\frac{\partial^2 N}{\partial \phi^2}\right|_{\phi}+\cdots\right\}\sum_m \frac{mP(m)}{N-m} \tag{7.31}$$

TABLE 7.1. Basic Trends of Macroevolutionary Patterns, Observed (from the Fossil Record, Mostly at Genera and Family Levels) and Predicted by the Network Model

PROPERTY	OBSERVED	NETWORK MODEL
Dynamics	Punctuated	Punctuated
Mass extinctions	Few events	Expected
Species decay	Exponential	Exponential
Extinction pattern, $N(E)$	Power law ($\alpha \approx 2$)	Power-law ($\alpha \approx 2$)
Hurst exponent, H	Persistence, $H > 1/2$	Persistence, $H > 1/2$
Genera lifetimes $N(T)$	Power law ($\gamma \approx 2$)	Power law ($\gamma \approx 2$)
Genera-species $N_g(S)$	Self-similar[a], ($\tau \approx 2$)	Self-similar, $\tau \approx 2$

[a] Data from plant groups indicates a different exponent, $\tau \approx 3/2$, thus suggesting an important difference in relation to animal communities.

and using a continuous approximation, it is easy to see that the previous equation reads:

$$\frac{1}{2}\int dm \int_{-m}^{m} \frac{P(m)}{2m}\left\{2\left[\frac{N(\phi+q)}{N(\phi)} - 1\right] + \left.\frac{\partial LnN}{\partial \phi}\right|_{\phi+q} + \cdots\right\}$$

$$+\frac{1}{2}\left\{2 + \left.\frac{\partial LnN}{\partial \phi}\right|_{\phi} + \cdots\right\}\int \frac{mP(m)}{N-m}dm = 0. \quad (7.32)$$

Assuming that $N(\phi)$ decays exponentially, $N(\phi) = \exp(-c\phi/N)$, we can integrate each part of the last equation, using $N(\phi+q)/N(\phi) = \exp(-cq/N) \approx 1 - cq/N$. The first term cancels exactly, the second gives $-2c/N$, and the third scales as $(1 - O(1/N))N^{1-\tau}$. So the previous equation leads to:

$$-\frac{2c}{N} + N^{1-\tau}G\left[1 - O\left(\frac{1}{N}\right)\right] = 0; \quad (7.33)$$

in order to satisfy this equality, we have $\tau = 2$, which gives us the scaling exponent for the extinction distribution.

Additionally, the taxonomy that emerges from this model displays fractal behavior (with an exponent $\alpha_b \approx 2$) in agreement with Burlando's analysis. Also, the lifetime distribution of genera can be computed and gives a power law with $\kappa \approx 2$, again consistent with data. Emergent genera are actually found to grow (on average) at a constant rate, independently of their size, and thus the exponents of extinction and lifetimes must be the same ($\alpha \approx \kappa$). The results obtained from these models are summarized in table 7.1 and compared with the fossil record.

Other models have successfully reproduced similar aspects by including a more comprehensive description of ecosystems. One particularly interesting example is the model introduced by Amaral and Meyer (1999) in which species are organized within layers, thus defining a trophic structure. The model considers L levels ($l = 0, 1, \ldots, L-1$) with N niches per level. The presence or absence of species at the i-th niche in the l-th level is indicated by a binary variable $S_i(l)$. The bottom species, belonging to the $l = 0$ layer, define the group of organisms that do not feed on others. From $l = 1$ to $l = L - 1$ the species of these layers feed on k or less species on the lower $(l-1)$-level.

The rules are very simple: (a) a new species is created at each niche at a rate μ. If $l > 0$, then k prey species are randomly chosen from the $(l-1)$-layer; (b) at a given rate p, species at the bottom layer are extinct. Then for species at $l > 0$ layers, extinction takes place if no input links are present. In other words, if $W(i,j; l-1 \to l)$ indicates the connection between species $i \in l-1$ and species $j \in l$ (here this connection is either one or zero), then

$$S_j = \Phi\left[\sum_{i=1}^{N} W(i,j; l-1 \to l)\right], \qquad (7.34)$$

where now $\Phi(z) = 1$ for positive z and zero for $z = 0$. Clearly, the link among connected species will generate avalanches of coextinction through different layers. From numerical simulations, Amaral and Meyer showed that in fact the distribution of avalanches is a power law with an exponent -2, also consistent with the fossil record data (Amaral and Meyer, 1999; Drossel, 1998; Camacho and Solé, 2000).

Decoupling Micro and Macroevolution

These models and their agreement with observed patterns can have important implications for evolutionary theory. There has been a long debate throughout the last decades concerning the basic mechanisms operating at small and large temporal scales. For some authors (especially among population geneticists), the rules operating at the small scale (the so-called microevolutionary events) can be directly translated into the process of macroevolution (Haldane, 1932; Dobzanski, 1951; Li and Graur, 1991). Others, such as Stephen Jay Gould, sustain that different processes are at work in evolution at different scales (Gould, 2003). Some criticism arose from evolutionary biologists since no well-defined mechanism for such decoupling was proposed. But in fact the network organization of ecologies immediately provides one possible source of decoupling.

Since species are not isolated entities, changes in the food web structure can propagate through it, eventually leading to large-scale extinction events (Solé et al., 1996a). The dynamics of these changes is not species-dependent, but network dependent. It is the global features of the food web that matter, not the specific properties of the species constituting the web. In this sense, at the small scale we observe species-level features dealing with individual adaptations as well as coevolution among pairs of species (such as predator-prey interactions). But in the long run, when extinction and replacement must be taken into account, it would be the network, and the emergent properties arising from network dynamics (together with external triggers), that are determinant for the collective behavior of communities.

ECOLOGY AS IT WOULD BE: ARTIFICIAL LIFE

One of the recurrent topics in this book is the presence of universals within ecology and evolutionary biology. Universality can be identified in different ways and can allow us to build simple models of complex systems. One obvious problem with universality in evolutionary ecology is the fact that most long-term evolutionary experiments cannot be reproduced. We intend to identify possible sources of universal patterns in the strong regularities exhibited by complex biological forms (Kauffman, 1993; Goodwin, 1994). In this context some basic universal principles of organization seem to be repeated again and again. Some types of structures in development, for example, are the result of convergent evolution: The same basic rules are used in different groups of organisms and play distinct functions. In general, to uncover a given universal rule, more than a single case is needed. As indicated by Maynard Smith, we have been able to study only one evolving system, and generalizations about evolving systems might need the building of new life forms.

One possible way of answering some key questions concerning universals in evolution is provided by the field of Artificial Life (AL) (Langton, 1989; Langton et al., 1992; Adami, 2000). AL deals with the emulation, simulation, and construction of living systems. Historically, one of the starting points of AL is, not surprisingly, based on cellular automata (see chapter 4). Actually, early work in this area involved the fundamental problem of the minimal requirements for reproduction by automata machines (Burks 1966; Siper and Reggia, 2001; see also Emmeche, 1994, chapter 3). In trying to define the minimal conditions for an automaton to reproduce itself, von Neumann found a solution that turned out to be extremely close to the one observed in cellular organization (despite the fact that no knowledge of the relevance of DNA was available at the time). Such matching between the abstract model and cellular reality is likely to be the result of fundamental constraints to the architecture of the replication process.

If an abstract, computational description of living structures is so close to the logic of living, perhaps universal principles of organization pervade other levels of complexity. Von Neumann's approach is a clear example of the AL approach to biocomplexity: By considering virtual, unconstrained scenarios of evolution, AL research intends to capture the fundamental principles of the organization of lifelike structures (Langton, 1989). *In silico* simulations allow one to rerun the tape of (virtual) evolution again and again and to keep track of any single phenomenon through the entire process. Such a privileged situation is only matched by long-term evolutionary experiments with microorganisms (Lenski and Travisano, 1994; Elena et al., 1996; Elena and Lenski, 2003).

Evolution in Tierra

What are the evolutionary questions that can be answered, or at least approached, by Artificial Life? One obvious one concerns the type of evolutionary patterns that can be expected to emerge *in silico* and how they relate to evolution. The first attempt in this direction was by the Italian-Norwegian mathematician Nils Baricelli, using von Neumann's computer in 1953 (Baricelli, 1957; Dyson 1998). A more recent example is provided by Tom Ray's Tierra model (Ray, 1991, 1994). The Tierra system involves a set of virtual organisms, each of which carrying a "genome" consisting of a sequence of instructions. These instructions can be thought of as strings of artificial DNA, and digital organisms replicate, mutate, and compete for "resources," here processing time from the computer CPU. Evolution thus takes place within the computer memory. The results emerging from these simulations are rather remarkable. Starting from an initial set of identical creatures consisting on eighty instructions, various easily recognizable patterns of evolution emerge. The first is the selection for shorter genomes. Smaller creatures replicated faster and occupied less CPU. This process of genome reduction was followed by the emergence of parasites, creatures with so few instructions that they could not self-replicate on their own. These parasites completed their reproduction by borrowing the host's replication code. The coevolving dynamics was complex, with some population catastrophes. Arms races involved the emergence of hosts able to make parasite replication more difficult. Some mutations allowed the host to hide from the attacking parasite. Although this "immunization" mechanism required extra instructions (and thus had

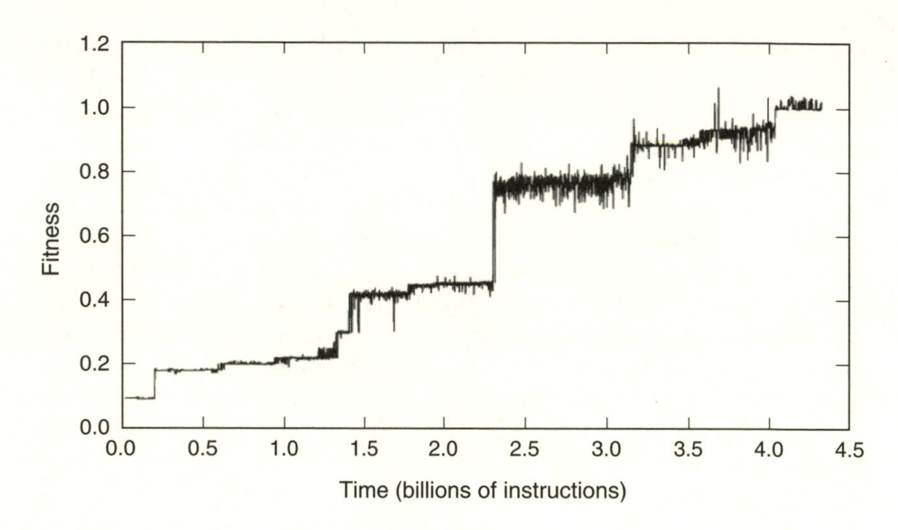

FIGURE 7.18. Typical example of fitness dynamics in a model of artificial life evolution. Here Adami's *Avida* program has been used. Fitness changes slowly over very long periods of evolutinary stasis, followed by sudden jumps in which fitness increases rapidly.

a higher cost), the gain in fitness in their response to parasite attacks was important.

Parasites in Tierra produced further mutations that modified their codes in sophisticated ways. Some of these changes allowed them to find mechanisms that rely in discovering programming tricks, some rather complicated (Ray, 1991; Levy, 1993). Other outcomes of this evolutionary experiment were hyperparasites and, at some point, cooperation among hyperparasites. The evolutionary changes in these artificial scenarios were typically punctuated by periods of stasis followed by rapid shifts in ecosystem composition (Ray, 1991; Adami, 1998). The ecology also exhibited a high diversity of creatures and of path dependence (see chapter 6). Other studies reached similar results independently (Hillis, 1990; Lindgren, 1992; Lindgren and Nordahl 1994; Rasmussen et al., 1992).

Two particularly interesting (and perhaps universal) features of these AL models are: (a) the punctuated nature of their evolutionary dynamics (fig. 7.18) and (b) the spontaneous emergence of parasites. In relation to the first, Gould (2003) correctly indicates that "the apparent ubiquity of punctuational patterns . . . may be telling us something about general properties of change itself, and about the nature of systems built of interacting components that propagate themselves through history."

Punctuated equilibrium might play a leading role in the evolution of complex systems, simply because new solutions are likely to emerge by crossing bifurcation points or reaching some new type of structural property through some discontinuous transition.

Concerning the emergence of parasites, AL experiments suggest that parasites might be the unavoidable consequence of evolution in a complex system, both organic and virtual. Additionally, as it occurs with real communities, artificial ecosystems remain more diverse when parasites are present (Ray, 1992). When parasites were deliberately forbidden, only a small number of host types were stabilized in Tierra. When allowed, they prevented any one species of host from reaching high densities.

The emergence and coevolution of these host-parasite systems is typically of Red Queen type: in many cases, the "environment" that acts on an organism (and defines its ecological context) is primarily defined by the other organisms present. The underlying landscape is thus changing and evolution can produce major leaps in biological complexity. An example, suggested by William Hamilton, is that (at least in some cases) sex would be an evolutionary response against parasites (Hamilton et al., 1990). Resistance against parasites requires constantly changing gene combinations, and sex provides a powerful mechanism of escaping to parasite's pressure by means of recombination. Actually, in Tierra too, some sort of sex also emerged: Some creatures spontaneously began to exchange parts of their coded instructions.

The emergence of sex and other qualitative features exhibited by Tierra raises the problem concerning how complexity is generated through evolution (Schuster, 1996). Once again, *in silico* models can help us understand the origins of complexity. An example is the analysis by Lenski and coworkers of the evolution of self-replicating, digital organisms growing and evolving in different selective environments (Lenski et al., 1999; Wilke et al., 2001; Wilke and Adami, 2002). In one case, simple organisms evolved under selection for fast replication. In a second scenaro, complex digital creatures were "rewarded" (with extra CPU time) for performing certain mathematical functions. One particularly interesting result was obtained by looking at the mutational robustness of the final, evolved organisms. The introduction of mutations into their programs and the proportion of those (single-point) mutations that were lethal was measured. The results are shown in figure 7.19. The essential message is that simple organisms are much more fragile than their complex counterparts. Most mutations have strong effects when acting on small-sized genomes, whereas increasing complexity leads to increased robustness against mutations. A detailed analysis reveals that epistatic interactions are more common in com-

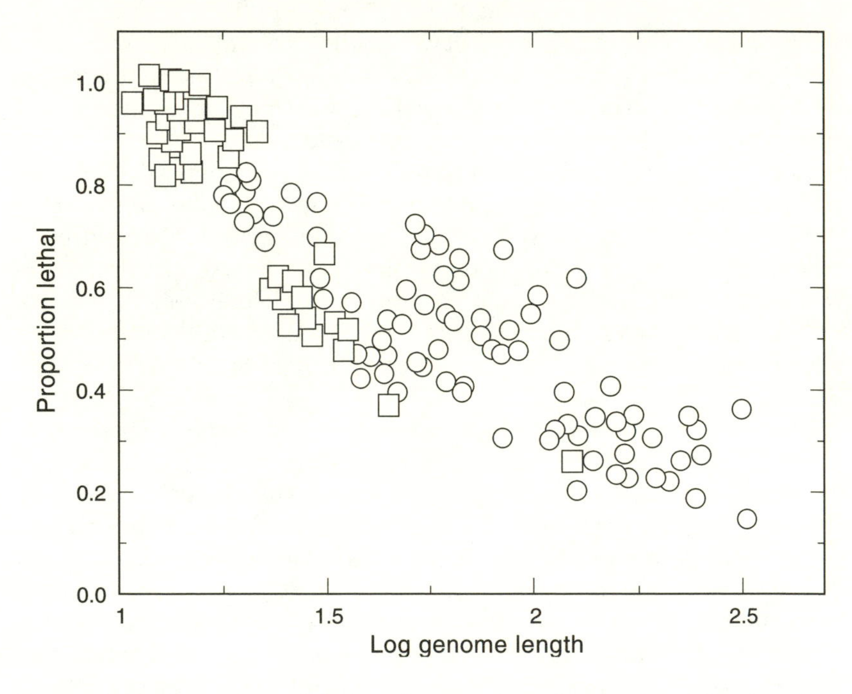

FIGURE 7.19. Proportion of single-point mutations that are lethal for digital organisms. Shown as a function of log_{10}-transformed genome length. Here circles indicate results for complex organisms, and squares are used for simple organisms (adapted from Lenski et al., 1999).

plex organisms, consistent with observations of real organisms (Elena and Lenski, 1997). Although care is needed when comparing real and virtual ecosystems (Bedau, 1999), these results encourage the view that general laws of organization might pervade any evolutionary system beyond the specific nature of its components.

RECOVERY AFTER MASS EXTINCTION

The quality and detail of the fossil record allows us to reach some understanding about how the biosphere is organized, and at some scales, the previous models are able to capture some fundamental aspects of macroevolutionary patterns. These models share the implicit assumption that the conditions under which evolution takes place are homogeneous through geological time. This would mean that the same basic rules can be applied once and once again. This is of course an extreme oversimplification: Throughout its geological history, our planet has undergone changes that vastly modified the shape of continents, the distribution and depth of seas and of course of the climate. Some of these changes happened only once or in an asymmetric way. An example is the rise of oxygen levels during the Proterozoic aeon. Somewhere between 2.4 and 2.8 billion years ago, our atmosphere reached around 1%–2% of the present-day oxygen levels (Knoll, 1992). This concentration favored the radiation of aerobic prokaryotes. The resulting physical environment prefaced the emergence and evolution of the first eukaryotic cells. In this context, it has been shown in geological and paleomagnetic studies that ice sheets may have reached the equator of our planet about 600 to 800 million years ago. Such a "snowball Earth" would have strongly constrained the early evolution of life (Hoffman et al., 1998; Hyde et al., 2000).

An important ingredient in the evolution of the biosphere through the Phanerozoic was the presence of irreversible qualitative changes in ecosystem function associated with the aftermath of mass extinctions (Erwin, 2001). All mass extinctions share a common trait: They reflect perturbations that stress ecosystems beyond their resilience. But the ecological and evolutionary impacts of different extinction events differ between extinctions, and the same happens with recovery patterns (Hallam and Wignall, 1997; Erwin, 1998a, 1998b). Some mass extinctions are known to involve little immediate ecological effect, or long-term evolutionary consequences (such as the end-Ordovician or the end-Triassic events). Others, such as the end-Permian event, dramatically shifted the course of evolution (Erwin, 1993, 1994). Recovery patterns provide a unique window through which to explore the structure and evolution of paleoecosystems. Ecological links do not fossilize, but the underlying structure of ancient food webs can be inferred, at least to some extent, from the fossil record. One example is provided from estimations of carbon fluxes (particularly carbon isotopic differences) for the ecological recovery in oceans after the Cretaceous-Terciary (K-T) event. This recovery involved the rebuilding of higher trophic levels (D'Hont et al., 1998). By tracking the evolution of carbon isotopes through time, it was shown that primary productivity quickly returned to previous levels over thousands of years but that the final recovery of the open-ocean ecosystem structure required the evolution of new species through a time window of 3 million years. Another example is the recovery of the marine food chain in the late Cenomanian event (Hart, 1996). Late Cenomanian ecosystems have been reconstructed prior to the start of the event, and followed through the subsequent changes. Changes in the oxygen minimum zone through the water column triggered the subsequent changes, which can be followed in their effects on the abundance of different taxa (over thousands of years).

The overall pattern of recovery after a large extinction has an interesting similarity to the recovery of an ecosystem subject to a transient shock. The resilience of ecosystems to perturbations, and how it relates to community structure has been a matter of extensive analysis (Yodzis, 1997; Wooton, 1998; Pimm, 1999). In a number of ways, the recovery pattern is similar to the process of ecological succession (Solé et al., 2002a). During ecological sucession (more precisely, secondary succession), ecological disturbances are followed by recolonization from surrounding areas or regions, depending on the magnitude of the disturbance. In contrast, postextinction rebounds are characterized by colonization, delayed increases in abundance, speciation, and for the largest events, evolutionary innovation.

One of the recovery patterns on evolutionary time scales is the common presence of opportunistic clades in the aftermath of the perturbation. They are later replaced by non-opportunistic clades. Examples include the lycopside *Isoetes* and the bivalve *Claraia* in the Early Triassic, and *Gumbelitria* in the earliest Tertiary. These taxa have been shown to be both locally and geographically widespread immediately following mass extinctions, perhaps indicating their adaptation to the unusual environmental conditions of the time. For example, *Claraia* is common in dysaerobic marine environments that might have been common through the early Triassic. We recognize priority effects and the replacement patterns that are also observed in ecological succession. In table 7.2 we summarize the most interesting aspects that define both succession and recoveries.

The remarkable similarities between the two processes seems to indicate that an ecological-based model of paleoecosystems is a good starting point for understanding the recovery process. Here we consider a model in which three layers of sites (which can be occupied by new species) are defined, involving primary producers, herbivores, and predators (Solé et al., 2002a). Instead of using a simpler model where links have no weights or signs (as in the Amaral-Meyer model), we introduce weighted interactions in order to gain some realism and provide the system an opportunity to self-organize with no constraint other than the layered structure. Besides, top-down and bottom-up control can emerge and extinctions can happen if prey in the bottom layer are gone, but also from the pressure from top predators (May, 1973; Pimm, 1991). Other sources of extinction can include overcompetition after a predator with a wide diet of prey is removed.

As in previous models, this one deals with species that are either present (1) or absent (0) and is organized into three layers (fig. 7.20, left). The time scale in our simulations is assumed to be very large. The

TABLE 7.2. Comparison between Different Trends Displayed by Ecological Succession and Those Observed in Recovery Dynamics

	ECOLOGICAL SUCCESSION	RECOVERY PATTERN
Initial Condition	Species-poor	Low-diverse
Initial Groups	Generalists, small	Opportunistic, small
Final Groups	Specialist, longer-lived	Lazarus sp., larger
Trophic Features	CPP, omnivory	Increase predation
Functions Developed	CPP + control of NC	Increase in PP
Niche Dynamics	Colonization	Colonization + construction
Final Pattern	Mature community	Functional ecology
Path Dependence	Present, moderate	Important
Predictability	Possible (community level)	?
Probability of Invasion	Decreasing with time	?
External Pool	Conserved	Evolving
Diversity Trends	Increasing	Increasing

Here PP: primary production, CPP: constant primary production. No information is available in relation to the degree of predictability or the probability of invasion for fossil record data (from Solé et al., 2002a).

state of the i-th species at the k-th layer at a given time t will be indicated as $S_i^k(t)$. Layers two and three (L_2, L_3) exploit species at lower layers and thus they do not compete directly. The interactions between layers L_1 and L_2 and layers L_2 and L_3 are indicated by C_{ij}^1 and C_{ij}^2, respectively. Producers at the lower level exploit some underlying, limited resource and thus compete among themselves. Their interactions are defined by a competition matrix $\beta = (\beta_{ij})$.

The state of each species at each layer is updated following simple rules. First, species from the bottom layer can become extinct with some probability p_d. Then the states are updated as follows:

$$S_i^1(t+1) = \Theta\left[1 - \sum_{j\in L_1} \beta_{ij} S_j^1 - \sum_{j\in L_2} C_{ij}^1 S_j^2\right] \tag{7.35}$$

$$S_i^2(t+1) = \Theta\left[\sum_{j\in L_1} C_{ij}^1 S_j^1 - \sum_{j\in L_3} C_{ij}^2 S_j^3\right] \tag{7.36}$$

$$S_i^3(t+1) = \Theta\left[\sum_{j\in L_2} C_{ij}^2 S_j^2\right], \tag{7.37}$$

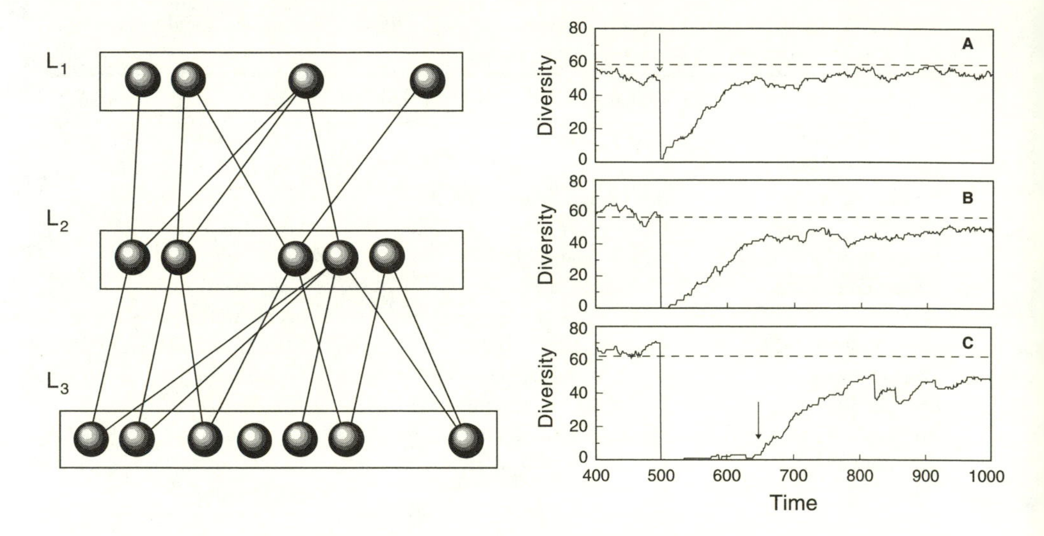

FIGURE 7.20. Left: Trophic structure of the evolutionary model of recovery. The basal level includes primary producers competing for some underlying resources. The second and third levels are connected to the lower layers through trophic links. Right: Recovery pattern displayed by the evolutionary assembly model. After a pulse extinction introduced at $T = 500$ (right), dynamics of recovery for different levels is shown. Here (A) primary producers, (B) herbivores, and (C) predators are shown. We can appreciate different recovery curves associated with each level. A clear lag occurs at the two upper levels.

where $i = 1, \ldots, N$, $\Phi(z) = 0$ if $z \leq 0$ and $\Phi(z) = 1$ otherwise. A maximum number of N sites per layer is considered.

In spite of their discrete character, we can easily recognize the basic features of (three-level) Lotka-Volterra models, although no population size is defined for each species, and species come and go. As we will see, however, some surprisingly common features are shared with the behavior of a standard ecological model.

The rules are completed by introducing speciation into the system. After the previous updating is applied (sequentially), new species can be created from existing ones. This rule is introduced as follows: At each layer, we look for empty niches (i.e., using vacant sites, such that $S_i = 0$), and with some probability of origination α, a new species can be generated from one of the already present species at the same layer. The new species is obtained by copying its ancestor's set of connections, adding a small amount of noise to the new connections. Assuming that the chosen species is S_k, the new connections of the new species (say from level L_2) are inherited from the parental one as:

$$C^1_{ij} = C^1_{kj} + \xi^1_{ij} \tag{7.38}$$

$$C^2_{ij} = C^2_{kj} + \xi^2_{ij} \tag{7.39}$$

and similarly for other levels. The copy process can include the addition or deletion of connections. Different variations around these rules gave the same basic results.

Extensive simulations of the recovery model were performed by introducing a pulse perturbation on the first layer (killing a fraction of the species present). Species removed are randomly chosen and the fraction removed is called the extinction size (E) at this level. Then, the system recovery is followed over other T steps. The model shows the observed delayed recoveries provided that the pulse perturbation is large enough (an example is shown in fig. 7.20a–c). A few predictions also emerge (Solé et al., 2002a). It was shown that there is a threshold of about 20% of removal below which recovery proceeds quickly (only the intermediate level seems to experience some delay). For $0.2 < E < 0.4$, top predators recover much faster than herbivores. This is due to the emergence of predators that have numerous, but weak connections (i.e., generalist species) preventing them from being removed and favoring further diversification. However, after $E_c \approx 0.5$, predators require longer time, reflecting the fact that their buffering against extinction is associated to broad trophic requirements and no longer enhances recovery. This type of model thus provides a powerful approach to recovery patterns and might help us understand the relevance of different components (ecology, innovation, history) in shaping the recovery process (Solé et al., 2005b).

IMPLICATIONS FOR CURRENT ECOLOGIES

The current, human-driven biodiversity extinction event is taking place in a catastrophically short time scale (Wilson, 1992; Lawton and May, 1995; Leakey and Lewin, 1995). Estimations based on extinction rates indicate that a large fraction of species will be extinct in a few decades. The scenario is thus not far from that of previous mass extinctions. An additional burden is introduced by the nonlinear character of many ecological responses, an inevitable consequence of ecological complexity. Together with direct extinction, secondary extinctions may be revealed to be more important in the long run.

Similar processes may also have been at work in the Late Pleistocene under extensive human hunting activities. This seems to be the case

TABLE 7.3. Comparison between Different Features of Past Mass Extinction Events (as Reported from the Fossil Record) and Present-Day, Human-Driven Mass Extinction

	FOSSIL RECORD	PRESENT DAY
Time Resolution	$\approx 10^5 - 10^6$ yr	≈ 1 yr
Most Affected Biotas	Tropical biotas	Coral reefs, rainforests
Selectivity on Size	Large-sized species	Large-sized species
Loss of Endemics	Not well known	Widespread
Effects of Habitat Loss	Widespread	Widespread
Time Scale of Recovery	$\approx 5 - 10$ Myr	Not known
Direct Effects on Food Webs	Very important	Very important
Indirect Effects on Food Webs	Not known	Very important
Extinction Rates	See 75–80%	$\approx 10^4$ species/year

From Solé and Newman, 2001. Here the effects of habitat loss are widespread but different mechanisms were at work: Changes in sea levels and continental breakup (fossil record) and human-driven habitat destruction/fragmentation (present day) respectively. Estimations of species loss are from Wilson (1992).

with Australia (Miller et al., 1999; Roberts et al., 2001) where all marsupials exceeding 100 kg, along with many reptiles and birds became extinct, presumably through a combination of hunting and ecosystem change resulting from burning practices. The same applies to the end-Pleistocene extinctions of mastodons and mammoths in North America (Owen-Smith, 1987; Martin and Klein, 1984; Alroy, 2001), which were associated with widespread changes in vegetation patterns and the disappearance of many other species. In this context, Norman Owen-Smith (1987) proposed a "keystone herbivore" hypothesis based on the ecology of extant African megaherbivores (i.e., animals exceeding 1,000 kg in body mass). The hypothesis suggests that, due to their invulnerability to nonhuman predation on adults, these species attain high (near-saturation) densities. Such large populations have huge effects on large-scale vegetation patterns: African elephants, for example, can convert closed woodland into open grassy savanna (Leakey and Lewin, 1995), and similar situations apply to other megahervibores. They play the role of keystone species (chapter 6), and their removal by human hunters promoted widespread changes in vegetation, eventually detrimental to other smaller herbivores dependent upon the vegetation diversity that is maintained by the constant perturbation introduced by large herbivores. One may well ask whether the current extinction of large herbivorous species in African ecosystems will result in similar concurrent extinctions. Evidence is mounting (from both data analysis and computer simulation) that supports the leading role played by humans in Quaternary extinctions (Whittington and Dyke, 1989; Alroy, 2001). Particularly interesting is the simulation model developed by Alroy for the end-Pleistocene megafaunal mass extinction, which included most available data and correctly predicted the extinction or survival of thirty-two out of forty-one prey species (Alroy, 2001).

Comparison between fossil and present-day extinction is not straightforward since, as mentioned above, the fossil record consists largely of marine organisms. Also, recently extinct or currently endangered species tend to be rare, whereas the fossil record primarily reflects the most abundant and numerous biotas. Still, there are a number of general patterns in the fossil record of extinction (Jablonski, 1999, 2001) that may help us in the conservation of modern-day biodiversity (see table 7.3).

APPENDIX ONE

Lyapunov Exponents for ID Maps

Starting from the definition of the distance between two nearest points belonging to two different, but initially close orbits,

$$\delta(\epsilon) = \epsilon e^{n\lambda(x_0)}, \tag{1}$$

an explicit form for the Lyapunov exponent can be derived.

Let us first consider the explicit form of the (euclidean) distance between close orbits, which, after n iterations is given by:

$$\delta(\epsilon) = |F^n_\mu(x_0 + \epsilon) - F^n_\mu(x_0)|. \tag{2}$$

Using (1) the Lyapunov exponent can be written as:

$$\lambda(x_0) = \frac{1}{n}\, log \left| \frac{F^n_\mu(x_0 + \epsilon) - F^n_\mu(x_0)}{\epsilon} \right|, \tag{3}$$

which in the limit can be written as:

$$\begin{aligned} \lambda(x_0) &= \lim_{n\to\infty} \lim_{\epsilon\to 0} \frac{1}{n}\, log \left| \frac{F^n_\mu(x_0 + \epsilon) - F^n_\mu(x_0)}{\epsilon} \right| \\ &= \lim_{n\to\infty} \frac{1}{n}\, log \left| \frac{\partial F^n_\mu(x_0)}{\partial x_0} \right|. \end{aligned} \tag{4}$$

Consider now the following property:

$$\begin{aligned} \frac{\partial F^2_\mu(x_0)}{\partial x_0} &= \frac{\partial}{\partial x_0}\left[F_\mu(F_\mu(x_0))\right] = \frac{\partial F_\mu(F_\mu(x_0))}{\partial x_0}\ \frac{\partial F_\mu(x_0)}{\partial x_0} \\ &= \frac{\partial F_\mu(x_1)}{\partial x_1}\ \frac{\partial F_\mu(x_0)}{\partial x_0}. \end{aligned} \tag{5}$$

By incorporating the generalization of the previous relation into (4), we can write the following definition for the Lyapunov exponent:

$$\lambda(x_0) = \lim_{n\to\infty} \frac{1}{n}\, log \left| \prod_{i=0}^{n-1} \frac{\partial F_\mu(x_i)}{\partial x_i} \right| = \lim_{n\to\infty} \frac{1}{n} \sum_{i=0}^{n-1} log \left| \frac{\partial F_\mu(x_i)}{\partial x_i} \right|. \tag{6}$$

For practical purposes, if we deal with a finite temporal series, we can write the following approximation:

$$\lambda(x_0) \approx \frac{1}{n} \sum_{i=0}^{n-1} log \left| \frac{\partial F_\mu(x_i)}{\partial x_i} \right|, \tag{7}$$

which has been used to perform the computations of the Lyapunov exponents shown in chapter 2.

APPENDIX TWO

Renormalization Group Analysis

Percolation thresholds can be estimated by using a powerful technique of statistical physics: the renormalization group (Binney et al., 1992; Yeomans, 1992). This is a fairly well-known technique that was early applied to a broad number of areas from particle physics to economics. Although in many cases it requires rather advanced mathematical background (Parisi, 1988), here a simple version known as real-space renormalization is considered.

Consider a two-dimensional lattice, with an occupation probability p. We might think of this lattice in terms of a two-dimensional landscape. In order to calculate the percolation threshold p_c, let us consider $m \times m$ boxes including four nearest sites and following Milne and Johnson, let us calculate the probability of percolation from left to right (Milne and Johnson, 1993). It can be shown that seven favorable configurations allow propagation to occur. Since we assume independence in the occupation probability, the probability of left-right propagation in the rescaled system is:

$$p(m = 2) = 2p^2(1 - p)^2 + 4p^3(1 - p) + p^4. \qquad (1)$$

This equation defines a recurrence relation $p_{n+1} = \phi(p_n)$ connecting two different scales: the first (n)-th the finer scale, and the second $(n + 1)$-th the coarser one. It is easy to find the percolation point by computing the fixed points of this recurrence relation and analyzing their stability. The set of (physically meaningful) fixed points $\pi = \{p^* | p^* = \phi(p^*)\}$ involves three elements. Two of them are the so-called trivial fixed points, $p^* = 0$ and $p^* = 1$, corresponding to a totally empty or a totally filled lattice, respectively. They are obvious fixed points since in both cases the probability is not altered when the block procedure is applied. But a third fixed point is found at $p^* = 0.618$ (very close to $p_c = 0.5928$). This fixed point is unstable, that is, $d\phi(p^*)/dp > 0$.

The connectivity of sites in a landscape involving percolation affects both species distribution patterns and the dynamics of whole ecosystems. As discussed by David Green, dispersal tends to produce clumped distributions, which promote species persistence and provide a possible mechanism for maintaining high species richness in tropical rain forests and other ecosystems. Simulations of multispecies systems show that, below a critical rate, disturbance regimes have little impact on species richness (Green, 1989, 1990). If supercritical rates of disturbance are introduced (such as might occur when large fires propagate through the entire forest), the rate of decrease in species richness depends on the balance between the rate of disturbance and dispersal range.

APPENDIX THREE

Stochastic Multispecies Model

The stochastic multispecies model introduced in chapter 6 has been shown to reproduce a broad number of basic, well-known field observations (Solé et al., 2000b; McKane et al., 2000). But the mechanisms of interaction used are too abstract and do not allow, in particular, the expansion of isolated species into empty sites. A second model involves again a set of S (possible) species plus empty sites (pseudospecies labeled 0). This set (the species pool) $\Sigma(S)$ is then a discrete set:

$$\Sigma(S) = \{0, 1, 2, \ldots, S\}, \tag{1}$$

where 0 indicates empty space, available to all species from the pool. Three basic types of transitions are allowed to occur. Let us call $A, B \in \Sigma(S)$ two given species present at a given step. These possible transitions are summarized as follows:

1. Immigration: An empty site is occupied by a species randomly chosen from the set of (nonempty) species, that is, $A \in \Sigma(S) - \{0\}$:

   $$0 \xrightarrow{\mu_i} A\,. \tag{2}$$

   This occurs with a probability of colonization (of empty sites) μ_i. Notice that this colonization depends on the particular species.

2. Death: All occupied sites can become empty with some fixed probability e_i:

   $$A \xrightarrow{e_i} 0. \tag{3}$$

3. Interaction: The same rule as in model A is used, but here the probability of success P_{ij} is weighed by the coefficients of the interaction matrix. Here we use:

   $$P_{ij} = \pi[\Omega_{ij} - \Omega_{ji}]\,, \tag{4}$$

   where $\pi[x] = x$ when $x > 0$ and zero otherwise. The probability of an interaction occurring in the system between species i and j is a measure of the interaction strength linking interacting species.

In other words, the parameters defining the interactions, colonization, and extinction of species in model B are given by two vectors:

$$\mu = (\mu_0, \mu_1, \ldots, \mu_S)\,, \tag{5}$$

$$e = (e_0, e_1, \ldots, e_S)\,, \tag{6}$$

the immigration and the extinction vectors respectively, where here $\mu_0 = e_0 = 0$, and an $(S+1) \times (S+1)$ matrix Ω, which is such that $\Omega_{0i} = 0$ for all $i \in \Sigma(S) - \{0\}$. Additional rules (such as the presence of trade-offs between colonization and competition) can be easily introduced. Several variations of the previous models have shown the same basic patterns.

This model allows the simulation (as particular cases) of a number of well-known problems. Let us mention five of them.

1. The predator-prey model (chapter 2): Here empty sites, prey (1), and predators (2) are considered. It would be represented as $\mu = (0, \mu_1, \mu_2)$, $e = (0, e_1, e_2)$, and the interaction matrix would be:

$$\Omega = \begin{pmatrix} 0 & 0 & 0 \\ \Omega_{10} & 0 & 0 \\ 0 & \Omega_{21} & 0 \end{pmatrix} . \tag{7}$$

2. The forest-fire model (chapter 4): Here the three states correspond to ashes (empty sites), green trees (1) and burning trees (2). Two different types of models have been considered: the Bak-Chen-Tang (BCT) version and the Drossel-Schwabl (DS) version. The only difference between them is that the second allows for the spontaneous burning of green trees. Now we have: $\mu = (0, \mu_1, \mu_2)$, $\mathbf{e} = (0, 0, 1)$, with $\mu_2 = 0$ in BCT and $\mu_2 \neq 0$ in the DS model. The interaction matrix is:

$$\Omega = \begin{pmatrix} 0 & 0 & 0 \\ 0 & 0 & 0 \\ 0 & 1 & 0 \end{pmatrix} . \tag{8}$$

3. Tilman's model (chapter 5). In this model, species i can colonize sites that are not occupied by superior competitors, but it is extinguished from any site invaded by superior competitors. The model involves the proportion of sites occupied by species i (p_i), species-specific colonization rates (c_i), and mortality rates (e_i). The equation for species i is:

$$\frac{dp_i}{dt} = c_i p_i \left(1 - \sum_{j=1}^{i} p_j \right) - e_i p_i - \sum_{j=1}^{i-1} c_j p_j p_i . \tag{9}$$

The stochastic counterpart of the deterministic model described by Eq (9) can also be rewritten as a particular case of the stochastic model B, assuming no external immigration and with the extinction vector being $\mathbf{e} = (0, e_1, \ldots, e_{S-1}, e_S)$. Notice how the interaction matrix captures the hierarchical structure of the competitive community:

$$\Omega = \begin{pmatrix} 0 & 0 & \cdots & \cdots & \cdots & \cdots & 0 \\ c_1 & 0 & c_1 & \cdots & \cdots & \cdots & c_1 \\ \vdots & \vdots & \ddots & \ddots & \cdots & \cdots & \vdots \\ c_i & 0 & \cdots & 0 & c_i & \cdots & c_i \\ \vdots & \vdots & \vdots & \vdots & \ddots & \ddots & \vdots \\ c_{S-1} & 0 & \cdots & 0 & \cdots & 0 & c_{S-1} \\ c_S & 0 & \cdots & 0 & \cdots & 0 & 0 \end{pmatrix}. \tag{10}$$

4. The contact process (chapter 4) would include only two types of particles, active and inactive (exposed and infected), with $\mu = (0, 0)$, $\mathbf{e} = (0, e_1)$ and the interaction matrix would be:
$$\Omega = \begin{pmatrix} 0 & 0 \\ \Omega_{10} & 0 \end{pmatrix}, \tag{11}$$
where Ω_{10} corresponds to the infection rate and e_1 to the recovery rate.

5. Hubbell's neutral theory of biodiversity (Hubbell, 2001) is also based on a stochastic model that can be formalized as a particular case of model B. The colonization vector must now be defined as $\mu = (\mu_0, \mu_1, \ldots, \mu_S)$, where μ_0 is the perturbation rate, that is, the probability of formation of a new gap in a lattice site per time unit. The matrix Ω takes into account only the ability of species to colonize empty space, and therefore has the same entries for all species. Species only differ in the immigration rate (μ_i) from an external biogeographical pool. Internal colonization of empty sites takes place in a way that is proportional to the relative abundance of any particular species in the lattice. Therefore, the matrix Ω is written as:
$$\Omega = \begin{pmatrix} 0 & \cdots & \cdots & 0 \\ c & 0 & \cdots & 0 \\ \vdots & \vdots & \ddots & \vdots \\ c & 0 & \cdots & 0 \end{pmatrix}. \tag{12}$$
In Hubbell's model, extinction takes place as a result of an external perturbation. There is no species-specific mortality rate. Therefore, the extinction vector $e = (e_0, e_1, \ldots, e_S)$ must be defined as a null vector.

The fact that these models exhibit a wide range of dynamical patterns (from spiral waves and chaos to self-organized critical behavior) is important here. This wide range of patterns validates our choice of a model in which mechanisms of different kinds (which can create conflicting constraints of different types) are explicitly included.

References

Abraham, E. R. 1998. "The Generation of Plankton Patchiness by Turbulent Stirring." *Nature* 391:577–80.

Abramowitz, M., and Stegun, I. A., 1965. *Handbook of Mathematical Functions*. Dover: New York.

Adami, C. 1995. "Self-Organized Criticality in Living Systems." *Physics Letters* A 203:29–32.

_____. 1998. *Introduction to Artificial Life*. Springer: New York.

Akhromeyeva, T. S., Kudryumov, S. P., Malinetskii, G. G., Samarskii, A. A. 1989. "Nonstationary Dissipative Structures and Diffusion-Induced Chaos in Nonlinear Media." *Physics Reports* 176 (5/6):189–372.

Albert, R., and Barabási, A. L. 2002. "Statistical Mechanics of Complex Networks." *Reviews of Modern Physics* 74:47–97.

Albert, R., Barabási, A-L., and Jeong, H. 2000. "The Internet's Achilles Heel: Error and Attack Tolerance in Complex Networks." *Nature* 406:378–82.

Allain C., Cloitre M. 1991. "Characterizing the Lacunarity of Random and Deterministic Fractal Sets." *Physical Review* A 44:3552–58.

Allee, W. C., Emerson, A. E., Park, O., Park, T., and Schmidt, K. P. 1949. *Principles of Animal Ecology*. W. B. Saunders: Philadelphia.

Allen, J. C., Schaffer, W. M., and Rosko, D. 1993. "Chaos Reduces Species Extinction by Amplifying Local Population Noise." *Nature* 364:229–32.

Alonso, D., Bartumeus, F., and Catalan, J. 2002. "Mutual Interference between Predators Can Give Rise to Turing Spatial Structures." *Ecology* 83:28–34.

Alonso, D., and Solé, R. V. 1998. "Random Walks, Fractals and the Origins of Rainforest Diversity." *Advances in Complex Systems* 1:203–20.

Alroy, J., 2001. "A Multi-Species Overkill Simulation of the End-Pleistocene Megafaunal Mass Extinction." *Science* 292:1893–1986.

_____. 2003. "Cenozoic Bolide Impacts and Biotic Change in North American Mammals." *Astrobiology* 3:119–32.

Alstrom, P., and Stassinopoulos, D. 1992. "Space-Time Renormalization at the Onset of Spatiotemporal Chaos in Coupled Maps." *Chaos* 2(3):301–12.

Amaral, L.A.N., Buldyrev, S. V., Havlin, S., Salinger, M. A., and Stanley, H. E. 1998. "Power Law Scaling for a System of Interacting Units with Complex Internal Structure." *Physical Review Letters* 80:1385–88.

Amaral, L.A.N., and Meyer, M. 1999. "Coextinction in a Simple Model of Evolution." *Physical Review Letters* 82:652–55.

Amarasekare, P. 1998. "Allee Effects in Metapopulation Dynamics." *American Naturalist* 152:298–302.

Anderson, R. M., and May, R. M. 1991. *Infectious Diseases of Humans. Dynamics and Control*. Oxford University Press: Oxford.

Andrén, H. 1994. "Effects of Habitat Fragmentation on Birds and Mammals in Landscapes with Different Proportions of Suitable Habitat: A Review." *Oikos* 71:355–66.

_____. 1996. "Population Responses to Habitat Fragmentation-Statistical Power and the Random Sample Hypothesis." *Oikos* 76:235–42.

Arnold, V. I. 1978. *Mathematical Methods in Classical Mechanics*. Springer Verlag: Berlin.

Arthur, B., Durlauf, S. N., and Lane, D. A. 1997. Introduction. In *The Economy as an Evolving Complex System II*, W. B. Arthur, S. N. Durlauf, and D. A. Lane, eds., *Santa Fe Institute Studies in the Science of Complexity*, Vol. 27. Addison-Wesley: Reading, MA.

Arthur, M. A., Zachos, J. C., and Jones, D. S. 1987. "Primary Productivity and the Cretaceous/Tertiary Boundary Event in the Oceans." *Cretaceous Research* 8:43–54.

Atmar, W., and Patterson, B. D. 1993. "The Measure of Order and Disorder in the Distribution of Species in Fragmented Habitat." *Oecologia* 96:373–82.

Bak, P. 1997. *How Nature Works*. Springer: New York.

Bak, P., Chen, K., and Tang, C. A. 1990. "The Forest-Fire Model and Some Thoughts on Turbulence." *Physics Letters* A. 147:297–300.

Bak, P., Flyvbjerg, H., and Lautrup, B. 1992. "Coevolution in a Rugged Fitness Landscape. *Physical Review* A 46:6724–30.

Bak, P., and Sneppen, K. 1993. "Punctuated Equilibrium and Criticality in a Simple Model of Evolution." *Physical Review Letters* 71:4083–86.

Bak, P., Tang, C., and Wiesenfeld, K. 1987. "Self-Organized Criticality: An Explanation of 1/f Noise." *Physical Review Letters* 59:381–84.

Ball, P. 1999. *The Self-Made Tapestry*. Oxford University Press: Oxford.

Barabási, A.-L., and Stanley, H. E. 1995. *Fractal Concepts in Surface Growth*. Cambridge University Press: New York.

Barbault, R., and Sastrapradja, S. D. 1995. "Generation, Maintenance, and Loss of Biodiversity." In *Global Biodiversity Assessment*, pp. 232–44. V. H. Heywood, ed. Cambridge University Press, Cambridge.

Baricelli, N. A., 1957. "Numerical Testing of Evolution Theories." *Acta Biotheoretica*, 16:69–126.

Bartumeus, F., Alonso, D., and Catalan, J. 2001. "Self-Organized Spatial Structures in a Ratio-Dependent Predator-Prey Model." *Physica* A 295:53–57.

Bartumeus, F., Catalan, J. Fulco, U. L., Lyra, M. L., and Viswanathan, G. M. 2002. "Optimizing the Encounter Rate in Biological Interactions: Levy versus Brownian Strategies." *Physical Review Letters* 88:097901.

Bartumeus, F., Peters, F., Pueyo, S., Marrase, C., and Catalan, J. 2003. "Helical Levy Walks: Adjusting Searching Statistics to Resource Availability in Microzooplankton." *Proceedings of the National Academy of Sciences, USA* 100:12771–75.

Bascompte, J. 2001. "Aggregate Statistical Measures and Metapopulation Dynamics." *Journal of Theoretical Biology* 209:373–79.

Bascompte, J. 2003. "Extinction Thresholds: Insights from Simple Models." *Annales Zoologici Fennici* 40:99–114.

Bascompte, J., Jordano, P., Melián, C. J., and Olesen, J. M. 2003. "The Nested Assembly of Plant-Animal Mutualistic Networks." *Proceedings of the National Academy of Sciences, USA* 100:9383–87.

Bascompte, J., Melián, C. J., and Sala, E. 2005. "Interaction Strength Combinations and the Overfishing of a Marine Food Web." *Proceedings of the National Academy of Sciences, USA* 102:5443–47.

Bascompte, J., Possingham, H., and Roughgarden, J. 2002. "Patchy Populations in Stochastic Environments: Critical Number of Patches for Presistence." *American Naturalist* 159:128–37.

Bascompte, J., and Rodríguez, M. A. 2000. "Self-Disturbance as a Source of Spatiotemporal Heterogeneity: The Case of the Tallgrass Prairie." *Journal of Theoretical Biology* 204:153–64.

Bascompte, J., and Solé, R. V. 1994. "Spatially Induced Bifurcations in Single-Species Population Dynamics." *Journal of Animal Ecology* 63:256–64.

_____. 1995. "Rethinking Complexity: Modelling Spatiotemporal Dynamics in Ecology." *Trends in Ecology and Evolution* 10:361–66.

_____. 1996. "Habitat Fragmentation and Extinction Thresholds in Spatially Explicit Metapopulation Models." *Journal of Animal Ecology* 65:465–73.

_____. 1998a. *Modeling Spatiotemporal Dynamics in Ecology*. Springer-Verlag: Berlin.

_____. 1998b. "Spatiotemporal Patterns in Nature." *Trends in Ecology and Evolution* 13:173–74.

_____. 1998c. "Effects of Habitat Destruction in a Prey-Predator Metapopulation Model." *Journal of Theoretical Biology* 195:383–93.

Bascompte, J., Solé, R. V., and Martinez, N. 1997. "Population Cycles and Spatial Patterns in Snowshoe Hares: An Individual-Oriented Simulation." *Journal of Theoretical Biology*, 187:213–22.

Bascompte, J. and Vilá, C. 1997. "Fractals and Search Paths in Mammals." *Landscape Ecology* 12:213–21.

Bassingthwaighte, J. B., Liebovitch, L. S., and West, B. J. 1994. *Fractal Physiology*. Oxford University Press: New York.

Bastolla, U., Lassig, M., Manrubia, S. C., and Valleriani, A. 2001. "Diversity Patterns in Ecological Models at Dynamical Equilibrium." *Journal of Theoretical Biology* 212:11–25.

Bazzaz, F. A. 1975. "Plant Species Diversity in Oldfield Successional Ecosystems in Southern Illinois." *Ecology* 56:485–88.

Bedau, M. A. 1999. "Can Unrealistic Computer Models Illuminate Theoretical Biology?" In *Proceedings of 1999 Genetic and Evolutionary Computation Conference Workshop Program*, A. Wu, ed., 20–23.

Beddington, J. R. 1975. "Mutual Interference between Parasites or Predators and Its Effect on Searching Efficiency." *Journal of Animal Ecology* 44:331–40.

Begon, M., and Mortimer, M. 1986. *Population Ecology*, 2nd ed. Sinauer Associates: Sunderland, MA.

Bell, G. 2000. "The Distribution of Abundance in Neutral Communities." *The American Naturalist* 155:606–17.

_____. 2001. "Neutral Macroecology." *Science* 293:2413–18.

Bengtsson, J. 1991. "Interspecific Competition in Metapopulations." *Biological Journal of the Linnean Society*. 42:219–37.

Benton, M. J., ed. 1993. *Fossil Record 2*. Chapman and Hall: London.

_____. 1996. "On the Nonprevalence of Competitive Replacement in the Evolution of Tetrapods." In *Evolutionary Paleoecology*, D. Jablonski, D. H. Erwin and J. Lipps, eds., 185–210. University of Chicago Press: Chicago.

_____. 1995a. "Diversification and Extinction in the History of Life." *Science* 268:52–58.

_____. 1995b. "Red Queen Hypothesis." In *Palaeobiology*, eds. D.E.G. Briggs and P. R. Crowther, Blackwell: Oxford.

Berryman, A. A., and Millstein, J. A. 1989a. "Are Ecological Systems Chaotic: And If Not Why Not?" *Trends in Ecology and Evolution* 4:26–28.

_____. 1989b. "Avoiding Chaos." *Trends in Ecology and Evolution* 4:240.

Binney, J. J., Dowrick, N. J., Fisher, A. J., and Newman, M.E.J. 1992. *The Theory of Critical Phenomena: An Introduction to the Renormalization Group*. Oxford Science Publications: Oxford.

Bjørnstad, O. N., and Bascompte, J. 2001. "Synchrony and Second-Order Spatial Correlation in Host-Parasitoid Systems." *Journal of Animal Ecology* 70:924–33.

Bjørnstad, O. N., and Bolker, B. 2000. "Canonical Functions for Dispersal-Induced Synchrony." *Proceedings of the Royal Society of London Series B* 267:1787–94.

Bjørnstad, O. N., Ims, R. A., and Lambin, X. 1999a. "Spatial Population Dynamics: Analyzing Patterns and Processes of Population Synchrony." *Trends in Ecology and Evolution* 14:427–32.

Bjørnstad, O. N., Peltonen, M., Liebhold, A. M., and Baltensweiler, W. 2002. "Waves of Larch Budmoth Outbreaks in the European Alps." *Science* 298:1020–23.

Bjørnstad, O. N., Stenseth, N. C., and Saitoh, T. 1999b. "Synchrony and Scaling in Dynamics of Voles and Mice in Northern Japan." *Ecology* 80:622–37.

Blasius, B., Huppert, A., and Stone, L. 1999. "Complex Dynamics and Phase Synchronization in Spatially Extended Ecological Systems." *Nature* 399:354–59.

Boerlijst, M. C. and Hogeweg, P. 1991. "Spiral Wave Structure in Pre-Biotic Evolution: Hypercycles Stable against Parasites." *Physica D* 48:17–28.

Boerlijst, M. C., Lamers, M. E., and Hogeweg, P. 1993. "Evolutionary Consequences of Spiral Waves in a Host-Parasitoid System." *Proceedings of the Royal Society of London Series B* 253:15–18.

Bolker, B. M., and Grenfell, B. T. 1996. "Impact of Vaccination on the Spatial Correlation and Persistence of Measles Dynamics. *Proceedings of the National Academy of Sciences, USA* 93:12648–53.

Bolker, B., and Pacala, S. W. 1999. "Spatial Moment Equations for Plant Competition: Understanding Spatial Strategies and the Advantages of Short Dispersal." *The American Naturalist* 153:575–602.

Bollobás, B. 1985. *Random Graphs*. Academic Press: London.

Bond, W. J. 1993. "Keystone species." In: *Biodiversity and Ecosystem Function*, E. D. Schulze and H. A. Mooney, eds., 237–53. Springer-Verlag: Berlin.

Borda-de-Agua, L., Hubbell, S. P., and McAllister, M. 2002. "Species-Area Curves, Diversity Indices, and Species Abundance Distributions: A Multifractal Analysis." *American Naturalist* 159:138–55.

Bornholdt, S., and Schuster, H. G., eds. 2002. *Handbook of Graphs and Networks from the Genome to the Internet*. Springer: Berlin.

Borvall, C., Ebenman, B., and Jonsson, T. 2000. "Biodiversity Lessens the Risk of Cascading Extinctions in Food Webs." *Ecology Letters* 3:131–36.

Boswell, G. P., Britton, N. F., and Franks, N. R. 1998. "Habitat Fragmentation, Percolation Theory and the Conservation of a Keystone Species." *Proceedings of the Royal Society of London Series B* 265:1921–25.

Briggs, C. J., Nisbet, R. M., and Murdoch, W. W. 1999. "Delayed Feedback and Multiple Attractors in a Host-Parasitoid System." *Journal of Mathematical Biology* 38:317–45.

Brown, J. H., Gupta, V. K., Li, B. L., Milne, B. T., Restrepo C., and West, G. B. 2002. "The Fractal Nature of Nature: Power Laws, Ecological Complexity and Biodiversity." *Philosophical Transactions of the Royal Society of London Series B* 357:619–26.

Brown, J. H., and Kodric-Brown, A. 1977. "Turnover Rates in Insular Biogeography: Effect of Immigration on Extinction." *Ecology* 58:445–49.

Brown, J. H., and West, G. B., eds. 2000. *Scaling in Biology*. Oxford University Press: New York.

Bunde, A. and Havlin, S., eds. 1994. *Fractals in Science*. Springer: Berlin.

Bunn, A. G., Urban, D. L., and Keitt, T. H. 2000. "Landscape Connectivity: A Conservation Application of Graph Theory." *Journal of Environmental Management*. 59:265–78.

Burks, A. W., ed. 1966. *Theory of Self-Reproducing Automata by John von Neumann*. University of Illinois Press: Urbana.

Burlando, B. 1990. "The Fractal Dimension of Taxonomic Systems." *Journal of Theoretical Biology* 146:99–114.

_____. 1993. "The Fractal Geometry of Evolution." *Journal of Theoretical Biology* 163:161–72.

Caldarelli, G., Frondoni, R., Gabrielli, A., Montuori, M., Retzlaff, R., and Ricotta, C. 2001. "Percolation in Real Wildfires." *Europhysics Letters* 56:510–16.

Caldarelli, G., Higgs, P., and McKane, A. 1998. "Modelling Coevolution in Multispecies Communities." *Journal of Theoretical Biology* 193:345–58.

Camacho, J., Guimera, R., and Amaral, L.A.N. 2002. "Analytical Solution of a Model for Complex Food Webs." *Physical Review* E 65: 030901.

Camacho, J., and Solé, R. V. 2000. "Scaling and Zipf's Law in Ecological Size Spectra." *Europhysics Letters* 55:774–80.

Camacho, J., and Solé, R. V. 2000. "Extinction and Taxonomy in a Trophic Model of Coevolution." *Physical Review* E 62:1119–23.

Camazine, S., Deneubourg, J., Franks, N. R., Sneyd, J., Theraula, G., and Bonabeau, E. 2001. *Self-Organization in Biological Systems*. Princeton University Press: Princeton.

Carpenter, S. R., and Kitchell, J. F. 1996. *The Trophic Cascade in Lakes*. Cambridge University Press: Cambridge.

Case, T. J. 1990. "Invasion Resistance Arises in Strongly Interacting Species-Rich Model Competition Communities." *Proceedings of the National Academy of Sciences, USA* 87:9610–14.

_____. 1991. "Invasion Resistance, Species Buildup and Community Collapse in Metapopulation Models with Interspecies Competition." *Biological Journal of the Linnean Society* 42:239–66.

_____. 2000. *An Illustrated Guide to Theoretical Ecology*. Oxford University Press: New York.

Caswell, H., 1976. "Community Structure: A Neutral Model Analysis." *Ecological Monographs* 46:327–54.

Caswell, H., and Johns, A. 1992. "From the Individual to the Population in Demographic Models." In *Individual-Based Models and Approaches in Ecology: Populations, Communities and Ecosystems*, D. L. De Angelis and L. J. Gross, eds., 36–61. Chapman and Hall: London.

Chacon, P. and Nuño, J. C. 1995. "Spatial Dynamics of a Model for Prebiotic Evolution." *Physica* D81:398–410.

Chaloner, W. G., and Hallam, A., eds. 1989. "Evolution and Extinction." *Philosophical Transactions of the Royal Society of London* B 325:239–488.

Chopard, B., and Droz, M. 1999. *Cellular Automata Modeling of Physical Systems*. Cambridge University Press: Cambridge.

Clar, S., Drossel, B., Schenk, K., and Schwabl, F. 1999. "Self-Organized Criticality in Forest-Fire Models." *Physica* A 266:153–59.

Cohen, J. E., Briand, F., and Newman, C. M. 1990. *Community Food Webs: Data and Theory*. Springer-Verlag: New York.

Cohen, J. E., and Newman, C. M. 1984. "The Stability of Large Random Matrices and Their Products." *Annals of Probability* 12:283–310.

_____. 1985. "When Will a Large Complex System be Stable?" *Journal of Theoretical Biology* 113:153–56.

Collins, S. L. 1992. "Fire Frequency and Community Heterogeneity in Tallgrass Prairie Vegetation." *Ecology* 73:2001–6.

Comins, H. N., Hassell, M. P., and May, R. M. 1992. "The Spatial Dynamics of Host-Parasitoid Systems." *Journal of Animal Ecology* 61:735–48.

Connell, J. H. 1978. "Diversity in Tropical Rainforests and Coral Reefs." *Science* 199:1302–10.

Conway Morris, S. 1986. "The Community Structure of the Middle Cambrian Phyllopod Bed (Burgess Shale)." *Paleontology* 28:423–67.

_____. 1998. *The Crucible of Creation*. Oxford University Press: Oxford.

Costantino, R. F., Cushing, J. M., Dennis, B., and Desharnais, R. A. 1995. "Experimentally Induced Transitions in the Dynamic Behaviour of Insect Populations." *Nature* 375:227–30.

Costantino, R. F., Desharnais, R. A., Cushing, J. M., and Dennis, B. 1997. "Chaotic Dynamics in an Insect Population." *Science* 275:389–91.

Courchamp, F., Clutton-Brock, T., and Grenfell, B. 1999. "Inverse Density Dependence and the Allee Effect." *Trends in Ecology and Evolution* 14:405–10.

Crawley, M. J., and May, R. M. 1987. "Population Dynamics and Plant Community Structure: Competition Between Annuals and Perennials." *Journal of Theoretical Biology* 125:475–89.

Creswick, R. J., Farach, H. A., and Poole, C. P., Jr. 1992. *Introduction to Renormalization Group Methods in Physics*. John Wiley: New York.

Crist, T. O., Guertin, D. S., Wiens, J. A. and Milne, B. T. 1992. "Animal Movements in Heterogeneous Landscapes: An Experiment with Eleodes Beetles in Shortgrass Prairie." *Functional Ecology* 6:536–44.

Cronhjort, M. 1995. Models and Computer Simulations of Origins of Life and Evolution. Doctoral dissertation, Stockolm University.

Cronhjort, M. B., and Blomberg, C. 1997. "Cluster Compartmentalization May Provide Resistance to Parasites for Catalytic Networks." *Physica* D101:289–98.

Cross, M. C., and Hohenberg, P. 1993. "Pattern Formation Out of Equilibrium." *Reviews of Modern Physics* 65:851–1112.

Crutchfield, J. P., and Kaneko, K. 1988. "Are Attractors Relevant to Turbulence?" *Physical Review Letters* 60:2715–18.

Cvitanovic, P., ed. 1984. *Universality in Chaos*. Adam Hilger: Bristol, UK.

Darwin, C. 1859. *On the Origin of Species by Means of Natural Selection*. John Murray: London.

DeAngelis, D. L. 1975. "Stability and Connectance in Food Web Models." *Ecology* 56:238–43.

DeAngelis, D. L., and Gross, L. J. 1992. *Individual-Based Models and Approaches in Ecology: Populations, Communities and Ecosystems*. Chapman and Hall: London.

DeAngelis, D. L., and Waterhouse, J. C. 1987. "Equilibrium and Nonequilibrium Concepts in Ecological Models." *Ecological Monographs* 57:1–21.

Debinski, D. M., and Holt, R. D. 2000. "A Survey and Overview of Habitat Fragmentation Experiments." *Conservation Biology* 14:342–55.

De Ruiter, P. C., Neutel, A. M., and Moore, J. C. 1995. "Energetics, Patterns of Interaction Strengths, and Stability in Real Ecosystems." *Science* 269:1257–60.

D'Hondt, S., Donaghay, P., Zachos, J. C., Luttenberg, D., and Lindinger, M. 1998. "Organic Carbon Fluxes and Ecological Recovery from the Cretaceous-Tertiary Mass Extinction." *Science* 282:276–79.

Dickman, R. 1986. "Kinetic Phase Transitions in a Surface-Reaction Model: Mean-Field Theory." *Physical Review* A 34:4246–50.

Dieckmann, U., and Doebeli, M. 1999. "On the Origin of Species by Sympatric Speciation." *Nature* 400:354–57.

Dimri, V. P., and Prakash, V. R., 2001. "Scaling of Power Spectrum of Extinction Events in the Fossil." *Earth and Planetary Science Letters* 186:363–70.

Dixon, P. A., Milicich, M. J., and Sugihara, G. 1999. "Episodic Fluctuations in Larval Supply." *Science* 283:1528–30.

Dobzhansky, T. 1951. *Genetics and the Origin of Species*, 3rd ed., Columbia University Press: New York.

Drake, J. A. 1991. "Community-Assembly Mechanics and the Structure of an Experimental Species Ensemble." *American Naturalist* 137:1–26.

Drake, J. B., and Weishampel, J. F. 2000. "Multifractal Analysis of Canopy Height Measures in a Longleaf Pine Savanna." *Forest Ecology and Management* 128:121–27.

Droser, M. L, Bottjer, D. J., and Sheehan, P. M. 1997. "Evaluating the Ecological Architecture of Major Events in the Phanerozoic History of Marine Invertebrate Life." *Geology* 25:167–70.

Drossel, B. 1998. "Extinction Events and Species Lifetimes in a Simple Ecological Model." *Physical Review Letters* 81:5011–14.

_____. 2001. "Biological Evolution and Statistical Physics." *Advances in Physics* 50:209–95.

Drossel, B., and Schwabl, F. 1992. "Self-Organized Critical Forest-Fire Model." *Physical Review Letters* 69:1629–32.

Dungan, M. L., Miller, T. E., and Thompson, D. 1982. "Catastrophic Decline of a Top Carnivore in the Gulf of California Rocky Intertidal Zone." *Science* 216:989–81.

Dunne, J. A., Williams, R. J., and Martinez, N. D. 2002. "Food-Web Structure and Network Theory: The Role of Connectance and Size." *Proceedings of the National Academy of Sciences, USA* 99:12917–22.

Durrett, R. 1999. "Stochastic Spatial Models." *SIAM Review* 41:677–718.

Durrett, R., and Levin, S. A. 1994. "Stochastic Spatial Models: A User's Guide to Ecological Applications." *Philosophical Transactions of the Royal Society of London* B 343:329–50.

Dyson, G. 1998. *Darwin among the Machines*. Perseus Books: New York.

Dytham, C. 1995a. "Competitive Coexistence and Empty Patches in Spatially Explicit Metapopulation Models." *Journal of Animal Ecology* 64:145–46.

_____. 1995b. "The Effect of Habitat Destruction Pattern on Species Persistence—A Cellular Model." *Oikos* 74:340–44.

Eble, G. J., 1998. "The Role of Development in Evolutionary Radiations." In *Biodiversity Dynamics*, M. L. McKinney and J. A. Drake, eds. Columbia University Press: New York.

Eigen, M., and Schuster, P. 1979. *The Hypercycle: A Principle of Natural Self-Organization*. Springer: Berlin.

Elena, S. F., Cooper, V., and Lenski, R. 1996. "Punctuated Evolution Caused by Selection of Rare Beneficial Mutations." *Science* 272:1802–04.

Elena, S. F., and Lenski, R.E. 1997. "Test of Synergistic Interactions between Deleterious Mutations in Bacteria." *Nature* 390:395–98.

_____. 2003. "Evolution Experiments with Microorganisms: The Dynamics and Genetic Bases of Adaptation." *Nature Reviews Genetics* 4:457–69.

Elliott, D. K., ed. 1986. *Dynamics of Extinction*. Wiley Interscience: New York.

Ellner, S. P., and Turchin, P. 1995. "Chaos in a Noisy World: New Methods and Evidence from Time-Series Analysis." *The American Naturalist* 145: 343–75.

Elton, C. S. 1942. *Voles, Mice, and Lemmings: Problems in Population Dynamics*. Clarendon Press: Oxford.

_____. 1958. *The Ecology of Invasions by Animals and Plants*. Methuen: London.

Emmeche, C. 1994. *The Garden in the Machine: The Emerging Science of Artificial Life*. Princeton University Press: Princeton.

Erlinge, S. 1987. "Predation and Noncyclity in a Microtine Population in Southern Sweden." *Oikos* 50:347–52.

Ermentrout, G. B., and Edelstein-Keshet, L. 1993. "Cellular Automata Approaches to Biological Modeling." *Journal of Theoretical Biology* 160:97–133.

Erwin, D. H. 1993. *The Great Paleozoic Crisis: Life and Death in the Permian*. Columbia University Press: New York.

_____. 1994. "The Permo-Triassic Extinction." *Nature* 367:231–36.

_____. 1998a. "The End and the Beginning: Recoveries from Mass Extinctions." *Trends in Ecology and Evolution* 13:344–49.

_____. 1998b. "After the End: Recovery from Mass Extinction." *Science* 279:1324–25.

_____. 2001. "Lessons from the Past: Biotic Recoveries from Mass Extinction." *Proceedings of the National Academy of Sciences, USA* 98:5399–403.

Estes, J. A., and Palmisano, J. F. 1974. "Sea Otters: Their Role in Structuring Nearshore Communities." *Science* 185:1058–60.

Estes, J. A., Tinker, M. T., Williams, T. M., and Doak, D. F. 1998. "Killer Whale Predation on Sea Otters Linking Coastal with Oceanic Ecosystems." *Science* 282:473–76.

Fahrig, L., and Merriam, G. 1994. "Conservation of Fragmented Populations." *Conservation Biology* 8:50–59.

Farmer, D., Toffoli, T., and Wolfram, S., eds. 1984. *Cellular Automata: Proceedings of an Interdisciplinary Workshop*. North-Holland Publishing Company: Amsterdam.

Feder, J. 1988. *Fractals*. Plenum Press: New York.

Feigenbaum, M. 1978. "Quantitative Universality for a Class of Nonlinear Transformations." *Journal of Statistical Physics* 19:25–52.

Food and Agriculture Organization of the United Nations. 1997. *State of the World's Forests*, 16. FAO: Rome.

Forster, M. A. 2003. "Self-Organised Instability and Megafaunal Extinctions in Australia." *Oikos* 103:235–38.

Fox, J. W., and McGrady-Steed, J. 2002. "Stability and Complexity in Microcosm Communities." *Journal of Animal Ecology* 71:749–56.

Frank, K., and Wissel, C. 2002. "A Formula for the Mean Lifetime of Metapopulations in Heterogeneous Landscapes." *American Naturalist* 159:530–52.

Frank, S. A. 1991. "Spatial Variation in Coevolutionary Dynamics." *Evolutionary Ecology* 5:193–217.

_____. 1997a. "The Price Equation, Fisher's Fundamental Theorem, Kin Selection and Causal Analysis." *Evolution* 51:1712–29.

_____. 1997b. "Developmental Selection and Self-Organization." *Biosystems* 40:237–43.

_____. 1998. *Foundations of Social Evolution*. Princeton University Press: Princeton.

Gaedke, U. 1992. "The Size Distribution of Plankton Biomass in a Large Lake and Its Seasonal Variability." *Limnology and Oceanography* 37:1202–20.

Gamarra, J., and Solé, R. V. 2001. "Bifurcations and Chaos in Ecology: Lynx Returns Revisited." *Ecology Letters* 3:114–21.

_____. 2002. "Biomass-Diversity Responses and Spatial Dependencies in Disturbed Tallgrass Prairies." *Journal of Theoretical Biology* 215:469–80.

Garlaschelli, D., Caldarelli, G., and Pietronero, L. 2003. "Universal Scaling Relations in Food Webs." *Nature* 423:165–68.

Gause, G. F. 1934. *The Struggle for Existence.* Williams and Wilkins: Baltimore.

Gavrilets, S. 1997. "Evolution and Speciation on Holey Adaptive Landscapes." *Trends in Ecology and Evolution* 12:307–12.

Gavrilets, S., and Gravner, J. 1997. "Percolation on the Fitness Hypercube and the Evolution of Reproductive Isolation." *Journal of Theoretical Biology* 184:51–64.

Gilinsky, N. L. 1991. "The Pace of Taxonomic Evolution." In *Analytical Paleobiology.* N. L. Gilinsky, P. W. Signor, eds., 157–74. University of Tennessee: Knoxville.

Gillespie, J. 1974. "Polymorphism in Patchy Environments." *American Naturalist* 108:145–51.

Gilpin, M. E., and Hanski, I. 1991. *Metapopulation Dynamics.* London: Academic Press.

Ginzburg, L. R., and Taneyhill, D. E. 1994. "Population Cycles of Forest Lepidoptera: A Maternal Effect Hypothesis." *Journal of Animal Ecology* 63:79–92.

Gisiger, T. 2001. "Scale Invariance in Biology: Coincidence or Footprint of a Universal Mechanism?" *Biological Reviews* 76:161–209.

Glass, L., and Mackey, M. C. 1988. *From Clocks to Chaos: The Rhythms of Life,* Princeton University Press: Princeton.

Goldberger, A. L., Rigney, D. R., and West, B. J. 1990. "Chaos and Fractals in Human Physiology." *Scientific American* 262:42–49.

Goodwin, B. C. 1994. *How the Leopard Changes Its Spots: The Evolution of Complexity.* Charles Scribner's Sons: New York.

Gotelli, N. J. 2001. "Research Frontiers in Null Model Analysis." *Global Ecology and Biogeography* 10:337–43.

Gould, S. J., 1989. *Wonderful Life: The Burgess Shale and the Nature of History* W. W. Norton: New York.

_____. 2003. *The Structure of Evolutionary Theory.* Harvard University Press: Cambridge, MA.

Gould, S. J., and Calloway, C. B. 1980. "Clams and Brachiopods: Ships That Pass in the Night." *Paleobiology* 6:383–96.

Gradstheyn, I. S. 2000. *Table of Integrals, Series and Products,* 6th ed. Academic Press: New York.

Grassberger, P., and Procaccia, I. 1983. "Characterization of Strange Attractors." *Physical Review Letters* 50:346–49.

Green, D. G. 1989. "Simulated Effects of Fire, Dispersal and Spatial Pattern on Competition within Vegetation Mosaics." *Vegetatio* 82:139–53.

_____. 1990. "Cellular Automata Models in Biology." *Mathematical and Computer Modelling* 13(6):69–74.

Grenfell, B. T., Bjørnstad, O. N., and Kappey, J. 2001. "Travelling Waves and Spatial Hierarchies in Measles Dynamics." *Nature* 414:716–23.

Groot, S., and Mazur, P. 1984. *Non-Equilibrium Thermodynamics*. Dover: New York.

Guckenheimer, J., and Holmes, P. 1983. *Nonlinear Oscillations, Dynamical Systems and Bifurcations of Vector Fields*. New York: Springer-Verlag.

Gurney, W.S.C., and Nisbet, R. M. 1978. "Single-Species Population Fluctuations in Patchy Environments." *American Naturalist* 112:1075–90.

Gustafson, E.J., and Parker, G. R. 1992. "Relationship between Landcover Proportion and Indices of Landscape Spatial Patterns." *Landscape Ecology* 7:101–10.

Gyllenber, M., and Hanski, I. 1997. "Habitat Deterioration, Habitat Destruction, and Metapopulation Persistence in a Heterogeneous Landscape." *Theoretical Population Biology* 52:198–215.

Hairston, Jr., N. G., and Hairston, Sr., N. G. 1993. "Cause-Effect Relationships in Energy Flow, Trophic Structure, and Interspecific Interactions." *American Naturalist* 142:379–411.

Haken, H. 1977. *Synergetics: An Introduction*. Springer-Verlag: Berlin.

_____. 1983. *Advanced Synergetics*. Springer-Verlag: Berlin.

Haldane J.B.S. 1932. *The Causes of Evolution*. Longmans and Green: London.

Hallam, A., and Wignall, P. B. 1997. *Mass Extinctions and Their Aftermath*. Oxford University Press: Oxford.

Halsey T. C., Jensen, M. H., Kadanoff, L. P., Procaccia, I., and Shraiman, B. I. 1986. "Fractal Measures and Their Singularities." *Physical Review* A33:1141–51.

Hamilton, W. D., Axelrod, R., and Tanese, R. 1990. "Sexual Reproduction as an Adaptation to Resist Parasites." *Proceedings of the National Academy of Sciences, USA* 87:3566–73.

Hanski, I. 1987. "Populations of Small Mammals Cycle—Unless They Don't." *Trends in Ecology and Evolution* 2:55–56.

_____. 1998. "Metapopulation Dynamics." *Nature* 396:41–49.

_____. 1999. *Metapopulation Ecology*. Oxford University Press: Oxford.

Hanski, I., and Gilpin, M. E. 1997. *Metapopulation Biology: Ecology, Genetics and Evolution*. Academic Press: San Diego.

Hanski, I. and Gyllenberg, M. 1993. "Two General Metapopulation Models and the Core-Satellite Species Hypothesis." *American Naturalist* 142:17–41.

Hanski, I., Hansson, L., and Henttonen, H. 1991. "Specialist Predators, Generalist Predators, and the Microtine Rodent Cycle." *Journal of Animal Ecology* 60:353–67.

Hanski, I., Moilanen, A., and Gyllenberg, M. 1996. "Minimum Viable Metapopulation Size." *American Naturalist* 147:527–41.

Hanski, I., and Ovaskainen, O. 2000. "The Metapopulation Capacity of a Fragmented Landscape." *Nature* 404:755–58.

Hanski, I., Turchin, P., Korpimaki, E., and Henttonen, H. 1993. "Population Oscillations of Boreal Rodents—Regulation by Mustelid Predators Leads to Chaos." *Nature* 364:232–35.

Hanski, I., and Woiwood, I. P. 1993. "Spatial Synchrony in the Dynamics of Moth and Aphid Populations." *Journal of Animal Ecology* 62:656–68.

Harada, Y., Ezoe, H., Iwasa, Y., Matsuda, H., and Sato, K. 1995. "Population Persistence and Spatially Limited Social Interaction." *Theoretical Population Biology* 48:65–91.

Harada, Y., and Iwasa, Y. 1994. "Lattice Population Dynamics for Plants with Dispersing Seeds and Vegetative Propagation." *Research on Population Ecology* 36:237–49.

Harris, T. E. 1963. *The Theory of Branching Processes*. Springer: Berlin.

Hart, M. B. 1996. "Recovery of the Food Chain after the Late Cenomanian Extinction Event." In *Biotic Recovery from Mass Extinction Events*, M. B. Hart, ed., 265–77. Geological Society Special Publication.

Hassell, M. P. 1975. "Density-Dependence in Single-Species Populations." *Journal of Animal Ecology* 44:283–95.

_____. 1985. "Insect Natural Enemies as Regulating Factors." *Journal of Animal Ecology* 54:323–34.

Hassell, M. P., Comins, H. N., and May, R. M. 1991. "Spatial Structure and Chaos in Insect Population Dynamics." *Nature* 353:255–58.

_____. 1994. "Species Coexistence and Self-Organizing Spatial Dynamics." *Nature* 370:290–92.

Hassell, M. P., Lawton, J. N., and May, R. M. 1976. "Patterns of Dynamics Behaviour in Single-Species Populations." *Journal of Animal Ecology* 45:471–86.

Hassell, M. P., Miramontes, O., Rohani, P., and May, R. M. 1995. "Appropriate Formulations for Dispersal in Spatially Structured Models." *Journal of Animal Ecology* 64:662–64.

Hastings, A. 1977. "Spatial Heterogeneity and the Stability of Predator-Prey Systems." *Theoretical Population Biology* 12:37–48.

_____. 1992. "Age-Dependent Dispersal Is Not a Simple Process: Density Dependence, Stability and Chaos." *Theoretical Population Biology* 41:388–400.

_____. 1993. "Complex Interactions between Dispersal and Dynamics: Lessons from Coupled Logistic Equations." *Ecology* 74:1362–72.

_____. 1998. "Transients in Spatial Ecological Models." In *Modeling Spatiotemporal Dynamics in Ecology*, J. Bascompte and R. V. Solé, eds., 189–98. Springer-Verlag: Berlin.

Hastings, A., and Higgins, K. 1994. "Persistence of Transients in Spatially Structured Ecological Models." *Science* 263:1133–36.

Hastings, A., Hom, C. L., Ellner, S., Turchin, P., and Godfray, H.C.J. 1993. "Chaos in Ecology: Is Mother Nature a Strange Attractor?" *Annual Review of Ecology and Systematics* 24:1–33.

Hastings, H. M. 1982. "The May-Wigner Stability Theorem." *Journal of Theoretical Biology* 97:155–66.

Hastings, H. M., and Sugihara, G. 1993. *Fractals: A User's Guide for the Natural Sciences*. Oxford University Press: New York.

Hawkins, B. A. 1992. "Parasitoid-Host Food Webs and Donor Control." *Oikos* 65:159–62.

_____. 1994. *Pattern and Process in Host-Parasitoid Interactions*. Cambridge University Press: Cambridge.

Hiebeler, D. 1997. "Stochastic Spatial Models: From Simulations to Mean Field and Local Structure Approximations." *Journal of Theoretical Biology* 187:307–19.

_____. 2000. "Populations on Fragmented Landscapes with Spatially Structured Heterogeneities: Landscape Generation and Local Dispersal." *Ecology* 81:1629–41.

Higgins, K., Hastings, A., and Botsford, L. W. 1997. "Density Dependence and Age Structure: Non-Linear Dynamics and Population Behavior." *The American Naturalist* 149:247–69.

Higgins, P.A.T., Mastrandrea, M. D., and Schneider, S. H. 2002. "Dynamics of Climate and Ecosystem Coupling: Abrupt Changes and Multiple Equilibria." *Philosophical Transactions of the Royal Society of London* B 357:647–55.

Hill, M. F., and Caswell, H. 1999. "Habitat Fragmentation and Extinction Thresholds on Fractal Landscapes." *Ecology Letters* 2:121–27.

Hillel, D. 1998. *Environmental Soil Physics*. San Diego: Academic Press.

Hillis, D. W. 1990. "Co-Evolving Parasites Improve Simulated Evolution as an Optimization Parameter." *Physica* D 42:228–34.

Hofbauer, J. and Sigmund, K. 1991. *The Theory of Evolution and Dynamical Systems*. Cambridge University Press: Cambridge.

Hoffman, P. F., Kaufman, A. J., Halverson, G. P., and Schrag, D. P. 1998. "A Neoproterozoic Snowball Earth." *Science* 281:1342–46.

Hogg, T., Huberman, B. A., and McGlade, J. M. 1989. "The Stability of Ecosystems." *Proceedings of the Royal Society of London* B 237:43–51.

Holldobler, B., and Wilson, E. O. 1990. *The Ants*. Springer-Verlag: New York.

Holt, R. D. 1977. "Predation, Apparent Competition, and the Structure of Prey Communities." *Theoretical Population Biology* 12:197–229.

_____. 1997. "From Metapopulation Dynamics to Community Structure: Some Consequences of Spatial Heterogeneity." In *Metapopulation Biology: Ecology, Genetics and Evolution*, I. Hanski and M. Gilpin, eds., 149–64. Academic Press: San Diego.

Holt, R. D., Robinson, G. R., and Gaines, M. S. 1995. "Vegetation Dynamics in an Experimentally Fragmented Landscape." *Ecology* 76:1610–24.

Hopfield, J. J. 1994. "Physics, Computation, and Why Biology Looks So Different." *Journal of Theoretical Biology* 171:53–60.

Hubbell, S., Foster, R. B., O'Brien, S. T., Harms, K. E., Condit, R., Wechsler, B., Wright, S. J., and Loo de Lao, S. 1999. "Light-Gap Disturbances, Recruitment Limitation, and Tree Diversity in a Neotropical Forest." *Science* 283:554–57.

Hubbell, S. P. 1979. "Tree Dispersion, Abundance, and Diversity in a Tropical Dry Forest." *Science* 203:1299–1309.

_____. 1997. "A Unified Theory of Biogeography and Relative Species Abundance and Its Application to Tropical Rain Forests and Coral Reefs." *Coral Reefs* 16 (Suppl.): S9–S21.

_____. 2001. *The Unified Theory of Biogeography and Biogeography*. Princeton University Press: Princeton.

Huffaker, C. B. 1958. "Experimental Studies on Predation: Dispersion Factors and Predator-Prey Oscillations." *Hilgardia* 27:343–83.

Huisman, J., and Weissing, F. J. 2001. "Biological Conditions for Oscillations and Chaos Generated by Multispecies Competition." *Ecology* 82:2682–95.

Hunt, H. W., Coleman, D. C., Ingham, E. R., Elliott, E. T., Moore, J. C., Reid, C.P.P., and Morley, C. R. 1987. "The Detrital Food Web in a Shortgrass Prairie." *Biology and Fertility of Soils* 3:57–68.

Hutchinson, E. 1965. *The Ecological Theater and the Evolutionary Play*. Yale University Press: New Haven.

Huxham, M., Beaney, S., and Raffaelli, D. 1996. "Do Parasites Reduce the Changes Triangulation in a Real Food Web?" *Oikos* 76:284–300.

Hyde, W. T., Crowley, T. J., Baum, S. K., and Peltier, W. R. 2000. "Neoproterozoic Snowball Earth Simulations with a Coupled Climate/Ice-Sheet Model." *Nature* 405:425–29.

Ikegami, T., and Kaneko, K. 1990. "Computer Symbiosis: Emergence of Symbiotic Behavior through Evolution." *Physica* D 42:235–43.

Ilachinski, A. 2001. *Cellular Automata: A Discrete Universe*. World Scientific: Singapore.

Ives, A. R., Turner, M. G., and Pearson, S. M. 1998. "Local Explanations of Landscape Patterns: Can Analytical Approaches Approximate Simulation Models of Spatial Processes?" *Ecosystems* 1:35–51.

Iwao, S. 1968. "A New Regression Method for Analyzing the Aggregation Pattern of Animal Populations." *Research in Population Ecology* 10:1–20.

Jablonski, D., 1998. "Geographic Variation in the Molluscan Recovery from the End-Cretaceous Extinction." *Science* 279:1327–30.

_____. 1999. "The Future of the Fossil Record." *Science* 284:2114–16.

_____. 2001. "Lessons from the Past: Evolutionary Impacts of Mass Extinctions." *Proceedings of the National Academy of Sciences, USA* 98:5393–98.

Jablonski, D., Erwin, D. H., and Lipps, J. H. 1996. *Evolutionary Paleobiology*. University of Chicago Press: Chicago.

Jackson, A. 1991. *Perspectives of Nonlinear Dynamics*, vol. 2. Cambridge University Press: Cambridge.

Jackson, J.B.C. 1997. "Reefs since Columbus." *Coral Reefs* 16:S23–S32.

Jackson, J.B.C., Budd, A., and Pandolfi, J. M. 1996. "The Shifting Balance of Natural Communities?" In *Evolutionary Paleobiology*, D. Jablonski, D. H. Erwin, and J. H. Lipps, eds., 89–122. University of Chicago Press: Chicago.

Jacob, F. 1977. "Evolution and Tinkering." *Science* 196:1161–66.

Jain, S., and Krishna, S. 2001. "Crashes, Recoveries, and 'Core-Shifts' in a Model of Evolving Networks." *Proceedings National Academy of Sciences, USA* 98:543–47.

_____. 2002. "Large Extinctions in an Evolutionary Model: The Role of Innovation and Keystone Species." *Proceedings National Academy of Sciences, USA* 99:2055–60.

Jansen, V.A.A., and Yoshimura, J. 1998. "Populations Can Persist in an Environment Consisting of Sink Habitats Only." *Proceedings of the National Academy of Sciences, USA* 95:3696–98.

Jensen, H. 1998. *Self-Organized Criticality*. Cambridge University Press: Cambridge.

Johnson, A. R., Milne, B. T., and Wiens, J. A. 1992. "Diffusion in Fractal Landscapes: Simulations and Experimental Studies of Tenebrionid Beetle Movements." *Ecology* 73:1968–83.

Jordano, P. 1987. "Patterns of Mutualistic Interactions in Pollination and Seed Dispersal: Connectance, Dependence Asymmetries, and Coevolution." *American Naturalist* 129:657–77.

Jordano, P., Bascompte, J., and Olesen, J. M. 2003. "Invariant Properties in Coevolutionary Networks of Plant-Animal Interactions." *Ecology Letters* 6:69–81.

Judson, O. 1994. "The Rise of the Individual-Based Model in Ecology." *Trends in Ecology and Evolution* 9:9–14.

Kaneko, K. 1984. "Period-Doubling of Kink-Antikink Patterns, Quasi-Periodicity in Antiferro-Like Structures and Spatial Intermittency in Coupled Map Lattices: Towards a Prelude to a 'Field Theory of Chaos'." *Progress in Theoretical Physics* 72:480–86.

_____. 1987. "Velocity Dependent Lyapunov Exponent as a Measure of Chaos for Open Flows." *Physics Letters* 119 A:397–402.

_____. 1990. "Supertransients, Spatiotemporal Intermittency, and Stability of Fully Developed Spatiotemporal Chaos." *Physics Letters* A 149:105–12.

_____, ed. 1992. "Special Issue on Coupled Map Lattices." *Chaos* 2, 279–313.

_____, ed. 1993. *Theory and Applications of Coupled Map Lattices*. John Wiley and Sons: Chichester.

_____. 1998. "Diversity, Stability and Metadynamics: Remarks from Coupled Map Studies." In *Modeling Spatiotemporal Dynamics in Ecology*, J. Bascompte and R. V. Solé, eds., 27–45. Springer-Verlag: Berlin.

Kaneko, K., and Ikegami, T. 1992. "Homeochaos: Dynamics Stability of a Symbiotic Dynamics and Evolving Mutation Rates." *Physica D* 56:406–29.

Kaneko, K., and Tsuda, I. 2000. *Complex Systems: Chaos and Beyond*. Springer: New York.

Kaplan, D., and Glass, L. 1995. *Understanding Nonlinear Dynamics*. Springer: New York.

Kapral, R. 1985. "Pattern Formation in Two-Dmensional Arrays of Coupled, Discrete-Time Oscillators." *Physical Review* A 31:3868–74.

Kareiva, P., and Wennergren, U. 1995. "Connecting Landscape Patterns to Ecosystem and Population Processes." *Nature* 373:299–302.

Katori, M., Kizaki, S., Terui, Y., and Kubo, T. 1998. "Forest Dynamics with Canopy Gap Expansion and Stochastic Ising Model." *Fractals* 6:81–86.

Kauffman, S. A. 1989. "Cambrian Explosion and Permian Quiescence: Implications of Rugged Fitness Landscapes." *Evolutionary Ecology* 3:274–81.

_____. 1993. *The Origins of Order*. Oxford University Press: Oxford.

Kauffman S. A., and Levin, S. 1987. "Towards a General Theory of Adaptive Walks on Rugged Landscapes." *Journal of Theoretical Biology* 128:11–45.

Kauffman, S. A., and Johnsen, J. 1991. "Coevolution to the Edge of Chaos: Coupled Fitness Landscapes, Poised States and Coevolutionary Avalanches." *Journal of Theoretical Biology* 149:467–505.

Keith, L. B. 1983. "Role of Food in Hare Population Cycles." *Oikos* 40:385–95.

Keitt, T. H., and Marquet, P. 1996. "The Introduced Hawaiian Avifauna Reconsidered: Evidence for Self-Organized Criticality?" *Journal of Theoretical Biology* 182:161–67.

Keitt, T. H., and Stanley, H. E. 1998. "Dynamics of North American Breeding Bird Populations." *Nature* 393:257–60.

Keitt, T. H., Urban, D. L., and Milne, B. T. 1997. "Detecting Critical Scales in Fragmented Landscapes." *Conservation Ecology* 1:4.

Kendall, B. E., Bjørnstad, O. N., Bascompte, J., Keitt, T. H., and Fagan, W. F. 2000. "Dispersal, Environmental Correlation, and Spatial Synchrony in Population Dynamics." *American Naturalist* 155:628–36.

Kephart, J. O., Hogg, T., and Huberman, B. A. 1989. "Dynamics of Computational Ecosystems." *Physical Review* A 40:404–21.

Kirchner, J. W., and Weil, A. 1998. "No Fractals in Fossil Extinction Statistics." *Nature* 395:337–38.

Kitching, R. L. 2000. *Food Webs and Container Habitats: The Natural History and Ecology of Phytotelmata*. Cambridge University Press: New York.

Klafter J., Schlesinger, M. F., and Zumofen, G. 1996. "Beyond Brownian Motion." *Physics Today* 49:33–39.

Klausmeier, C. A. 1999. "Regular and Irregular Patterns in Semiarid Vegetation." *Science* 284:1826–28.

______. 2001. "Regular and Irregular Patterns in Semiarid Vegetation." *Science* 284:1826–28.

Klinkhamer, P.G.I., de Jong, T. J., and Metz, J.A.J. 1983. "An Explanation for Low Dispersal Rates: A Simulation Experiment." *Netherland Journal of Zoology* 33:532–41.

Knoll, A. H. 1992. "The Early Evolution of Eukaryotes: A Geological Perspective." *Science* 256:622–27.

Korvin, G. 1992. *Fractal Models in the Earth Sciences*. Elsevier: New York.

Kot, M., Sayler, G. S., and Schultz, T. W. 1992. "Complex Dynamics in a Model Microbial Ecosystem." *Bulletin of Mathematical Biology* 54:619–48.

Krapivsky, P. L., and Redner, S. 2001. "Organization of Growing Random Networks." *Physical Review* E 63:066123.

Krapivsky, P. L., Rodgers, G. J., and Redner, S. 2001. "Degree Distributions of Growing Random Networks." *Phyical Review Letters* 86:5401–4.

Krause, A. E., Frank, K. A., Mason, D. M., Ulanowicz, R. E., and Taylor, W. W. 2003. "Compartments Revealed in Food-Web Structure." *Nature* 426:282–85.

Krebs, C. J., Boutin, S., Boonstra, R., Sinclair, A.R.E., Smith, J.N.M., Dale, M.R.T., and Turkington, K.M.R. 1995. "Impact of Food and Predation on the Snowshoe Hare Cycle." *Science* 269:1112–15.

Krebs, C. J., Gaines, M. S., Keller, B. L., Myres, J. H., and Tamarin, R. H. 1973. "Population Cycles in Small Rodents." *Science* 175:35–41.

Krues, A., and Tscharntke, T. 1994. "Habitat Fragmentation, Species Loss and Biological Control." *Science* 264:1581–84.

Kuno, E. 1981. "Dispersal and the Persistence of Populations in Unstable Habitats: A Theoretical Note." *Oecologia* 49:123–26.

Kuramoto, Y. 1984. *Chemical Oscillations, Chaos and Turbulence*. Springer-Verlag: Berlin.

Kuztnesov, Y. A. 1995. *Elements of Applied Bifurcation Theory*. Springer: Berlin.

Lande, R. 1987. "Extinction Thresholds in Demographic Models of Territorial Populations." *American Naturalist* 130:624–35.

Lande, R. 1988. "Genetics and Demography in Biological Conservation." *Science* 241:1455–60.

_____. 1993. "Risks of Population Extinction from Demographic and Environmental Stochasticity and Random Catastrophes." *American Naturalist* 142:911–27.

Lande, R., Engen, E., and Saether, B. E. 1998. "Extinction Times in Finite Metapopulation Models with Stochastic Local Dynamics." *Oikos* 83:383–89.

_____. 1999. "Spatial Scale of Population Synchrony: Environmental Correlation Versus Dispersal and Density Regulation." *American Naturalist* 154:271–81.

Langton, C. G., ed. 1989. *Artificial Life*. Addison-Wesley: Redwood City, CA.

_____, ed. 1994. *Artificial Life III*. Addison-Wesley: Reading, MA.

Langton, C. G., Taylor, C. E., Doyne Farmer, J., and Rasmussen, S., eds. 1992. *Artificial Life II*. Addison-Wesley: Redwood City, CA.

Laurance, W. F., and Bierregaard, R. O., Jr., eds. 1997. *Tropical Forest Remnants: Ecology, Management, and Conservation of Fragmented Communities*. University of Chicago Press: Chicago.

Law, R., and Morton, R. D. 1993. "Alternative Permanent States of Ecological Communities." *Ecology* 74:1347–61.

Lawton, J. H., and May, R. M. 1995. *Extinction Rates*. Oxford University Press: Oxford.

Langton, C. G., ed. 1989. *Artificial Life*. Santa Fe Institute Studies in the Sciences of Complexity, vol. 6. Addison-Wesley: Reading, MA.

Lavorel, S., Gardner, R. H., and O'Neill, R. V. 1993. "Analysis of Patterns in Hierarchically Structured Landscapes." *Oikos* 67:521–28.

Lawler, S. P. 1993. "Species Richness, Species Composition and Population Dynamics of Protists in Experimental Microcosms." *Journal of Animal Ecology* 62:711–19.

Lawler, S. P., and Morin, P. J. 1993. "Food Web Architecture and Population Dynamics in Laboratory Microcosms of Protists." *American Naturalist* 141:675–86.

Lawton, J. H., and May, R. M., eds. 1995. *Extinction Rates*. Oxford University Press: Oxford.

Leakey, R., and Lewin, R. 1995. *The Sixth Extinction*. Doubleday: New York.

Lefever, R., and Lejeune, O. 1997. "On the Origin of Tiger Bush." *Bulletin of Mathematical Biology* 59:263–94.

Lehman, C. L., and Tilman, D. 1997. "Competition in Spatial Habitats." In *Spatial Ecology: The Role of Space in Population Dynamics and Interspecific Interactions*, D. Tilman and P. Kareiva, eds., 185–203. Princeton University Press: Princeton.

Lenski, R. E., Ofria, C., Collier, T. C., and Adami, C. 1999. "Genome Complexity, Robustness, and Genetic Interactions in Digital Organisms." *Nature* 400:661–64.

Lenski, R. M., and Travisano, M. 1994. "Dynamics of Adaptation and Diversification: A 10,000-Generation Experiment with Bacterial Populations." *Proceedings of the National Academy of Sciences, USA* 91:6808–14.

Lenton, T. M. 1998. "Gaia and Natural Selection." *Nature* 394:439–47.

Lenton, T. M., and van Oijen, M. 2002. "Gaia as a Complex Adaptive System." *Philosophical Transactions of the Royal Society of London* B 357:683–95.

Levin, S. A. 1974. "Dispersion and Population Interactions." *American Naturalist* 108:207–28.

_____. 1992. "The Problem of Pattern and Scale in Ecology." *Ecology* 73:1943–67.

_____. 1998. "Ecosystems and the Biosphere as Complex Adaptive Systems." *Ecosystems* 1:431–36.

_____. 1999. *Fragile Dominion*. Perseus Books: New York.

Levin, S. A., and Durrett, R. 1996. "From Individuals to Epidemics." *Philosophical Transactions of the Royal Society of London*, Series B 351:1615–21.

Levin, S. A., and Pacala, S. W. 1998. "Theories of Simplification and Scaling of Spatially Distributed Processes." In *Spatial Ecology. The Role of Space in Population Dynamics and Interspecific Interactions*, D. Tilman and P. Kareiva, eds., 271–95. Princeton University Press: Princeton.

Levin, S. A., and Segel, L. A. 1976. "Hypothesis for the Origin of Planktonic Patchiness." *Nature* 259:659.

Levins, R. 1966. "The Strategy of Model Building in Population Biology." *American Scientist* 54:421–31.

_____. 1968. *Evolution in Changing Environments*. Princeton University Press: Princeton.

_____. 1969a. "Some Demographic and Genetic Consequences of Environmental Heterogeneity for Biological Control." *Bulletin of the Entomological Society of America* 15:237–40.

_____. 1969b. "The Effect of Random Variations of Different Types on Population Growth." *Proceedings of the National Academy of Sciences, USA* 62:1061–65.

Lévy, P. 1965. *Processus Stochastiques et Mouvement Brownien*, 2nd ed. Gauthier-Villars: Paris.

Levy, S. 1993. *Artificial Life*. Vintage Books: New York.

Lewontin, R. C., and Cohen, D. 1969. "On Population Growth in a Randomly Varying Environment." *Proceedings of the National Academy of Sciences, USA* 62:1056–60.

Li, T. Y., and Yorke, J. A. 1975. "Period Three Implies Chaos." *American Mathematical Monthly* 82:985–92.

Li, W. H., and Graur, D. 1991. *Fundamentals of Molecular Evolution*. Sinauer Associates: Sunderland, MA.

Lindgren, K. 1992. "Evolutionary Phenomena in Simple Dynamics." In *Artificial Life II*, C. Langton et al., eds., 295–312. Addison-Wesley: Redwood City, CA.

Lindgren, K., and Nordahl, M. G. 1994. "Evolutionary Dynamics of Spatial Games." *Physica* D 75:292–309.

Lloyd, M. 1967. "Mean Crowding." *Journal of Animal Ecology* 36:1–30.

Loehle, C., and Li, B. L. 1996. "Habitat Destruction and the Extinction Debt Revisited." *Ecological Applications* 6:784–89.

Loehle, C., Li, B. L., and Sundell, R. C. 1996. "Forest Spread and Phase Transitions at Forest-Prairie Ecotones in Kansas, USA." *Landscape Ecology* 11:225–35.

Lomnicki, A. 1989. "Avoiding Chaos." *Trends in Ecology and Evolution* 4:239–40.

Lorenz, E. 1963. "Deterministic Nonperiodic Flow." *Journal of Atmospheric Sciences* 20:130–41.

Lotka, A. J. 1925. *Elements of Physical Biology*. Williams and Wilkins: Baltimore.

Lovelock, J. 1988. *The Ages of Gaia*. W. W. Norton: New York.

Lundberg, P., Ranta, E., and Kaitala, V. 2000. "Species Loss Leads to Community Closure." *Ecology Letters* 3:465–68.

MacArthur, R. 1955. "Fluctuations of Animal Populations, and a Measure of Community Stability." *Ecology* 36:533–36.

Maddox, J. 1994. "Punctuated Equilibrium by Computer." *Nature* 371.

Malamud, B. D., Morein, G., and Turcotte, D. L. 1989. "Forest Fires: An Example of Self-Organized Critical Behavior." *Science* 281:1840–42.

Mandelbrot, B. 1977. *Fractals: Form, Chance, and Dimension*. Freeman: San Francisco.

_____. 1983. *The Fractal Geometry of Nature*. Freeman: New York.

Mani, G. S. 1989. "Avoiding Chaos." *Trends in Ecology and Evolution* 4:239–40.

Manrubia, S. C., and Paczuski, M. 1998. "A Simple Model of Large-Scale Organization in Evolution." *International Journal of Modern Physics* C 9:1025–32.

Mantegna, R. N., and Stanley, H. E. 2000. *An Introduction to Econophysics: Correlations and Complexity in Finance*. Cambridge University Press: Cambridge.

Margalef, R. 1968. *Perspectives in Ecological Theory*. University of Chicago Press: Chicago.

_____. 1969. "Diversity and Stability: A Practical Proposal and a Model of Interdependence." *Diversity and Stability in Ecological Systems, Brookhaven Symposia in Biology* 22:25–37.

_____. 1986. "Reset Successions and Suspected Chaos in Models of Marine Populations." *International Symposium in Long-Term Changes of Marine Fish Populations. Vigo*: 321–43.

Maron, J. L., and Harrison, S. 1997. "Spatial Pattern Formation in an Insect Host-Parasitoid System." *Science* 278:1619–21.

Marro, J., and Dickman, R. 1999. *Nonequilibrium Phase Transitions in Lattice Models*. Cambridge University Press: Cambridge.

Martin, P., and Klein, R., eds. 1984. *Quaternary Extinctions*. University of Arizona Press: Tuscon.

Martin, P. S., and Klein, R. G., eds. 1984. *Quaternary Extinctions: A Prehistoric Revolution*. University of Arizona Press: Tucson.

Martinez, N. 1991. "Artifacts or Attributes? Effect of Resolution on the Little Rock Lake Food Web." *Ecological Monographs* 61:367–92.

Maslov, S., and Sneppen, K. 2002. "Specificity and Stability in Topology of Protein Networks." *Science* 296:910–13.

Matsuda, H., Ogita, N., Sasaki, A., and Sato, W. 1992. "Statistical Mechanics of Population—The Lattice Lotka-Volterra Model." *Progress in Theoretical Physics* 88:1035–49.

Maurer, B. A. 1999. *Untangling Ecological Complexity: The Macroscopic Perspective*. University of Chicago Press: Chicago.

May, R. M. 1972. "Will a Large Complex System Be Stable?" *Nature* 238:413–14.

_____. 1973. *Stability and Complexity in Model Ecosystems*. Princeton University Press: Princeton.

_____. 1974. "Biological Populations with Non-Overlapping Generations: Stable Points, Stable Cycles, and Chaos." *Science* 186:645–47.

_____. 1975. "Patterns of Species Abundance and Diversity." In *Ecology and Evolution of Communities*, M. L. Cody and J. M. Diamond, eds., 197–227. Harvard University Press: Cambridge, MA.

_____. 1976. "Simple Mathematical Models with Very Complicated Dynamics." *Nature* 261:459–67.

_____. 1977. "Thresholds and Breakpoints in Ecosystems with a Multiplicity of Stable States." *Nature* 269:471–77.

_____. 1994. "The Effects of Spatial Scale on Ecological Questions and Answers." In *Large Scale Ecology and Conservation Biology*, P. J. Edwards, R. M. May, and N. R. Webb, eds., 1–17. Blackwell: Oxford.

_____. 2004. "Uses and Abuses of Mathematics in Biology." *Science* 303:790–93.

May, R. M., and Oster, G. F. 1976. "Bifurcations and Dynamic Complexity in Simple Ecological Models." *The American Naturalist* 110:573–99.

Maynard Smith, J., 1989. "The Causes of Extinction." *Philosophical Transactions of the Royal Society of London* B 325:241–52.

Maynard Smith, J., and Szathmáry, E. 1995. *The Major Transitions in Evolution*. W. H. Freeman: Oxford.

McCann, K. S. 2000. "The Diversity-Stability Debate." *Nature* 405: 228–33.

McCann, K. S., and Hastings, A. 1997. "Re-evaluating the Omnivory-Stability Relationship in Food Webs." *Proceedings of the Royal Society London* B 264:1249–54.

McGhee, G. R. 1999. *Theoretical Morphology*. Columbia University Press: New York.

McKane, A., Alonso, D., and Solé, R. V. 2000. "A Mean Field Stochastic Theory for Species-Rich Assembled Communities." *Physical Review* E62:8466–84.

McKinney, M. L., and Drake, J. A., eds. 1998. *Biodiversity Dynamics: Turnover of Populations, Taxa and Communities*. Columbia University Press: New York.

McLaughlin, S. P. 1986. "Floristic Analysis of the Southwestern United States." *Great Basin Naturalist* 46:46–65.

Meinhardt, H. 1982. *Models of Biological Pattern Formation*. Academic Press: London.

_____. 1995. *The Algorithmic Beauty of Sea Shells*. Springer: Berlin.

Melián, C. J., and Bascompte, J. 2002a. "Food Web Structure and Habitat Loss." *Ecology Letters* 5:37–46.

_____. 2002b. "Complex Networks: Two Ways to Be Robust?" *Ecology Letters* 5:705–708.

_____. 2004. "Food Web Cohesion." *Ecology* 85:352–58.

Menge, B. A. 1995. "Indirect Effects in Marine Rocky Intertidal Interactive Webs: Patterns and Importance." *Ecological Monographs* 65:21–74.

Meron, E., Gilad, E., von Hardenberg, J., Shachak, M., and Zarmi, Y. 2004. "Vegetation Patterns Along a Rainfall Gradient." *Chaos, Solitons and Fractals* 19:367–76.

Metz, J.A.J., de Jong, T. J., and Klinkhamer, P.G.L. 1983. "What Are the Advantages of Dispersing: A Paper by Kuno Explained and Extended." *Oecologia* 57:166–69.

Metz, J.A.J., and de Roos, A. M. 1992. "The Role of Physiologically Structured Population Models within a General Individual-Based Modelling Perspective." In *Individual-Based Models and Approaches in Ecology: Populations, Communities and Ecosystems*, D. L. De Angelis and L. J. Gross, eds., 36–61. Chapman and Hall: London.

Mezard, M., Parisi, G., and Virasoro, M. 1999. *Sping Glass Theory and Beyond*. World Scientific: Singapore.

Miller, A. I. 2000. "Conversations about Phanerozoic Global Diversity." In *Deep Time: Paleobiology's Perspective* (Special Twenty-Fifth Anniversary Issue of *Paleobiology*), D. J. Erwin and S. L. Wing, eds. *Paleobiology* 26 (supplement to No. 4): 53–73.

Miller, G. H., Magee, J. W., Johnson, B. J., Fogel, M., Spooner, N. A., McCulloch, M. T., and Ayliffe, L. K. 1999. "Pleistocene Extinction of *Genyornis Newtoni*: Human Impact on Australian Megafauna." *Science* 283:205–208.

Miller, K. D., Keller, J. B., and Stryker, M. P. 1989. "Ocular Dominance Column Development: Analysis and Simulation." *Science* 245:605–15.

Milne, B. T. 1998. "Motivation and Benefits of Complex Systems Approaches in Ecology." *Ecosystems* 1:449–56.

Milne, B. T., and Johnson, A. R. 1993. "Renormalization Relations for Scale Transformation in Ecology." In *Some Mathematical Questions in Biology: Predicting Spatial Effects in Ecological Systems*, R. H. Gardner, ed., 109–28. American Mathematical Society: Providence, RI.

Milne, B. T., Johnson, A. R., Keitt, T. H., Hatfield, C. A., David, J., and Hraber, P. 1996. "Detection of Critical Densities Associated with Piñon-Juniper Woodland Ecotones." *Ecology* 77:805–21.

Milo, R., Shen-Orr, S., Itzkovitz, S., Kashtan, N., Chklovskii, D., and Alton, U. 2002. "Network Motifs: Simple Building Blocks of Complex Networks." *Science* 298:824–27.

Moilanen, A., and Hanski, I. 1995. "Habitat Destruction and Coexistence of Competitors in a Spatially Realistic Metapopulation Model." *Journal of Animal Ecology* 64:141–44.

Montoya, J. M., and Solé, R. V. 2002. "Small World Patterns in Food Webs." *Journal of Theoretical Biology* 214:405–12.

Moran, P.A.P. 1953. "The Statistical Analysis of the Canadian Lynx Cycle, II: Synchronization and Metereology." *Australian Journal of Zoology* 1:291–98.

Morgan Ernest, S.K.M., and Brown, J. H. 2001. "Delayed Compensation for Missing Keystone Species by Colonization." *Science* 292:101–104.

Morin, P. J. 1999. *Community Ecology*. Blackwell: Oxford.

Morris, W. F. 1990. "Problems in Detecting Chaotic Behavior in Natural Populations by Fitting Simple Discrete Models." *Ecology* 71:1849–62.

Morse, D. R., Lawton, J. H., Dodson, J. H., and Williamson, M. M. 1985. "Fractal Dimension of Vegetation and the Distribution of Arthropod Body Lengths." *Nature* 314:731–33.

Mossa, S., Barthelemy, M., Stanley, H. E., and Amaral, L.A.N. 2002. "Truncation of Power Law Behavior in Scale-Free Network Models Due to Information Filtering." *Physical Review Letters* 88:138701.

Mueller, L. D., and Ayala, F. J. 1981. "Dynamics of Single-Species Population Growth: Stability or Chaos?" *Ecology* 62:1148–54.

Murray, J. D. 1989. *Mathematical Biology*. Springer: Heidelberg.

Myers, J. H., and Rothman, L. D. 1995. "Field Experiments to Study Regulation of Fluctuating Populations." In *Population Dynamics: New Approaches and Synthesis*, N. Cappuccino and P. Price, eds., 229–51. Academic Press: New York.

Myers, R. A., Mertz, G., and Barrowman, N. J. 1995. "Spatial Scales of Variability in Cod Recruitment in the North Atlantic." *Canadian Journal of Fisheries and Aquatic Science* 52:1849–62.

Myers, R. A., Mertz, G., and Bridson, J. 1997. "Spatial Scales of Interannual Recruitment Variations of Marine, Anadromous, and Freshwater Fish." *Canadian Journal of Fisheries and Aquatic Science* 54:1400–07.

Naeem, S., and Li, S. 1997. "Biodiversity Enhances Ecosystem Reliability." *Nature* 390:507–9.

Nee, S. 1994. "How Populations Persist." *Nature* 367:123–24.

Nee, S., and May, R. M. 1992. "Dynamics of Metapopulations: Habitat Destruction and Competitive Coexistence." *Journal of Animal Ecology* 61:37–40.

Nee, S., May, R. M., and Hassell, M. P. 1997. "Two Species Metapopulation Models." In *Metapopulation Biology: Ecology, Genetics and Evolution*, I. Hanski and M. Gilpin, eds., 123–47. Academic Press: San Diego.

Neutel, A. M., Heesterbeek, J.A.P., and de Ruiter, P. C. 2002. "Stability in Real Food Webs: Weak Links in Long Loops." *Science* 296:1120–23.

Newman, M.E.J. 1996. "Self-Organized Criticality, Evolution, and the Fossil Extinction Record." *Proceedings of the Royal Society of London* B 263:1605–10.

_____. 1997. "A Model of Mass Extinction." *Journal of Theoretical Biology* 189:235–52.

Newman, M.E.J., and Sneppen, K. 1996. "Avalanches, Scaling, and Coherent Noise." *Physical Review* E 54:6226–31.

Newman, M.E.J., and Eble, G. J. 1999. "Decline in Extinction Rates and Scale Invariance in the Fossil Record." *Paleobiology* 25:434–39.

Newman, M.E.J., and Palmer, R. G. 2003. *Modeling Extinction*. Oxford University Press: New York.

Nicholson, A. J., and Bailey, V. A. 1935. "The Balance of Animal Populations I." *Proceedings of the Zoological Society of London* 3:551–98.

Nicolis, G. 1995. *Introduction to Nonlinear Science*. Cambridge University Press: New York.

Nicolis, G., and Prigogine, I. 1977. *Self-Organization in Nonequilibrium Systems*. Wiley: New York.

_____. 1990. *Exploring Complexity*. Freeman: New York.

Nijhout, F. H. 1991. *The Development and Evolution of Butterfly Wing Patterns*. Smithsonian Institution: Washington D.C.

Niklas, K. J. 1994. "Morphological Evolution through Complex Domains of Fitness." *Proceedings of the National Academy of Science, USA* 91:6772–79.

_____. 1997. *The Evolutionary Biology of Plants*. University of Chicago Press: Chicago.

Niklas, K. J., and Kerchner, V. 1984. "Mechanical and Photosynthetic Constraints on the Evolution of Plant Shape." *Paleobiology* 10:79–101.

Nikolaus, P. M., Martin Gonzalez, J. M., and Solé, R. V. 2002. "Spatial Forecasting: Detecting Determinism from Single Snapshots." *International Journal of Bifurcations and Chaos* 12:369–76.

Nisbet, R., Blythe, S., Gurney, B., Metz, H., and Stockes, K. 1989. "Avoiding Chaos." *Trends in Ecology and Evolution* 4:238–39.

Nobel, P. S. 1983. *Biophysical Plant Physiology and Ecology*. Freeman: New York.

Odum, E. P. 1977. "The Emergence of Ecology as a New Integrative Discipline." *Science* 195:1289–93.

Odum, H. T. 1983. *Systems Ecology: An Introduction*. John Wiley: New York.

Okubo, A. 1980. *Diffusion and Ecological Problems: Mathematical Models*. Springer: Berlin.

Olsen, L. F., Truty, G. L., and Schaffer, W. M. 1988. "Oscillations and Chaos in Epidemics: A Nonlinear Dynamics Study of Six Childhood Diseases in Copenhagen, Denmark." *Theoretical Population Biology* 33:344–70.

Omohundro, S. M. 1984. "Modelling Cellular Automata with Partial, Differential Equations." *Physica* D 10:128–34.

Ovaskainen, O., and Hanski, I. 2002. "Transient Dynamics in Metapopulation Response to Perturbation." *Theoretical Population Biology* 61:285–95.

Ovaskainen, O., Sato, K., Bascompte, J., and Hanski, I. 2002. "Metapopulation Models for Extinction in Spatially Correlated Landscapes." *Journal of Theoretical Biology* 215:95–108.

Owen-Smith, N. 1987. "Pleistocene Extinctions: The Pivotal Role of Megaherbivores." *Paleobiology* 13:351–62.

Pacala, S. W., and Levin, S. A. 1998. "Biologically Generated Spatial Pattern and the Coexistence of Competing Species." In *Spatial Ecology: The Role of Space in Population Dynamics and Interspecific Interactions*, D. Tilman and P. Kareiva, eds., 204–32. Princeton University Press: Princeton.

Pace, M. L., Cole, J. J., Carpenter, S. R., and Kitchell, J. F. 1999. "Trophic Cascades Revealed in Diverse Ecosystems." *Trends in Ecology and Evolution* 14:459–503.

Paine, R. T. 1966. "Food Web Complexity and Species Diversity." *American Naturalist* 100:65–75.

_____. 1988. "Food Webs: Road Maps of Interactions or Grist for Theoretical Development?" *Ecology* 69:1648–54.

_____. 1992. "Food-Web Analysis through Field Measurements of Per Capita Interaction Strength." *Nature* 355:73–75.

Parisi, G. 1988. *Statistical Field Theory*. Perseus Books: Reading, MA.

Partridge, L. W., Britton, N. F., and Franks, N. R. 1996. "Army Ant Population Dynamics: The Effects of Habitat Quality and Reserve Size on Population Size and Time to Extinction." *Proceedings of the Royal Society London* B 263:735–41.

Pascual, M. 1993. "Diffusion-Induced Chaos in a Spatial Predator-Prey System." *Proceedings of the Royal Society of London* B 251:1–7.

Pascual, M., Roy, M., Guichard, F., and Flierl, G. 2002. "Cluster Size Distributions: Signature of Self-Organization in Spatial Ecologies." *Philisophical Transactions of the Royal Society of London* B 357:657–66.

Pastor-Satorras, R., Vázquez, A., and Vespignani, A. 2001. "Dynamical and Correlation Properties of the Internet." *Physical Review Letters* 87:258701.

Peitgen, H.-O., Jurgens, H., and Saupe, D. 1992. *Chaos and Fractals: New Frontiers of Science*. Springer-Verlag: New York.

Peters, S. E., and Foote, M. 2002. "Determinants of Extinction in the Fossil Record." *Nature* 416:420–24.

Pielou, E. C. 1969. *Ecological Diversity*. John Wiley: New York.

Pietronero, L. 1995. "Theoretical Concepts for Fractal Growth and Self-Organized Criticality." *Fractals* 3:405–12.

Pimm, S. L. 1979. "Complexity and Stability: Another Look at MacArthur's Original Hypothesis." *Oikos* 33:357.

_____. 1979. "The Structure of Food Webs." *Theoretical Population Biology* 16:144–58.

_____. 1980. "Species Deletion and the Design of Food Webs." *Oikos* 35:139–49.

_____. 1982. *Food Webs*. Chapman and Hall: London.

_____. 1991. *The Balance of Nature? Ecological Issues in the Conservation of Species and Communities*. University of Chicago Press: Chicago.

_____. 1999. "The Dynamics of the Flows of Matter and Energy." In *Advanced Ecological Theory*, J. McGlade, ed., 172–93. Blackwell Science: Oxford.

Pimm, S. L., and Kitching, R. L. 1987. "The Determinants of Food Chain Lengths." *Oikos* 50:302–307.

Pimm, S. L., and Lawton, J. H. 1977. "Number of Trophic Levels in Ecological Communities." *Nature* 268:329–31.

_____. 1980. "Are Food Webs Divided into Compartments?" *Journal of Animal Ecology* 49:879–98.

Pimm S. L., and Raven, P. H. 2000. "Biodiversity: Extinction by Numbers." *Nature* 403:843–45.

Plotnick, R., Gardner, R. H., Hargrove, W. W., Prestegaard, K., and Perlmutter, M. 1996. "Lacunarity Analysis: A General Technique for the Analysis of Spatial Patterns." *Physical Review* E 53:5461–68.

Plotnick, R., and McKinney, M. 1993. "Ecosystem Organization and Extinction Dynamics." *Palaios* 8:202–12.

Plotnick, R., and Sepkoski, J.J., Jr. 2001. "A Multifractal Model for Macroevolution." *Paleobiology* 27:126–39.

Poag, C. W. 1997. "Roadblocks on the Kill Curve: Testing the Raup Hypothesis." *Palaios* 12:582–90.

Polis, G. A. 1991. "Complex Desert Food Webs: An Empirical Critique of Food Web Theory." *American Naturalist* 138:123–55.

_____. 1994. "Food Webs, Trophic Cascades and Community Structure." *Australian Journal of Ecology* 19:121–36.

_____. 1998. "Stability is Woven by Complex Webs." *Nature* 395:744–45.

Pollard, E. 1991. "Synchrony of Population Fluctuations: The Dominant Influence of Widespread Factors on Local Butterfly Populations." *Oikos* 60:7–10.

Post, D. M. 2002. "The Long and Short of Food-Chain Length." *Trends in Ecology and Evolution* 17:269–77.

Poston, T., and Stewart, I. 1977. *Catastrophe Theory and Its Applications*. Dover: New York.

Powell, T. M., and Okubo, A. 1994. "Turbulence, Diffusion and Patchiness in the Sea." *Philosophical Transactions of the Royal Society of London* B 343:11–18.

Price, G. R. 1970. "Selection and Covariance." *Nature* 227:520–21.

Raffaelli, D., and Hall, S. J. 1992. "Compartments and Predation in an Esturine Food Web." *Journal of Animal Ecology* 61:551–60.

Rahmstorf, S. 1996. "On the Freshwater Forcing and Transport of the Atlantic Thermohaline Circulation." *Climate Dynamics* 12:799–811.

Rand, D. A., Wilson, H. B., and McGlade, J. M. 1994. "Dynamics and Evolution: Evolutionarily Stable Attractors, Invasion Exponents and Phenotype Dynamics." *Philosophical Transactions of the Royal Society of London* 243:261–83.

Ranta, E., and Kaitala, V. 1997. "Travelling Waves in Vole Population Dynamics." *Nature* 390:456.

Ranta, E., Kaitala, V., and Lindstrom, J. 1998. "Spatial Dynamics of Populations." In *Modeling Spatiotemporal Dynamics in Ecology*, J. Bascompte and R.V. Solé, eds., 47–62. Springer-Verlag: Berlin.

Ranta, E., Kaitala, V., Lindstrom, J., and H. Lindén. 1995. "Synchrony in Population Dynamics." *Proceedings of the Royal Society of London* B 262:113–18.

Ranta, E., Lindström, J., Kaitala, V., Kokko, H., Lindén, H., and Helle, E. 1997a. "Solar Activity and Hare Dynamics: A Cross-Continental Comparison." *American Naturalist* 149:765–75.

Ranta, E., Kaitala, V., and Lundberg, P. 1997b. "The Spatial Dimension in Population Fluctuations." *Science* 278:1621–23.

Rasmussen, D. R., and Bohr, T. 1987. "Temporal Chaos and Spatial Disorder." *Physics Letters* A 125:107–110.

Rasmussen, S., Knudsen, C., and Feldberg, R. 1991. "Dynamics of Programmable Matter." In *Artificial Life II, SFI Studies in the Sciences of Complexity*,

Volume X, C. Langton, C. Taylor, J. D. Farmer, and S. Rasmussen, eds., 211–54. Addison-Wesley: Redwood City, CA.

Raup, D. M. 1985. "Mathematical Models of Cladogenesis." *Paleobiology* 11:42–52.

_____. 1986. "Biological Extinction and Earth History." *Science* 231:1528–33.

_____. 1991. "A Kill Curve for Phanerozoic Marine Species." *Paleobiology* 17:37–48.

_____. 1993. *Extinction: Bad Genes or Bad Luck?* Oxford University Press: Oxford.

Raup, D. M., Gould, S. J., Schopf, T.J.M., and Simberloff, D. S. 1973. "Stochastic Models of Phylogeny and the Evolution of Diversity." *Journal of Geology* 81:525–42.

Raup, D. M., and Sepkoski, J. J., Jr. 1984. "Periodicity of Extinctions in the Geologic Past." *Proceedings of the National Academy of Science, USA* 81:801–805.

_____. 1986. "Periodic Extinction of Families and Genera." *Science* 231:833–36.

Ravasz, E., Somera, A. L., Mongru, D. A., Oltavi, Z. N., and Barabási, A.-L. 2002. "Hierarchical Organization of Modularity in Metabolic Networks." *Science* 297:1551–55.

Ray, T. S. 1992. "An Approach to the Synthesis of Life." In *Artificial Life II*, C. Langton, J. Taylor, D. Farmer, and S. Rasmussen, eds., 371–408. Addison-Wesley: Redwood City, CA.

_____. 1994. "Evolution, Complexity, Entropy, and Artificial Reality." *Physica* D 75:239–63.

Reagan, D. P., and Waide, R. B. 1996. *The Food Web of a Tropical Rain Forest*. University of Chicago Press: Chicago.

Rejmánek, M., and Stary, P. 1979. "Connectance in Real Biotic Communities and Critical Values for Stability of Model Ecosystems." *Nature* 280:311–13.

Renshaw, E. 1995. *Modelling Biological Populations in Space and Time*. Cambridge University Press: London.

Rhodes, C. J., and Anderson, R. M. 1996. "Power Laws Governing Epidemics in Isolated Populations." *Nature* 381:600–602.

Ricotta, C., Avena, G. C., and Marchetti, M. 1999. "The Flaming Sandpile: Self-Organized Criticality and Wildfires." *Ecological Modeling* 119:73–77.

Ripa, J. 2000. "Analysing the Moran Effect and Dispersal: Their Significance and Interaction in Synchronous Population Dynamics." *Oikos* 89:175–87.

Roberts, R. G., Flannery, T. F., Ayliffe, L. K., Yoshida, H., Olley, J. M., Prideaux, G. J., Laslett, G. M., Baynes, A., Smith, M. A., Jones, R., and Smith, B. L. 2001. "New Ages for the Last Australian Megafauna: Continent-Wide Extinction about 46,000 years ago." *Science* 292:1888–92.

Robinson, G. R., Holt, R. D., Gaines, M. S., Hamburg, S. P., Johnson, M. L., Fitch, H. S., and Martinko, E. A. 1992. "Diverse and Contrasting Effects of Habitat Fragmentation." *Science* 257:524–26.

Rodriguez-Iturbe, I., and Rinaldo, A. 1997. *Fractal River Basins: Chance and Self-Organization*. Cambridge University Press: New York.

Rohani, P., Lewis, T. J., Grunbaum, D., and Ruxton, G. D. 1997. "Spatial Self-Organization in Ecology: Pretty Patterns or Robust Reality?" *Trends in Ecology and Evolution* 12:70–74.

Rohani, P., and Ruxton, G. 1999. "Dispersal-Induced Instabilities in Host-Parasitoid Metapopulations." *Theoretical Population Biology* 55:23–36.

Roughgarden, J. 1979. *Theory of Population Genetics and Evolutionary Ecology: An Introduction*. Macmillan Publishing Company: New York.

Royama, T. 1992. *Analytical Population Dynamics*. Chapman and Hall: London.

Ruxton, G. D. 1994. "Local and Ensemble Dynamics of Linked Populations." *Journal of Animal Ecology* 63:1002.

Saether, B-E., Engen, S., and Lande, R. 1999. "Finite Metapopulation Models with Density-Dependent Migration and Stochastic Local Dynamics." *Proceedings of the Royal Society of London* B 266:113–18.

Saravia, L. A., Coviella, C. E., and Ruxton, G. D. 2000. "The Importance of Transients' Dynamics in Spatially Extended Populations." *Proceedings of the Royal Society* B 267:1781–85.

Sato, K., Matsuda, H., and Sasaki, A. 1994. "Pathogen Invasion and Host Extinction in Lattice Structured Populations." *Journal of Mathematical Biology* 32:251–68.

Schaffer, W. M. 1984. "Stretching and Folding in Lynx Fur Returns: Evidence for a Strange Attractor in Nature?" *The American Naturalist* 124:796–820.

Schaffer, W. M., and Kot, M. 1986a. "Chaos in Ecological Systems: The Coals That Newcastle Forgot." *Trends in Ecology and Evolution* 1:58–63.

_____. 1986b. "Differential Systems in Ecology and Evolution." In *Chaos*, A. V. Holden, ed. Manchester University Press: Manchester, UK.

Scheffer, M., Carpenter, S., Foley, J., Folke, C., and Walker, B. 2001. "Catastrophic Shifts in Ecosystems." *Nature* 413:591–96.

Schroeder, M. 1991. *Fractals, Chaos, Power Laws*. Freeman: New York.

Schuster, H. G. 1984. *Deterministic Chaos*. Physik-Verlag: Berlin.

Schuster, P. 1996. "How Does Complexity Arise in Evolution?" *Complexity* 2:22–30.

Schulze, E.-D., and Mooney, H. A. 1994. "Ecosystem Function of Biodiversity: A Summary." In *Biodiversity and Ecosystem Function*, E.-D. Shulze and H. A. Mooney, eds., 497–510. Springer-Verlag: Berlin.

Segel, L. A., and Jackson, J. 1972. "Dissipative Structure: An Explanantion and an Ecological Example." *Journal of Theoretical Biology* 37:545–59.

Sepkoski, J. J. 1978. "A Kinetic Model of Phanerozoic Taxonomic Diversity I: Analysis of Marine Orders." *Paleobiology* 4:223–51.

_____. 1979. "A Kinetic Model of Phanerozoic Taxonomic Diversity II. Early Phanerozoic Families and Multiple Equilibria." *Paleobiology* 5:222–51.

_____. 1984. "A Kinetic Model of Phanerozoic Taxonomic Diversity III. Post-Paleozoic Families and Mass Extinction." *Paleobiology* 10:246–67.

_____. 1989. "Periodicity in Extinction and the Problem of Catastrophism in the History of Life." *Journal of the Geological Society London* 146:7–19.

_____. 1996. "Competition in Macroevolution: The Double Wedge Revisited." In *Evolutionary Paleoecology*, D. Jablonski, D. H. Erwin, and J. Lipps, eds., 211–55. University of Chicago Press: Chicago.

Sheffer, M. 1991. "Should We Expect Strange Attractors Behind Plankton Dynamics: And If So, Should We Bother?" *Journal of Plankton Research* 13:1291–1306.

Shnerb, N. M., Sarah, P., Lavee, H., and Solomon, S. 2003. "Reactive Glass and Vegetation Patterns." *Physical Review Letters* 90:0381011.

Shorrocks, B. 1991. "A Need for Niches?" *Trends in Ecology and Evolution* 6:262–63.

Sibani, P., Brandt, M., Alström, P. 1998. "Evolution and Extinction Dynamics in Rugged Fitness Landscapes." *International Journal of Modern Physics* B 12:361–91.

Sinclair, A.R.E., Gosline, J. M., Holdsworth, G., Krebs, C. J., Boutin, S., Smith, J.N.M., Boonstra, R., and Dale, M. 1993. "Can the Solar Cycle and Climate Synchronize the Snowshoe Hare Cycle in Canada? Evidence from Tree Rings and Ice Cores." *American Naturalist* 141:173–98.

Sinclair, A.R.E., Krebs, C. J., Smith, J.N.M., and Boutin, S.. 1988. "Population Biology of Snowshoe Hares. III. Nutrition, Plant Secondary Compounds and Food Limitation." *Journal of Animal Ecology* 57:787–806.

Sipper, M., and Reggia, J. A. 2001. "Go Forth and Replicate." *Scientific American* 285:34–43.

Slatkin, M. 1974. "Competition and Regional Coexistence." *Ecology* 55:128–34.

Smith, C. H. 1983. "Spatial Trends in Canadian Snowshoe Hare *Lepus Americanus* Population Cycles." *Canadian Field-Naturalist* 97:151–60.

Sneppen, K., Bak, P., Flyvbjerg, H., and Jensen, M. H. 1995. "Evolution as a Self-Organized Critical Phenomenon." *Proceedings National Academy of Sciences, USA* 92:5209–13.

Snyder, R. E., and Nisbet, R. E. 2000. "Spatial Structure and Fluctuations in the Contact Process and Related Models." *Bulletin of Mathematical Biology* 62:959–75.

Solé, R. V. 1996. "On Macroevolution, Extinctions and Critical Phenomena." *Complexity* 1(4):40–46.

_____. 2001. "Modelling Macroevolutionary Patterns: An Ecological Perspective." In *Biological Evolution and Statistical Physics*, M. Lassig and A. Valleriani, eds., 318–44. Springer: Berlin.

Solé., R. V., and Alonso, D. 2000. "The DivGame Simulator: A Stochastic Cellular Automata Model of Rainforest Dynamics." *Ecological Modelling* 133:131–41.

Solé., R. V., Alonso, D., and McKane, A. 2002. "Connectivity and Scaling in S-Species Model Ecosystems." *Physica* A 286:337–44.

_____. 2002b. "Self-Organized Instability in Complex Ecosystems." *Philosophical Transactions of the Royal Society of London* B 357:667–81.

Solé, R. V., Alonso, D., and Saldanya, J. 2004. "Habitat Fragmentation and Biodiversity Collapse in Neutral Communities." *Ecological Complexity* 1:65–75.

Solé, R. V., Bartumeus, F., and Gawarra, J.P.G. 2005. "Gap Percolation in Rainforests." *Oikos* 110:177–85.

Solé, R. V., and Bascompte, J. 1993. "Chaotic Turing Structures." *Physics Letters* A 179:325–31.

_____. 1995. "Measuring Chaos from Spatial Information." *Journal of Theoretical Biology* 175:139–47.

_____. 1996. "Are Critical Phenomena Relevant to Large-Scale Evolution?" *Proceedings of the Royal Society of London* B 263:161–68.

Solé, R. V., Bascompte, J., and Manrubia, S. C. 1996a. "Extinctions: Bad Genes or Weak Chaos?" *Proceedings of the Royal Society of London* B 263:1407–13.

Solé, R. V., Bascompte, J., and Valls, J. 1992c. "Nonequilibrium Dynamics in Lattice Ecosystems: Chaotic Stability and Dissipative Structures." *Chaos* 2:387–95.

Solé, R. V., Bascompte, J., and Valls, J. 1992a. "Stability and Complexity of Spatially Extended Two-Species Competition." *Journal of Theoretical Biology* 159:469–80.

Solé, R. V., Ferrer, R., Montoya, J. M., and Valverde, S. 2002b. "Selection, Tinkering and Emergence in Complex Networks." *Complexity* 8:20–33.

Solé, R. V., and Goodwin, B. C. 2001. *Signs of Life: How Complexity Pervades Biology*. Basic Books: New York.

Solé, R. V., and Manrubia, S. C. 1995a. "Are Rainforests Self-Organized in a Critical State?" *Journal of Theoretical Biology* 173:31–40.

_____. 1995b. "Self-Similarity in Rain Forests: Evidence for a Critical State." *Physical Review* E 51:6250–53.

_____. 1996. "Extinction and Self-Organized Criticality in a Model of Large-Scale Evolution." *Physical Review* E 54:R42–R46.

_____. 1997. "Criticality and Unpredictability in Macroevolution." *Physical Review* E 55:4500–4508.

Solé, R. V., Manrubia, S. C., Kauffman, S. A., Benton, M., and Bak, P., 1999. "Criticality and Scaling in Evolutionary Ecology." *Trends in Ecology and Evolution* 14:156–60.

Solé, R. V., Manrubia, S. C., Luque, B., Delgado, J., and Bascompte, J. 1996b. "Phase Transitions and Complex Systems." *Complexity* 1:13–26.

Solé, R. V., Manrubia, S. C., Mercader, J. P., Benton, M., and Bak, P. 1998. "Long-Range Correlations in the Fossil Record and the Fractal Nature of Macroevolution." *Advances in Complex Systems* 1:255–66.

Solé, R. V., and Montoya, J. M. 2001. "Complexity and Fragility in Ecological Networks." *Proceedings of the Royal Society of London* B 268:2039–45.

Solé, R. V., Montoya, J. M., and Erwin, D. H. 2002a. "Recovery after Mass Extinction: Evolutionary Assembly in Large-Scale Biosphere Dynamics." *Philosophical Transactions of the Royal Society of London* B 357:697–707.

Solé, R. V., and Newman, M.E.J. 2001. "Paleontologic Patterns of Extinction and Biodiversity." In *Encyclopaedia of Global Environmental Change*, vol. 2, 297–301. John Wiley: New York.

Solé, R. V., Saldonya, J., Montoya, J. M., and Erwin, D. M. 2005b. "A Simple Model of Recovery Dynamics after Mass Extinction." Santa Fe Institute Working Paper 05-06-028.

Solé, R. V., and Valls, J. 1991. "Order and Chaos in a Two-Dimensional Lotka-Volterra Coupled Map Lattice." *Physics Letters* A 153:330–36.

_____. 1992. "Nonlinear Phenomena and Chaos in a Monte Carlo Simulated Microbial Ecosystem." *Bulletin of Mathematical Biology* 54:939–55.

Solé, R. V., Valls, J., and Bascompte, J. 1992b. "Spiral Waves, Chaos and Multiple Attractors in Lattice Models of Interacting Populations." *Physics Letters* A 166:123–28.

Sornette, D. 2000. *Critical Phenomena in the Natural Sciences*. Springer: Berlin.

Stanley, H. E. 1971. *Introduction to Phase Transitions and Critical Phenomena*. Oxford University Press: New York.

Stanley H. E., Amaral, L.A.N., Buldyrev, S. V., Goldberger, A. L., Havlin, S., Leschhorn, H., Maass, P., Makse, H. A., Peng, C. K., Salinger, M. A., Stanley, M.H.R., and Viswanathan, G. M. 2000. "Scaling and Universality in Animate and Inanimate Systems." *Physica* A 231:20–48.

Stauffer, D., and Aharony, A. 1985. *Introduction to Percolation Theory*. Taylor and Francis: London.

Steele, J. H., and Henderson, E. W. 1981. "A Simple Plankton Model." *American Naturalist* 117:676–91.

Steen, H., Yoccoz, N. G., and Ims, R. A. 1990. "Predators and Small Rodent Cycles: An Analysis of a 29-Year Time Series of Small Rodent Population Fluctuations." *Oikos* 59:115–20.

Stenseth, N. C., and Maynard Smith, J. 1984. "Coevolution in Ecosystems: Red Queen Evolution or Stasis?" *Evolution* 38:870–80.

Stephens, P. A., and Sutherland, W. J. 1999. "Consequences of the Allee Effect for Behaviour, Ecology, and Conservation." *Trends in Ecology and Evolution* 14:401-5.

Stone, L. 1995. "Biodiversity and Habitat Destruction: A Comparative Study of Model Forest and Coral Reef Ecosystems." *Proceedings of the Royal Society of London* B 261:381–88.

Stone, L., and Roberts, A. 1991. "Conditions for a Species to Gain Advantage from the Presence of Competitors." *Ecology* 72:1964–72.

Strogatz, S. 1994. *Nonlinear Dynamics and Chaos*. Addison-Wesley: New York.

_____. 2001. "Exploring Complex Networks." *Nature* 410:268–76.

Sugihara, G. 1982. *Niche Hierarchy: Structure, Organization and Assembly in Natural Communities*. Ph.D. Thesis. Princeton University: Princeton.

Sugihara, G., and May, R. M. 1990a. "Nonlinear Forecasting as a Way of Distinguishing Chaos from Measurement Error in Time Series." *Nature* 344:734–41.

_____. 1990b. "Application of Fractals in Ecology." *Trends in Ecology and Evolution* 5(3):79–86.

Sugihara, G., Schoenly, K., and Trombla, A. 1989. "Scale Invariance in Food Web Properties." *Science* 245:48–52.

Swihart, R. K., Feng, Z., Slade, N. A., Mason, D. M., and Gehring, T. M. 2001. "Effects of Habitat Destruction and Resource Supplementation in a Predator-Prey Metapopulation Model." *Journal of Theoretical Biology* 210:287–303.

Takayasu, H. 1990. *Fractals in the Physical Sciences*. Manchester University Press: New York.

Takens, F. 1981. "Detecting Strange Attractors in Turbulence." In *Lecture Notes in Mathematics*, vol. 898, D. A. Rand and L. S. Young, eds., 366–81. Springer: Berlin.

Taylor, K. 1999. "Rapid Climate Change." *American Scientist* 87:320–27.

Terborgh, J. 1988. "The Big Things That Run the World." *Conservation Biology* 2:402–403.

Terborgh, J., et al. 2001. "Ecological Meltdown in Predator-Free Forest Fragments." *Science* 294:1923–26.

Theraulaz, G., Bonabeau, E., Nicolis, S. C., Solé, R. V., Fourcassié, V., Blanco, S., Fournier, R., Joly, J.-L., Fernández, P., Grimal, A., Dalle, A., and Deneubourg, J.-L. 2002. "Spatial Patterns in Ants Colonies." *Proceedings of the National Academy of Sciences, USA* 99:9645–49.

Thomas, W. R., Pomeranz, M. J., and Gilpin, M. E. 1980. "Chaos, Asymmetric Growth and Group Selection for Dynamical Stability." *Ecology* 61:1312–20.

Tilman, D., and Kareiva, P., eds. 1997. *Spatial Ecology. The Role of Space in Population Dynamics and Interspecific Interactions*. Princeton University Press: Princeton.

Tilman, D., Lehman, C. L., and Chengjun, Y. 1997. "Habitat Destruction, Dispersal, and Deterministic Extinction in Competitive Communities." *American Naturalist* 149:407–35.

Tilman, D., May, R. M., Lehman, C. L., and Nowak, M. A. 1994. "Habitat Destruction and the Extinction Debt." *Nature* 371:65–66.

Tilman, D., and Wedin, D. 1991. "Oscillations and Chaos in the Dynamics of a Perennial Grass." *Nature* 353:653–55.

Turchin, P. 1990. "Rarity of Density Dependence or Population Regulation with Lags?" *Nature* 344:660–62.

Turchin, P., and Ellner, S. P. 2000. "Living on the Edge of Chaos: Population Dynamics of Fennoscandian Voles." *Ecology* 81:3099–3116.

Turchin, P., and Hanski, I. 1997. "An Empirically Based Model for Latitudinal Gradient in Vole Population Dynamics." *American Naturalist* 149:842–74.

Turchin, P., Oksanen, L., Ekerholm, P., Oksanen, T., and Henttonen, H. 2000. "Are Lemmings Prey or Predators?" *Nature* 405:562–65.

Turchin, P., Reeve, J. D., Cronin, J. T., and Wilkens, R. T. 1998. "Spatial Pattern Formation in Ecological Systems: Bridging Theoretical and Empirical Approaches." In *Modeling Spatiotemporal Dynamics in Ecology*, J. Bascompte and R. V. Solé, eds., 199–213. Springer-Verlag: Berlin.

Turcotte, D. L. 1992. *Fractals and Chaos in Geology and Geophysics*. Cambridge University Press: Cambridge.

Turing, A. 1952. "On the Chemical Basis of Morphogenesis." *Philosophical Transactions of the Royal Society* B 237:37–72.

Turner, M. G. 1989. "Landscape Ecology: The Effect of Pattern on Process." *Annual Review of Ecology and Systematics* 29:171–97.

Turner, M. G., Gardner, R. H., Dale, V. H., and O'Neill, R. V. 1989. "Predicting the Spread of Disturbances Across Heterogeneous Landscapes." *Oikos* 55:121–29.

Ulanowicz, R. E. 1997. *Ecology, the Ascendent Perspective*. Columbia University Press: New York.

Van de Koppel, J., Rietkerk, M., and Weissing, F. J. 1997. "Catastrophic Vegetation Shifts and Soil Degradation in Terrestrial Grazing Systems." *Trends in Ecology and Evolution* 12(9):352–56.

Van der Laan, J. D., and Hogeweg, P. 1995. "Predator-Prey Coevolution: Interaction Across Different Time Scales." *Proceedings of the Royal Society of London* B 259:35–42.

Vandermeer, J. H. 1990. "Indirect and Diffuse Interactions." *Journal of Theoretical Biology* 142:429–42.

Van Kampen, N. G., 1981. *Stochastic Processes in Physics and Chemistry*. Elsevier: Amsterdam.

Van Valen, L. 1973. "A New Evolutionary Law." *Evolutionary Theory* 1:1–30.

Vermeij, G. J., 1987. *Evolution and Escalation: An Ecological History of Life*. Princeton University Press: Princeton.

Vicsek, T. 1992. *Fractal Growth Phenomena*, 2nd ed. World Scientific: Singapore.

Vilar, J.M.G., Solé, R. V., and Rubi, J. M. 2003. "On the Origin of Plankton Patchiness," *Physica* A 317:239–46.

Viswanathan, G. M., Buldyrev, S. V., Havlin, S., et al. 1999. "Optimizing the Success of Random Searches." *Nature* 401:911–14.

Volterra, V. 1926. "Variations and Fluctuations of the Number of Individuals in Animal Species Living Together." Translation in Chapman, R. N., 1931, *Animal Ecology*, 409–48. Wiley: New York.

Von Bertalanffy, L. 1951. *General System Theory: A New Approach to Unity of Science*. Johns Hopkins University Press: Baltimore.

———. 1968. *General System Theory: Foundations, Development, Applications*. Braziller: New York.

Von Hardenberg, J., Meron, E., Shachak, M., and Zarmi, Y. 2001. "Diversity of Vegetation Patterns and Desertification." *Physical Review Letters* 87:198101.

Vrba, E. S., 1985. "African Bovidae: Evolutionary Events Since the Miocene." *South African Journal of Science* 81:263–66.

Waller, I., and Kapral, R. 1984. "Spatial and Temporal Structure in Systems of Coupled Nonlinear Oscillators." *Physical Review* A30:2047–55.

Ward, P. D. 1992. *On Methusela's Trail: Living Fossils and the Great Extinctions*. Freeman: New York.

Warren, P. H. 1989. "Spatial and Temporal Variation in the Structure of a Freshwater Food Web." *Oikos* 55:299–311.

———. 1996. "Structural Constraints on Food Web Assembly." In *Aspects of the Genesis and Maintenance of Biological Diversity*, M. E. Hochberg, J. Clobert, and R. Barbault, eds., 142–61. Oxford University Press: Oxford.

Watts, D. J. 2000. *Small Worlds*. Princeton University Press: Princeton.

Watts, D. J., and Strogatz, S. H. 1998. "Collective Dynamics of 'Small-World' Networks." *Nature* 393:440–42.

Weinberger, E. D. 1989. "A More Rigorous Derivation of Some Properties of Uncorrelated Fitness Landscapes." *Journal of Theoretical Biology* 134:125–29.

———. 1990. "Correlated and Uncorrelated Fitness Landscapes and How to Tell the Difference." *Biological Cybernetics* 63:325–26.

_____. 1991. "Local Properties of Kauffman's NK Model: A Tunably Rugged Energy Landscape." *Physical Review* A 44:6399–6413.

Weitz, J. S., and Rothman, D. H. 2004. "Scale-Dependence of Resource-Biodiversity Relationships." *Journal of Theoretical Biology* 225(2):205–14.

Welden, C. W., Hewett, S. W., Hubbell, S. P., and Foster, R. B. 1991. "Sapling Survival, Growth, and Recruitment: Relationship to Canopy Height in a Neotropical Forest." *Ecology* 72:35–50.

West, B. J., and Goldberger, A. L. 1987. "Physiology in Fractal Dimensions." *American Scientist* 75:354–65.

White, A., Begon, M., and Bowers, R. G. 1996. "Host-Pathogen System in a Spatially Patchy Environment." *Proceedings of the Royal Society of London* B 263:325–32.

Whittington, S. L., and Dyke, B. 1989. "Simulating Overkill: Experiment with the Mossiman and Martin Model." In *Quaternary Extinctions: A Prehistoric Revolution*, P. S. Martin and R. G. Klein, eds. University of Arizona Press: Tucson.

Wigner, E. P. 1958. "On the Distribution of the Roots of Certain Symmetric Matrices." *Annals of Mathematics* 67:325–27.

Wilcove, D. S. 1987. "From Fragmentation to Extinction." *Natural Areas Journal* 7:23–29.

Wilcox, B. A., and Murphy, D. D. 1985. "Conservation Strategy: The Effects of Fragmentation on Extinction." *American Naturalist* 125:879–87.

Wilke, C. O., and Adami, C. 2002. "The Biology of Digital Organisms." *Trends in Ecology and Evolution* 17:528–32.

Wilke, C. O., Wang, J. L., Ofria, C., Lenski, R. E., and Adami, C. 2001. "Evolution of Digital Organisms at High Mutation Rate Leads to Survival of the Flattest." *Nature* 412:331–33.

Williams, C. B. 1944. "Some Applications of the Log Series and the Index of Diversity to Ecological Problems." *Journal of Biology* 32:1–44.

Williams, D. W., and Liebhold, A.M. 1995. "Influence of Weather on the Synchrony of Gypsy Moth (Lepidoptera: Lymantriidae) Outbreaks in New England." *Environmental Entomology* 24:987–95.

Williams, R. J., Berlow, E. R., Dunne, J. A., Barabási, A.-L., and Martinez, N. D. 2002. "Two Degrees of Separation in Complex Food Webs." *Proceedings of the National Academy of Sciences, USA* 99:12913–16.

Willliams, R. J., and Martinez, N. D. 2000. "Simple Rules Yield Complex Food Webs." *Nature* 404:180–83.

Williamson, M. H., and Lawton, J. H. 1991. "Measuring Habitat Structure with Fractal Geometry." In *Habitat Structure*. S. Bells, E. D. McCoy, and H. R. Mushinsky, eds., 69–86. Chapman and Hall: London.

Willis, E. O., and Oniki, Y. 1978. "Birds and Army Ants." *Annual Reviews in Ecology and Systematics* 9:243–63.

Willis, J. C. 1922. *Age and Area*. Cambridge University Press: Cambridge.

Wilson, E. O. 1992. *The Diversity of Life*. Harvard University Press: Cambridge, MA.

_____. 1998. *Consilience: The Unity of Science*. Knopf: New York.

Wilson, W. G., de Roos, A. M., and McCauley, E. 1993. "Spatial Instabilities within the Diffuse Lotka-Volterra System: An Individual-Based Simulation Result." *Theoretical Population Biology* 43:91–127.

Winemiller, K. O., Pianka, E. P., Vitt, L. J., and Joern, A. 2001. "Food Web Laws or Niche Theory? Six Independent Empirical Tests." *American Naturalist* 158:193–99.

Winfree, A. 1972. "Spiral Waves of Chemical Activity." *Science* 175:634–36.

With, K. A., and King, A. W. 1999. "Extinction Thresholds for Species in Fractal Landscapes." *Conservation Biology* 13:314–26.

Wolf, A., Swift, J. B., Swinney, H. L., and Vastano, J. A. 1985. "Determining Lyapunov Exponents from Time Series." *Physica* D 16:285–317.

Wolf-Gladrow, D. A. 2000. *Lattice-Gas Cellular Automata and Lattice Boltzmann Models: An Introduction*. Springer: Berlin.

Wolff, J. O. 1980. "The Role of Habitat Patchiness in the Population Dynamics of Snowshoe Hares." *Ecological Monographs* 50:111–30.

Wolfram, S. 1984. "Cellular Automata as Models of Complexity." *Nature* 311:419–24.

_____. 1986. *Cellular Automata and Complexity: Collected Papers*. Addison-Wesley: Reading, MA.

Wootton, J. T. 1998. "Effects of Disturbance on Species Diversity: A Multitrophic Perspective." *American Naturalist* 152:803–25.

_____. 2001. "Local Interactions Predict Large-Scale Pattern in Empirically Derived Cellular Automata." *Nature* 413:841–44.

Wuensche, A., and Lesser, M. 1992. *The Global Dynamics of Celullar Automata*. Discrete Dynamics Incorporated: Santa Fe, NM.

Yeomans, J. M. 1992. *Statistical Mechanics of Phase Transitions*. Oxford University Press: Oxford.

Yodzis, P. 1980. "The Connectance of Real Ecosystems." *Nature* 284:544–45.

_____. 1988. "The Indeterminacy of Ecological Interactions, as Perceived through Perturbation Experiments." *Ecology* 69:508–15.

_____. 1997. "Food Webs and Perturbation Experiments: Theory and Practice." In *Food Webs: Integration of Patterns and Dynamics* G. A. Polis and K. O. Winemiller, eds., 192–200. Chapman and Hall: New York.

Yoshimura, J., and Jansen, V.A.A. 1996. "Evolution and Population Dynamics in Stochastic Environments." *Research in Population Ecology* 38:165–82.

Zeeman, E. C. 1976. "Catastrophe Theory." *Scientific American* 4:65–83.

Index

Page numbers in italics indicate figures and tables.

MONOGRAPHS IN POPULATION BIOLOGY

EDITED BY SIMON A. LEVIN AND HENRY S. HORN

1. *The Theory of Island Biogeography*, by Robert H. MacArthur and Edward O. Wilson
2. *Evolution in Changing Environments: Some Theoretical Explorations*, by Richard Levins
3. *Adaptive Geometry of Trees*, by Henry S. Horn
4. *Theoretical Aspects of Population Genetics*, by Motoo Kimura and Tomoko Ohta
5. *Populations in a Seasonal Environment*, by Steven D. Fretwell
6. *Stability and Complexity in Model Ecosystems*, by Robert M. May
7. *Competition and the Structure of Bird Communities*, by Martin L. Cody
8. *Sex and Evolution*, by George C. Williams
9. *Group Selection in Predator-Prey Communities*, by Michael E. Gilpin
10. *Geographic Variation, Speciation, and Clines*, by John A. Endler
11. *Food Webs and Niche Space*, by Joel E. Cohen
12. *Caste and Ecology in the Social Insects*, by George F. Oster and Edward O. Wilson
13. *The Dynamics of Arthropod Predator-Prey Systems*, by Michael P. Hassel
14. *Some Adaptations of Marsh-Nesting Blackbirds*, by Gordon H. Orians
15. *Evolutionary Biology of Parasites*, by Peter W. Price
16. *Cultural Transmission and Evolution: A Quantitative Approach*, by L. L. Cavalli-Sforza and M. W. Feldman
17. *Resource Competition and Community Structure*, by David Tilman
18. *The Theory of Sex Allocation*, by Eric L. Charnov
19. *Mate Choice in Plants: Tactics, Mechanisms, and Consequences*, by Nancy Burley and Mary F. Wilson
20. *The Florida Scrub Jay: Demography of a Cooperative-Breeding Bird*, by Glen E. Woolfenden and John W. Fitzpatrick
21. *Natural Selection in the Wild*, by John A. Endler
22. *Theoretical Studies on Sex Ratio Evolution*, by Samuel Karlin and Sabin Lessard
23. *A Hierarchical Concept of Ecosystems*, by R. V. O'Neill, D. L. DeAngelis, J. B. Waide, and T.F.H. Allen
24. *Population Ecology of the Cooperatively Breeding Acorn Woodpecker*, by Walter D. Koenig and Ronald L. Mumme
25. *Population Ecology of Individuals*, by Adam Lomnicki
26. *Plant Strategies and the Dynamics and Structure of Plant Communities*, by David Tilman
27. *Population Harvesting: Demographic Models of Fish, Forest, and Animal Resources*, by Wayne M. Getz and Robert G. Haight
28. *The Ecological Detective: Confronting Models with Data*, by Ray Hilborn and Marc Mangel

29. *Evolutionary Ecology across Three Trophic Levels: Goldenrods, Gallmakers, and Natural Enemies*, by Warren G. Abrahamson and Arthur E. Weis
30. *Spatial Ecology: The Role of Space in Population Dynamics and Interspecific Interactions*, edited by David Tilman and Peter Kareiva
31. *Stability in Model Populations*, by Laurence D. Mueller and Amitabh Joshi
32. *The Unified Neutral Theory of Biodiversity and Biogeography*, by Stephen P. Hubbell
33. *The Functional Consequences of Biodiversity: Empirical Progress and Theoretical Extensions*, edited by Ann P. Kinzig, Stephen J. Pacala, and David Tilman
34. *Communities and Ecosystems: Linking the Aboveground and Belowground Components*, by David Wardle
35. *Complex Population Dynamics: A Theoretical/Empirical Synthesis*, by Peter Turchin
36. *Consumer-Resource Dynamics*, by William W. Murdoch, Cheryl J. Briggs, and Roger M. Nisbet
37. *Niche Construction: The Neglected Process in Evolution*, by F. John Odling-Smee, Kevin N. Laland, and Marcus W. Feldman
38. *Geographical Genetics*, by Bryan K. Epperson
39. *Consanguinity, Inbreeding, and Genetic Drift in Italy*, by Luigi Luca Cavalli-Sforza, Antonio Moroni, and Gianna Zei
40. *Genetic Structure and Selection in Subdivided Populations*, by François Rousset
41. *Fitness Landscapes and the Origin of Species*, by Sergey Gavrilets
42. *Self-Organization in Complex Ecosystems*, by Ricard V. Solé and Jordi Bascompte